TEACH YOURSELF BOOKS

CALCULUS

TEACH YOURSELF BOOKS

CALCULUS

P. ABBOTT

TEACH YOURSELF BOOKS
ST. PAUL'S HOUSE WARWICK LANE LONDON EC4

First printed 1940
This Edition 1970

ISBN 0 340 05536 7

Printed in Great Britain for The English Universities Press Ltd., by Richard Clay (The Chaucer Press) Ltd., Bungay, Suffolk.

INTRODUCTION

SOME years ago an excellent little book was published bearing the title, "The Calculus Made Easy." The author adopted as his motto, "What one fool can do another can," intending thereby to encourage a diffident student. As the author, however, disclosed the fact that he was a "Fellow of the Royal Society" it is doubtful whether the words would bring much comfort to those who were proposing to study the subject.

In those days the calculus was looked upon by many as abstruse and lying beyond the boundaries of elementary mathematics. But the increasing use of the subject in engineering and science, and consequently the desirability of bringing such a powerful mathematical instrument within the reach of a wider circle of students, led to the gradual simplification of its presentation.

The present volume is in the line of this development. It aims at making it easier for the private student, who is unable to obtain the guidance and help of a teacher, to acquire a working knowledge of the calculus. Like other books in the series, it attempts, within the inevitable limitations of space, to provide something of the presentation and illustrations employed by a teacher of the subject, especially in the earlier stages when the student is trying to discover what it is all about.

Those who propose to use the book will naturally want to know what previous knowledge of other branches of mathematics are necessary. It is assumed that the readers possess an elementary knowledge of algebra, trigonometry and the fundamental principles of geometry such as is contained, for example, in the companion books on these subjects in the same series. To assist the student, cross references to the relevant parts of these books are given wherever they may be of assistance to him.

Perhaps the greatest difficulty in writing a book of this character is to determine what to include and what to omit. The calculus is so wide and deep in its ramifications and applications, that the temptation is continually present to include much that the limitations imposed by the available space, make impossible. The author, therefore, has been guided by the policy of including what seems to him to be necessary to enable and encourage the student to proceed further in his study of the subject or to utilise it in its application to science and engineering. It was only after much hesitation that the book was lengthened by the inclusion of the last three chapters. They were inserted in the hope that they would convey to the student some idea of the possibilities of the calculus and lead him to continue his study of it.

As far as possible the " proofs " of many of the theorems have been simplified and curtailed. In consequence of this simplification they may frequently be lacking in the mathematical rigidity and exactitude which are possible in a larger and more ambitious volume. It is hoped, however, that they will supply the student with a sufficiently logical basis for an intelligent study of the subject.

The majority of the tables at the end of this book are taken from Mr. Abbott's *Mathematical Tables and Formulae*, by courtesy of the publishers, Messrs. Longmans, Green & Co., Ltd.

P. ABBOTT.

CONTENTS

CHAPTER I

FUNCTIONS

1. What is the calculus?

THE word "calculus" is the Latin name for a stone which was employed by the Romans for reckoning—*i.e.*, for "calculation". When used as in the title of this book, it is an abbreviation for "Infinitesimal Calculus", which implies a reckoning, or calculation, with numbers which are infinitesimally small. This, in all probability, will not convey much to the beginner, and the real meaning of it will in many cases not be understood until the student has made some headway with his study of the subject. The following example may help to throw a little light on it.

Consider the growth of a small plant. In the ordinary way we know that it grows gradually and continuously. If it be examined after an interval of a few days, the growth will be obvious and readily measured. But if it be observed after an interval of a few minutes, although growth has taken place the amount is too small to be distinguished. If observation takes place after a still smaller interval of time, say a few seconds, although no change can be detected, we know that there has been growth, which, to use a mathematical term, can be regarded as **infinitesimally small**, or infinitesimal.

The process of gradual and continuous growth or increase may be observed in innumerable other instances, of which the case of a living organism referred to above is but one. What is of real importance in most cases is not necessarily the actual amount of growth or increase, but the **rate of growth or increase.** It is this problem, closely connected as it is with infinitesimal increases, that is the basis of the **Infinitesimal Calculus,** and more especially that part of it which is called the **Differential Calculus.** The meaning of differential will be apparent later.

Historical Note. The calculus is the most powerful mathematical invention of modern times. The credit for its discovery has been claimed for both Sir Isaac Newton and Leibnitz, the great German mathematician, and a controversy raged for years in England and Germany as to who was the first to invent it. Leibnitz was the first to publish an account of it, in 1684, though his notebooks showed that he used the method for the first time in 1675. Newton published his book on the subject in 1693, but he communicated his discovery of it to friends in 1669. It is generally agreed now that the fundamental basis of the invention was reached independently by the two mathematicians.

2. Functions.

The student will realise, from his knowledge of Algebra, that the example cited above of the growth of a plant is an instance of a functional relation. It may be affected by variations in temperature, moisture, sunlight, etc., but if these remain constant the **growth is a function of time,** although we are not able to express it in mathematical form.

It is desirable, therefore, that we should begin the study of calculus by clarifying our ideas about the meaning of a function, since this is fundamental in understanding the subject. The student will have become acquainted with the meaning of "function" in his Algebra (*Algebra*, Chaps. XIII and XVIII), but a brief revision is given below for the benefit of those who may not be quite clear on this very important matter.

3. Variables and constants.

Of the letters and symbols used to represent quantities or numbers in an algebraical expression or formula, some represent **variable** quantities, others represent **constants.**

Thus in the formula for the volume of a sphere, viz.

$$V = \frac{4}{3}\pi r^3$$

where **V** represents the volume and **r** represents the radius of the sphere,

(1) **V** and **r** vary with different spheres and are called **variables.**

(2) π and $\frac{4}{3}$ are **constants** whatever the size of the sphere.

Again, in the formula for a falling body, viz.:

$$s = \frac{1}{2}gt^2$$

in which s represents the distance fallen in time t,

s and t are **variables.**
$\frac{1}{2}$ and g are **constants.**

4. Dependent and independent variables.

It will be seen that in each of the above examples the variables are of two kinds.

Thus in $V = \frac{4}{3}\pi r^3$ if the radius (r) be increased or decreased, the volume (V) will increase or decrease in consequence.

i.e., **the variation of V depends upon the variation of r.**

Similarly in $s = \frac{1}{2}gt^2$, the distance (s) fallen depends on the time (t).

So, generally, it will be found that in all such formulae and mathematical expressions there are **two kinds of variables : dependent and independent.**

(1) That variable whose value depends upon the value assigned to the other is called a **dependent variable,** as V and s above.

(2) The variable in which changes in value produce corresponding changes in the other is called the **independent variable,** as r and t in the above formulæ.

In a general form of an expression of the second degree such as

$$y = ax^2 + bx + c$$

a, b and c represent constants, and the value of y depends on the value of x. Consequently x is an independent variable, and y a dependent variable. The constants a, b, and c are used to indicate the relation which exists between the two variables.

5. Functions.

This connection between two variables—viz. that the value of one is dependent upon the value of the other—is expressed by the statement that the **dependent variable is a function of the independent variable.** When the variables represent quantities we say that one quantity is a function of the other. Thus in the examples above

(1) The **volume** of a sphere is a function of its **radius.**
(2) The **distance** moved by a falling body is a function of **time.**

Note.—For the use of the word "quantity" see *Algebra*, § 6.
Innumerable examples might be given of the functional relation between quantities. Here are a few common examples:

The **logarithm** of a number is a function of the **number.**
The **volume** of a fixed mass of gas is a function of the **temperature** while the pressure remains constant.
The **sines, cosines** and **tangents** of angles are functions of the **angle.**
The **time** of beat is a function of the **length** of the pendulum.
The **range** of a gun, with a constant propelling force, is a function of the **angle of projection.**

Definition of a function.

Generally if two variable quantities X and Y are so related that, when any value is assigned to X there is thus determined a corresponding value of Y, then Y is termed a function of X.

6. Expression of functions.

When treating generally of functional relations letters such as x and y are commonly employed to represent variable quantities. Thus, in the expression $y = x^2 + 3x$ if, when any value be assigned to x there is always a corresponding value of y, then y is said to be expressed as a function of x. Similar examples are:

$$y = \sqrt{x^3 + 5}$$
$$y = \log_{10}x$$
$$y = \sin x + \cos x.$$

It is usual, when dealing generally with functions in this way, to employ letters at the end of the alphabet to represent the variables; when x and y are so employed the independent variable is generally expressed by x and the dependent by y.

For constants, other than actual numbers, letters at the beginning or middle of the alphabet are usually selected.

Thus in the equation of the straight line in general form

$$y = mx + b,$$

x and y are variables, m and b are constants.

When expressing functions of angles, the Greek letters θ (*theta*) or ϕ (*phi*) as well as x are often employed to represent the angle.

7. General notation for functions.

When it is necessary to denote a function of x in general, without specifying the form of the function, the notation f(x) is employed. In this notation the letter "f" is used as being the first letter of "function", while the letter "x" or other letter which might be employed indicates the independent variables. Thus $f(\theta)$ would be a general method of indicating a function of "θ".

Other forms of this notation are $F(x)$, $\phi(x)$, $\psi(x)$.

A statement such as $f(x) = x^2 - 7x + 8$

or $f(\theta) = \sin^2\theta - \cos^2\theta$

defines the specific function of the variable concerned.

This convenient notation is employed when it is desired to indicate that in a particular function, which has been defined, a **numerical value is to be substituted.**

Thus if $f(x) = x^2 - 4x + 3$, $f(1)$ would stand for the numerical value of the function when "1" is substituted for x.

Thus

$$\begin{aligned} f(1) &= 1^2 - (4 \times 1) + 3 = 0. \\ f(2) &= 2^2 - (4 \times 2) + 3 = -1. \\ f(0) &= 0 - 0 + 3 = 3. \\ f(a) &= a^2 - 4a + 3. \\ f(a+h) &= (a+h)^2 - 4(a+h) + 3. \end{aligned}$$

Again, if

$$\begin{aligned} \phi(\theta) &= 2\sin\theta. \\ \phi\left(\frac{\pi}{2}\right) &= 2\sin\frac{\pi}{2} = 2. \\ \phi(0) &= 2\sin 0 = 0. \\ \phi\left(\frac{\pi}{3}\right) &= 2\sin\frac{\pi}{3} = 2 \times \frac{\sqrt{3}}{2} = \sqrt{3}. \end{aligned}$$

8. Notation for increases in functions.

If x be any variable, the symbol δx (sometimes Δx) is used to denote an increase in the value of x. A similar notation is employed for any other variable. The symbol " δ " is the Greek small " d ", and is pronounced " delta ". Contrary to the ordinary usage of Algebra, δx does not mean $(\delta \times x)$. The letters should not be separated. Thus " δx " means " an increment of x ".

In accordance with the definition of a function, if y be a function of x, and if x be increased by δx, then y will be increased in consequence and its increment will be denoted by δy.

Accordingly, if $y = f(x)$

then $y + \delta y = f(x + \delta x)$

whence $\delta y = f(x + \delta x) - f(x)$.

If for example $y = x^3 - 7x^2 + 8x$

and x receives the increment δx, y will receive the increment δy. Then

$$y + \delta y = (x + \delta x)^3 - 7(x + \delta x)^2 + 8(x + \delta x).$$

Again, if $s = ut + \frac{1}{2}at^2$

and t receive an increment δt, then s will receive the increment δs.

Then $s + \delta s = u(t + \delta t) + \frac{1}{2}a(t + \delta t)^2$,

Single letters are sometimes employed to denote increments instead of the above method. For example

Let $y = f(x)$.

Let x receive the increment h and k be the corresponding increment of y.

Then $y + k = f(x + h)$

whence $k = f(x + h) - f(x)$.

9. Graphic representation of functions.

Let $f(x)$ be a function of x.

Then by the definition of a function (§ 5), for every value assigned to x there is a corresponding value of $f(x)$. Thus by giving a series of values to x a corresponding set of values of $f(x)$ is obtained. If these pairs of values of x and $f(x)$ are plotted as shown in *Algebra*, § 108, a graphical representation of $f(x)$ may be drawn.

Consider the example of $f(x) = x^2$, or $y = x^2$.

Assigning to x the values 0, 1, 2, 3, -0, -1, -2, -3, . . . we obtain the corresponding values of $f(x^2)$ or y.

Thus
$$f(0) = 0$$
$$f(1) = 1, f(-1) = 1$$
$$f(2) = 4, f(-2) = 4$$
$$f(3) = 9, f(-3) = 9.$$

From these values we deduce the fact that $f(-a)$ has the same value as $f(a)$. Hence the curve must be symmetrical about the axis of y. It is a parabola (*Algebra*, § 108), and is shown in Fig. 1. At the points on the

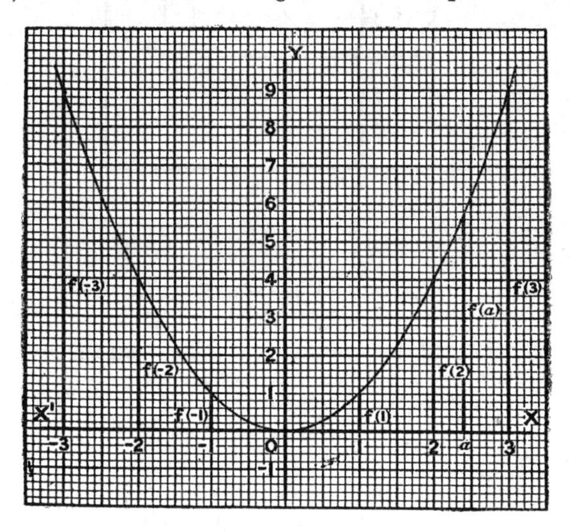

FIG. 1.—CURVE OF $f(x) = x^2$.

x-axis where $x = 1, 2, 3$ . . . the corresponding ordinates are drawn, the lengths of these represent $f(1), f(2), f(3)$ . . . and the ordinate drawn where $x = a$, represents $f(a)$.

In Fig. 2, which represents part of the curve of $f(x) = x^2$ or $y = x^2$,

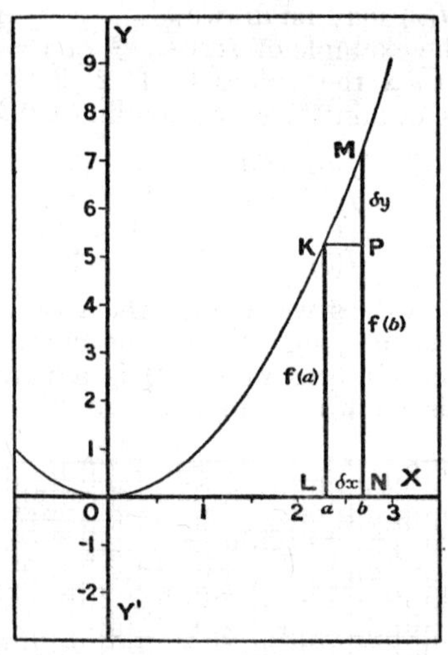

FIG. 2.

points L and N are taken on OX, so that

$$OL = a, \ ON = b.$$

Drawing the corresponding ordinates KL, MN, then

$$KL = f(a), \ MN = f(b).$$

In general, if L be any point on OX so that $OL = x$, let x be increased by LN where $LN = \delta x$.

MP represents the corresponding increase in $f(x)$ or y.

$$\therefore \quad MP = \delta y.$$

Since

$$KL = f(x)$$

$$MN = f(x + \delta x).$$

$$\therefore \quad MP = f(x + \delta x) - f(x)$$

or

$$\delta y = f(x + \delta x) - f(x).$$

10. Inverse functions.

Let $y = x^2$; then $x = \sqrt{y}$.

In the first equation y is expressed in terms of x and is a function of x. In the second x is expressed in terms of y—that is, as a function of y. The two functions—*i.e.*, $y = x^2$, and $x = \sqrt{y}$—are called **Inverse functions.** Similar examples will occur to the student, as for example:

If $y = a^x$, then $x = \log_a y$.
If $y = \sin x$, ,, $x = \sin^{-1} y$.

11. Implicit functions.

If an equation such as

$$x^2 - 2xy - 3y = 4$$

can be satisfied by values of x and y, but x and y are together on the same side of the equation, *i.e.*, y is not defined directly in terms of x, **y is said to be an implicit function of x.** In this particular case it is possible to solve for y in terms of x, giving $y = -\frac{4 - x^2}{2x + 3}$, which is an **explicit** function of y. But the solution is not always possible. Further examples of implicit functions are:

$$x^3 - 3x^2y + 5y^3 - 7 = 0$$
$$x \log y + y^2 = 4xy.$$

12. Functions of more than one variable.

We have been dealing with quantities which are functions of a single variable, but there are also quantities which are functions of two or more variables.

For example, the **area of a triangle** is a function of both base and height; the **volume of a fixed mass of gas** is a function of both pressure and temperature; the **volume of a rectangular-shaped room** is a function of three variables, the length, breadth and height of the room; the **resistance of a wire to electrical current** is a function of both the length of the wire and its sectional area.

In this book, however, we shall confine ourselves in the main to functions of a single variable.

Exercise I.

1. If $f(x) = 2x^2 - 4x + 1$, find the values of

$$f(1), f(0), f(2), f(-2), f(a), f(x + \delta x).$$

2. If $f(x) = (x - 1)(x + 5)$, find the values of

$$f(2), f(1), f(0), f(a + 1), f\left(\frac{1}{a}\right), f(-5).$$

3. If $f(\theta) = \cos\theta$, find the values of

$$f\left(\frac{\pi}{2}\right), f(0), f\left(\frac{\pi}{3}\right), f\left(\frac{\pi}{6}\right), f(\pi).$$

4. If $f(x) = x^2$, find the values of

$$f(3), f(3{\cdot}1), f(3{\cdot}01), f(3{\cdot}001).$$

Also find the value of $\dfrac{f(3{\cdot}001) - f(3)}{0{\cdot}001}$.

5. If $\phi(x) = 2^x$, find the values of $\phi(0)$, $\phi(1)$, $\phi(3)$, $\phi(0{\cdot}5)$.
6. If $F(x) = x^3 - 5x^2 - 3x + 7$, find the values of

$$F(0), F(1), F(2), F(-x).$$

7. If $f(t) = 3t^2 + 5t - 1$, find an expression for $f(t + \delta t)$.
8. If $f(x) = x^2 + 2x + 1$, find an expression for

$$f(x + \delta x) - f(x).$$

9. If $f(x) = x^3$, find expressions for:

(1) $f(x + \delta x)$.
(2) $f(x + \delta x) - f(x)$.
(3) $\dfrac{f(x + \delta x) - f(x)}{\delta x}$.

10. If $f(x) = 2x^2$, find expressions for:

(1) $f(x + h)$.
(2) $f(x + h) - f(x)$.
(3) $\dfrac{f(x + h) - f(x)}{h}$.

CHAPTER II

VARIATIONS IN FUNCTIONS. LIMITS

13. Variations in functions.

FROM the definition of a function we learn that when the independent variable changes in value the function changes its value in consequence. We now proceed to examine in a few examples **how** the function changes. We shall consider its variations as the independent variable changes through a range of numerical values. The graph of a function provides a revealing way of observing these changes.

As our first example we will consider the familiar function:

$$f(x) = x^2 \quad \text{or} \quad y = x^2$$

and refer to the graph as shown in Fig. 1. It shows within the limits of the values plotted how the function changes as x changes. In the conventional way **x is represented as increasing through the complete number scale** which is marked on the x axis OX (see *Algebra*, §§ 35, 36, 67). The values of the function x^2 are similarly shown on another complete number scale on the y axis (OY).

Remembering that the values of x are shown as **continously increasing from left to right,** we see, from examination of the curve, that

(1) **As x increases continuously through negative values to zero, values of y are positive and decrease to zero,** at the origin.

(2) **As x increases through positive values, y also increases and is positive.**

(3) **At the origin y ceases to decrease and begins to increase.** This is called a **turning point** on the curve.

(4) **If x be increased without limit, y will also increase without limit.** For values of x which are negative, but numerically very great, y is also very great and positive.

14. Variations in the function $y = \frac{1}{x}$.

In considering this function we recall the effect on a fraction of changes in the value of the denominator. It is seen that if the numerator of a fraction remains constant:

(1) **When the denominator increases, the fraction decreases.**
(2) **When the denominator decreases, the fraction increases.**

Thus in the function $y = \frac{1}{x}$:

(1) If x is very large, say, 10^{10}, y is a very small number.
(2) If $x = (10^{10})^{10}$, $y = \frac{1}{(10^{10})^{10}}$, an exceedingly small number.

These numbers, both very large and very small, are numbers which can be specified in arithmetical form. They are **finite** numbers.

If, however, we conceive of x being increased so that it is greater than any number which can be specified or expressed in arithmetical form, then we speak of it as being **increased without limit.** It is said to **approach infinity,** and is expressed by the symbol ∞.

This is not a number with which we can operate. Multiplication or division of it by any finite number leaves it still infinite.

It is evident from the above reasoning that when x becomes infinitely large the function $\frac{1}{x}$, which can now be represented by $\frac{1}{\infty}$, becomes an infinitely small magnitude, smaller than any finite number which can be specified or represented in arithmetical terms.

This is denoted by zero—*i.e.*, 0.

We must therefore in this connection conceive of zero, not as a number, but as an infinitely small magnitude. Multiplication or division by any finite number does not alter it; it remains zero.

If, however, a finite number be divided by zero—*e.g.*, the

above function becomes $\frac{1}{0}$—then by the converse of the above reasoning, the result will be infinitely large.

These conclusions can be expressed as follows, using the notation employed in Algebra (*Algebra*, § 201).

When $x \longrightarrow \infty, \quad \frac{1}{x} \longrightarrow 0$

" $x \longrightarrow 0, \quad \frac{1}{x} \longrightarrow \infty.$

It may be noted that the same conclusions will be reached if the numerator is any finite number—*e.g.*, $\frac{a}{x}$.

The above conclusions can be illustrated by drawing the graph of $y = \frac{1}{x}$.

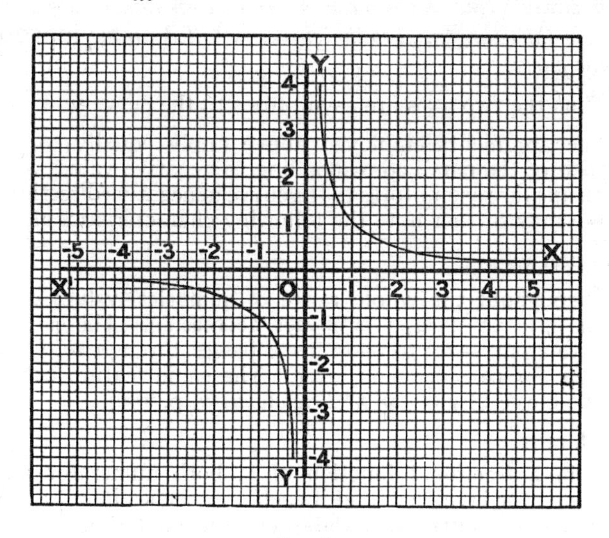

Fig. 3.

Plotting the curve from the usual table of values, *Algebra*, § 173, we obtain the curve shown in Fig. 3.

The curve is known as a hyperbola, and consists of two

branches of the same shape, corresponding to positive and negative values of x.

Considering the positive branch, we note the graphical expression of the conclusions reached above.

(1) As x increases, y decreases and the curve approaches the x-axis. Clearly as x approaches infinity, the distance between the curve and OX becomes infinitely small and the curve approaches coincidence with OX at an infinite distance. In geometrical terms the x-axis is tangential to the curve at infinity.

(2) For values between 0 and 1 it will be noted that the curve is approaching coincidence with OY at an infinite distance—*i.e.*, the y-axis is also tangential to the curve at infinity.

A straight line which meets a curve at an infinite distance, and is thus tangential to the curve, is called an **asymptote** *to the curve.*

Thus the two axes are asymptotes to the curve $y = \frac{1}{x}$.

The arguments employed above apply equally to the branch of the curve corresponding to negative values of x. Both axes are asymptotes to the curve in negative directions.

We may further note the following characteristics of the function $y = \frac{1}{x}$.

Throughout the whole range of numerical values of x, from $-\infty$ to $+\infty$, y is always decreasing. The sudden change from $-\infty$ to $+\infty$ as x passes through zero is a matter for consideration later. The same feature occurs in the curve of $y = \tan x$ (*Trigonometry*, p. 160).

15. Limits.

If in a fractional function of x, both numerator and denominator involve x, and if each approaches infinity as x approaches infinity, then the fraction ultimately takes the form $\frac{\infty}{\infty}$.

For example, if $$f(x) = \frac{2x}{x+1},$$

both numerator and denominator become infinite when x becomes infinite. The question then arises can any meaning be given to the fraction when it assumes the form $\frac{\infty}{\infty}$? In this case a meaning can be found as follows.

Dividing both numerator and denominator by x

$$f(x) = \frac{2x}{x+1}$$
$$= \frac{2}{1 + \frac{1}{x}}.$$

If now $\quad x \longrightarrow \infty, \quad$ then $\frac{1}{x} \longrightarrow 0.$

Consequently in the limit the fraction approaches $\frac{2}{1+0}$ or 2, but clearly it cannot exceed this number—*i.e.*,

$\frac{2x}{x+1}$ approaches the limiting value 2 as x approaches infinity.

Thus 2 is said to be the **limit** which $\frac{2x}{x+1}$ approaches as x approaches infinity; it is called the **limiting value,** or the **limit of the function.**

The following notation is employed to denote a "limit" of a function:

$$\underset{x \to \infty}{Lt} \frac{2x}{x+1} = 2.$$

The value towards which x approaches when a limit is approached is indicated by $x \longrightarrow \infty$, placed beneath *Lt*.

The idea of a limit is one of very great importance not only in the Differential Calculus, but in all advanced forms of mathematics.

16. Limit of a function of the form $\frac{0}{0}$.

Let us examine the function

$$f(x) = \frac{x^2 - 4}{x - 2}.$$

The value of this function for any value of x is readily found. But if the value assigned to x is 2, both numerator and denominator become zero, and the fraction takes the form of $\frac{0}{0}$. This form is said to be indeterminate, and it would be a mistake to suppose that its value is 0.

The form $\frac{0}{0}$ is of great importance, and we must carefully investigate it further.

Let us begin by assigning to x a number of values which are slightly greater or slightly less than that which produces the indeterminate form—viz., 2:

(1) Let $x = 2{\cdot}1$.

$$\text{Then}\quad \frac{x^2 - 4}{x - 2} = \frac{4{\cdot}41 - 4}{2{\cdot}1 - 2} = \frac{0{\cdot}41}{0{\cdot}1} = 4{\cdot}1.$$

(2) Let $x = 2{\cdot}01$.

$$\text{Then}\quad \frac{x^2 - 4}{x - 2} = \frac{4{\cdot}0401 - 4}{2{\cdot}01 - 2} = \frac{0{\cdot}0401}{0{\cdot}01} = 4{\cdot}01.$$

(3) Let $x = 2{\cdot}001$.

$$\text{Then}\quad \frac{x^2 - 4}{x - 2} = \frac{4{\cdot}004001 - 4}{2{\cdot}001 - 2} = \frac{0{\cdot}004001}{0{\cdot}001} = 4{\cdot}001.$$

Or, taking values less than 2:

(4) Let $x = 1{\cdot}9$.

$$\text{Then}\quad \frac{x^2 - 4}{x - 2} = \frac{3{\cdot}61 - 4}{1{\cdot}9 - 2} = \frac{-0{\cdot}39}{-0{\cdot}1} = 3{\cdot}9.$$

(5) Let $x = 1{\cdot}99$.

$$\text{Then}\quad \frac{x^2 - 4}{x - 2} = \frac{3{\cdot}9601 - 4}{1{\cdot}99 - 2} = \frac{-0{\cdot}0399}{-0{\cdot}01} = 3{\cdot}99.$$

A comparison of these results leads to the conclusion that, **as the value of x approaches 2 the value of the fraction approaches 4**, and that ultimately when the value of x differs from 2 by an infinitely small number, the value of the fraction also differs from 4 by an infinitely small number. This might be expressed in the form employed previously—viz.:

$$\text{as}\qquad x \longrightarrow 2,\ \frac{x^2 - 4}{x - 2} \longrightarrow 4.$$

It will thus be seen that the function $\frac{x^2 - 4}{x - 2}$ has a limiting value as x approaches 2, or with the notation for limits.

$$\underset{x \to 2}{Lt} \frac{x^2 - 4}{x - 2} = 4.$$

17. Let us next investigate the problem in a more general form, taking as our example the fraction $\frac{x^2 - a^2}{x - a}$ and find its value when $x = a$, for which value of x the fraction takes the form $\frac{0}{0}$.

Following the method employed above, but in a general form:

Let $$x = a + h,$$

i.e., h is the variable amount by which x differs from a for any assigned value of x.

Substituting in the fraction

$$\frac{x^2 - a^2}{x - a} = \frac{(a + h)^2 - a^2}{(a + h) - a}$$

$$= \frac{2ah + h^2}{h}.$$

Dividing numerator and denominator by h which is not zero,

$$\frac{x^2 - a^2}{x - a} = 2a + h.$$

As h decreases, x approaches in value to a, or when x approaches infinitely near in value to a, h approaches zero.

then $$2a + h \text{ approaches } 2a,$$

i.e., as x approaches in value to a, $\frac{x^2 - a^2}{x - a}$ approaches $2a$.

Or, using the symbols previously employed,

when $$x \longrightarrow a,\ h \longrightarrow 0,$$

and $$\frac{x^2 - a^2}{x - a} \longrightarrow 2a,$$

i.e., $2a$ is the limiting value of the function.

With the notation employed above:

$$\underset{x \to a}{Lt} \frac{x^2 - a^2}{x - a} = 2a.$$

It will be seen, therefore, that the expression $\frac{0}{0}$, as used in the above examples, can be regarded as representing the ratio of two infinitely small magnitudes. The value of this ratio approaches a finite limit as the numerator and denominator approach zero.

18. Limit of a series.

In the foregoing Sections we have considered a simple example of the limit of a function. But the student will have learned from *Algebra* that the term "*limit*" is also applied in certain cases to the sum of a series. In a Geometrical Progression, if the common ratio is a proper fraction, the sum of the terms of the series, as the number of them becomes great, approaches a finite number, which is called the limit of the sum. A more detailed examination of this will be found in *Algebra*, §§ 201–205. In this chapter, however, we will confine ourselves only to the expression for this limit as deduced from the general formula for the sum of n terms.

If a be the first term of the series,
n be the number of terms,
r be the common ratio,
S_n be the sum of n terms,

then
$$S_n = \frac{a(1 - r^n)}{1 - r}$$

or
$$S_n = \frac{a}{1 - r} - \frac{ar^n}{1 - r} \quad . \quad . \quad . \quad . \quad . \quad . \quad \text{(A)}$$

If r be a proper fraction, the value of r^n decreases as n increases. Using the notation employed above

as
$$n \longrightarrow \infty, \ r^n \longrightarrow 0$$
and
$$ar^n \longrightarrow 0.$$

Hence
$$\underset{n \to \infty}{Lt} \left(\frac{ar^n}{1 - r}\right) = 0.$$

Consequently it is evident from (A) that S_n approaches $\frac{a}{1 - r}$ as a limit as n becomes infinitely great.

Thus $\frac{a}{1-r}$ becomes the **limit of the series** as n becomes infinitely great, and is called the **sum to infinity.**

If r is numerically greater than unity, the magnitude of the terms increases as n increases; and if n approaches infinity, so also does the sum.

As the student extends his knowledge of Mathematics he will be concerned with many series of different kinds and he will find that it is important to know the following about the sum of n terms, when n becomes infinitely great.

(1) Does it approach a finite limit?
or (2) Does it become infinite?

If the sum of the series approaches a finite limit it is called **Convergent,** but if its sum becomes infinite it is called **Divergent.**

With a limited number of exceptions most series are either Convergent or Divergent, and we will return to the matter in Chap XIX.

19. A trigonometrical limit, $\underset{\theta \to 0}{Lt} \frac{\sin \theta}{\theta} = 1.$

Note.—Throughout this volume it will be assumed, unless specified to the contrary, that angles are measured in **radians**—*i.e.*, in circular measure.

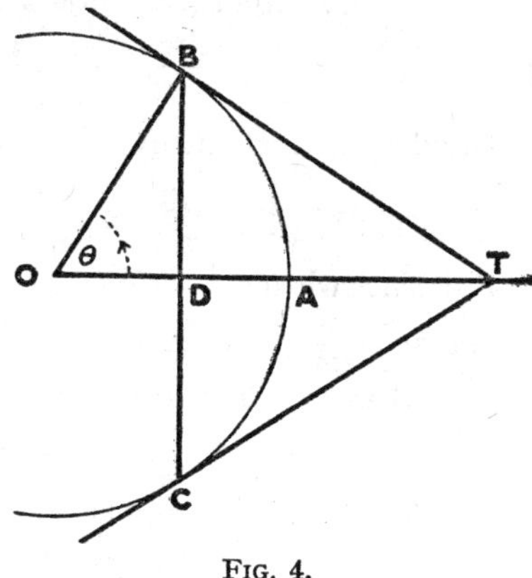

Fig. 4.

It is clear that as θ becomes very small, so also does sin θ, so that ultimately when θ and consequently sin θ approach zero, the ratio $\frac{\sin \theta}{\theta}$ approaches the form $\frac{0}{0}$. The limit of this ratio can be found as follows. In Fig. 4 let O be the centre of a circle of unit radius. Let BAC be an arc of this circle and BC its chord. Let OA be the radius which bisects BC at right angles and consequently bisects the arc BC. From B and C, draw BT, CT tangents to the circle. They will meet on OA produced.

Let $\angle AOB$ be θ radians.

Then $\qquad TB + TC > \text{arc } BAC,$

also $\qquad \text{arc } BAC > \text{chord } BC.$

Considering halves of these

$$BT > \text{arc } BA > BD \qquad . \quad . \quad . \quad \text{(A)}$$

Now, $\tan \theta = \dfrac{BT}{OB} = BT$, since OB is of unit length.

similarly $\theta = \dfrac{\text{arc } BA}{OB} = \text{arc } BA$, since OB is of unit length.

and $\sin \theta = \dfrac{BD}{OB} = BD$, since OB is of unit length.

$$\therefore \text{ from (A), } \tan \theta > \theta > \sin \theta$$

or

$$\frac{\sin \theta}{\cos \theta} > \theta > \sin \theta$$

Dividing throughout by $\sin \theta$.

$$\therefore \quad \frac{1}{\cos \theta} > \frac{\theta}{\sin \theta} > 1.$$

But, since when $\theta \longrightarrow 0$, $\cos \theta \longrightarrow 1$, and $\therefore \dfrac{1}{\cos \theta} \longrightarrow 1$

and as $\dfrac{\theta}{\sin \theta}$ always lies between $\dfrac{1}{\cos \theta}$ and 1

$\therefore$ when $\theta \longrightarrow 0$, and $\dfrac{1}{\cos \theta} \longrightarrow 1$

$$\frac{\theta}{\sin \theta} \longrightarrow 1,$$

i.e., as $\theta \longrightarrow 0$, $\dfrac{\sin \theta}{\theta}$ approaches unity as a limit,

or

$$\underset{\theta \to 0}{Lt} \frac{\sin \theta}{\theta} = 1.$$

It is left as an exercise to the student to prove, using the above, that as $\theta \longrightarrow 0$, $\dfrac{\tan \theta}{\theta}$ approaches unity as a limit.

20. A geometrical illustration of a limit.

Let OAB be a circle.

Let OB be a chord intersecting the circumference at O and B.

Suppose the chord OB to rotate in a clockwise direction about O. The point of intersection B will move along the circumference towards O. Consequently the arc OB and chord OB decrease.

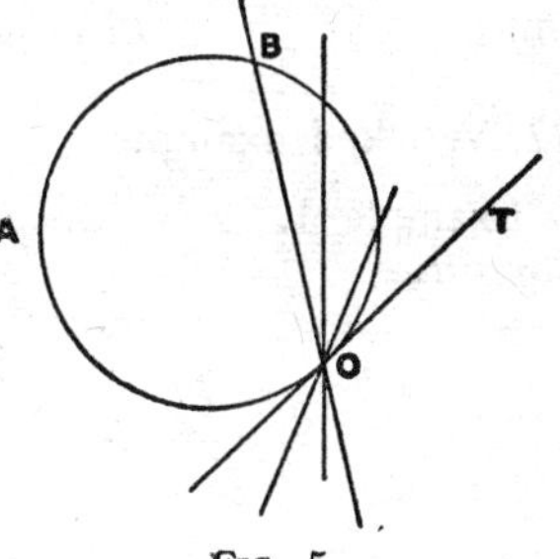

FIG. 5.

Let the rotation continue until B is infinitely close to O and the chord and arc become infinitely small.

It can be conceived that in the limiting position when B moves to coincidence with O—*i.e.*, the two points of intersection coincide—the straight line does not cut the circumference in a second point. **Therefore in the limiting position the chord becomes a tangent to the circle at O.**

21. Theorems on limits.

We now state, without proofs, four theorems on limits, to which reference will be made later.

[This can be omitted, if desired, on a first reading.]

(1) *If two variables are always equal, their limits are equal.*

(2) **Limit of a sum.**

The limit of the sum of any number of functions is equal to the sum of the limits of the separate functions.

Let u and v be functions of the same variable x.

Then $Lt(u + v) = Lt(u) + Lt(v)$.

(3) **Limit of a product.**

The limit of the product of any number of functions is equal to the product of the limits of the separate functions.

u and v standing for functions as above

$$Lt(u \times v) = Lt(u) \times Lt(v).$$

(4) **Limit of a quotient.**

The limit of the quotient of the functions is equal to the quotient of the limits of the functions provided that the limit of the divisor is not zero.

Thus $$Lt\left(\frac{u}{v}\right) = Lt(u) \div Lt(v)$$

unless $$Lt(v) = 0.$$

22. Worked examples.

Example 1. *Find the limit of* $\frac{x^2 + 3x}{2x^2 - 5}$ *when* x *becomes infinite.*

$$\underset{x \to \infty}{Lt} \frac{x^2 + 3x}{2x^2 - 5} = \underset{x \to \infty}{Lt} \frac{1 + \frac{3}{x}}{2 - \frac{5}{x^2}}$$

$$= Lt\left\{1 + \underset{x \to \infty}{Lt} \frac{3}{x}\right\} \div Lt\left\{2 - \underset{x \to \infty}{Lt} \frac{5}{x^2}\right\} \quad \text{(Th. 4)}$$

$$= \frac{1 + 0}{2 - 0} = \frac{1}{2}.$$

Example 2. *Find the value of* $\underset{x \to a}{Lt} \frac{x^n - a^n}{x - a}$.

When $x = a$, the function is of the form $\frac{0}{0}$, and therefore indeterminate.

Let $x = a + h$, where h is small.

Then $$\frac{x^n - a^n}{x - a} = \frac{(a + h)^n - a^n}{(a + h) - a}.$$

Expanding $(a + h)^n$ by the Binomial Theorem (*Algebra*, p. 281)

$$\frac{x^n - a^n}{x - a} = \frac{\left\{a^n + na^{n-1}h + \frac{n(n-1)}{\lfloor 2} a^{n-2}h^2 + \ldots\right\} - a^n}{h}$$

$$= \left\{na^{n-1} + \frac{n(n-1)}{\lfloor 2} a^{n-2}h + \ldots\right\}$$

But since $x = a + h$

when $x \longrightarrow a,\ h \longrightarrow 0.$

$\therefore$ Limit becomes

$$\underset{x \to a}{Lt} \frac{x^n - a^n}{x - a} = \underset{h \to 0}{Lt} \left\{ na^{n-1} + \frac{n(n-1)}{\underline{|2}} a^{n-2}h + \ldots \right\}$$

$$= na^{n-1}.$$

since all other terms have a power of h as a factor and therefore vanish when $h \longrightarrow 0$.

Example 3. *Find the limit of* $\dfrac{x-3}{\sqrt{x-2} - \sqrt{4-x}}$ *when* $x = 3$.

Both numerator and denominator vanish when $x = 3$.
Then the function takes the form $\frac{0}{0}$.
Rationalising the denominator (*Algebra*, p. 252),

$$\frac{x-3}{\sqrt{x-2} - \sqrt{4-x}} = \frac{(x-3)\{\sqrt{x-2} + \sqrt{4-x}\}}{(x-2) - (4-x)}$$

$$= \frac{(x-3)\{\sqrt{x-2} + \sqrt{4-x}\}}{2x - 6}$$

$$= \frac{\sqrt{x-2} + \sqrt{4-x}}{2}$$

and limit when $x = 3$ becomes

$$\frac{\sqrt{1} + \sqrt{1}}{2} = 1.$$

Exercise 2.

1. (*a*) What number does the function $\dfrac{1}{x-1}$ approach as x becomes infinitely large?
   (*b*) For what values of x is the function negative?
   (*c*) What are values of the function when the values of x are 2, 1·8, 1·5, 1·2, 1·1, 0·5, 0, -1, -2?
   (*d*) What limit is approached by the function as x approaches unity?
   (*e*) Using the values of the function found in (*c*) draw its curve.

2. (*a*) Find the values of the function $\frac{3x + 1}{x}$ when x has the values 10, 100, 1000, 1,000,000.
   (*b*) What limit does the function approach as x becomes very great?
   Find the limit of the function by using the method of § 17.
3. (*a*) Find $\underset{x \to \infty}{Lt} \frac{5x + 2}{x - 1}$.
   (*b*) Find the limit of the function as x approaches $+ 1$.
4. (*a*) Find the values of the function $\frac{x^2 - 1}{x - 1}$ when x has the values 10, 4, 2, 1·5, 1·1, 1·01.
   (*b*) Find the limit of $\frac{x^2 - 1}{x - 1}$ as x approaches unity.
5. Find the limit of the function $\frac{x^2 + 1}{x^2 - 1}$ as x approaches infinity.
6. Find $\underset{x \to 2}{Lt} \frac{x^2 - 4}{x^2 - 2x}$.
7. Find the limit of $\frac{(x + h)^3 - x^3}{h}$ as $h \to 0$.
8. Find the limit of the function $\frac{x}{2x + 1}$ as x approaches ∞.
9. Find the limit $\underset{x \to \infty}{Lt} \frac{4x^2 + x - 1}{3x^2 + 2x + 1}$.
10. Show from the proof given in § 19 that $\underset{\theta \to 0}{Lt} \frac{\tan \theta}{\theta} = 1$.

CHAPTER III

RATE OF CHANGE OF A FUNCTION. GRADIENTS

23. Rate of change of a function.

We have seen that a function changes in value when the variable upon which it depends is changed. The important question which next arises is, how to determine the **rate of change**?

In the Calculus we are fundamentally concerned with the rate of variation of a function with respect to the change in the variable on which it depends.

We will illustrate the problems which arise by examining a few simple cases, and in doing so will make use of the graph of a function, since the graph makes visible the changes in the function.

24. Uniform motion.

When a body moves so that it covers **equal distances in equal intervals of time** it is said to move uniformly. The distance is a function of the time, and from the above definition the rate of change of the function must be constant. This will appear in what follows.

Let s be the distance moved, and
t be the time taken.

Then it is shown in books on Mechanics that

$$s = vt$$

where v, the velocity, is a constant, and is the distance moved in each second.

The ratio of the two variables—viz. $\frac{s}{t}$—is constant for all corresponding values of them.

Consider the following graphical example:

A motor car travels distances in times as shown in the following table:

Time (t) (*in s*) . .	1	2	3	4	5
Distance (s) (*in m*) . .	20	40	60	80	100

These quantities are reckoned from a fixed point in the motion.

Plotting these points and joining them, they are seen to lie on a straight line. This is shown in Fig. 6, which represents the graph.

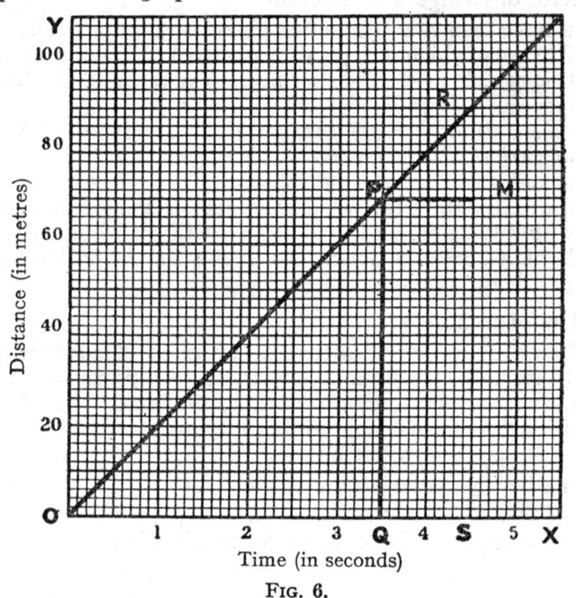

FIG. 6.

Let OQ, OS, represent two intervals of time (t). Then PQ, RS represent the corresponding distances (s). From the above general statement it follows that

$$\frac{PQ}{OQ} = \frac{RS}{OS}.$$

This is true for any positions of P and R, and therefore the graph must be a straight line.

Let θ be the angle made by this line with OX—*i.e.*, $\angle POQ$.

Then $$\frac{PQ}{OQ} = \frac{RS}{OS} = \tan \theta.$$

$\therefore \frac{PQ}{OQ}$ represents the gradient of the line (*Algebra*, § 72).

Let PM be drawn parallel to OX.

Then, between the time intervals represented by OQ and OS

PM represent the increase in time. Let this be δt.

RM represent the increase in distance. Let this be δs.

$$\therefore \text{ ratio of } \frac{\text{increase in distance}}{\text{increase in time}} = \frac{\delta s}{\delta t}$$
$$= \tan \theta$$
$$= \text{gradient of the line.}$$

Hence—for any corresponding values of s and t the ratio of increase of the distance with respect to the increase in time is constant and equal to the gradient of the line.

In the example above of uniform motion, this gradient is seen to be 20 ms^{-1}. This is the velocity of the car.

25. Gradient of a linear function.

Generalising the above:

Let y be a function of x. The straight line representing the function may be of two forms:

(1) The function $y = mx$.

The graph is a straight line passing through the origin. Comparing with the above example, if δy and δx be increments of y and x, $\frac{\delta y}{\delta x}$ is a constant and represents the gradient of the line. But this is represented by m (*Trigonometry*, § 67).

$$\therefore \quad \frac{\delta y}{\delta x} = m.$$

$\therefore$ m represents the rate of increase of y with respect to x.

(2) The function $y = mx + b$.

This straight line does not pass through the origin, but has an intercept b on the y axis.

In Fig. 7 let CPQ be the line whose equation is $y = mx + b$.

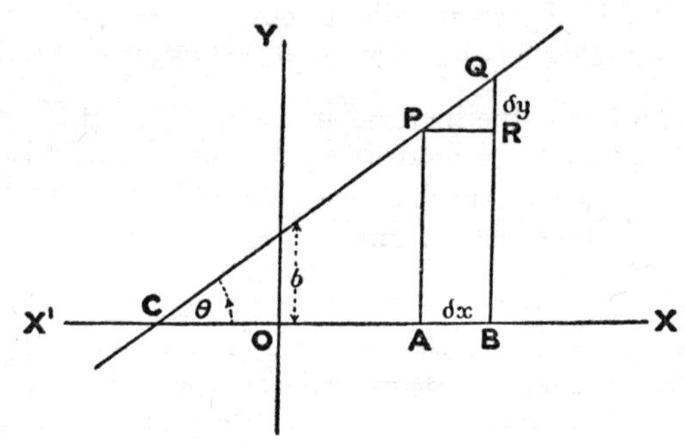

FIG. 7.

Let θ be the angle made with OX.

Let P be any point on the line, its co-ordinates being (x, y).

Then $OA = x$, $PA = y$.

Let x be increased by δx from OA to OB.

Let y be increased correspondingly by δy, from AP to BQ.

Drawing PR parallel to OX, $QR = \delta y$.

$\therefore$ co-ordinates of Q are

$$(x + \delta x, y + \delta y).$$

i.e., $$OB = x + \delta x, \quad QB = y + \delta y.$$

Substituting their values in the equation,

$$y = mx + b \quad \ldots \ldots \quad (1)$$
$$y + \delta y = m(x + \delta x) + b \quad \ldots \quad (2)$$

Subtracting (1) from (2)

$$\delta y = m(\delta x)$$

$$\therefore \quad m = \frac{\delta y}{\delta x} = \tan QPR = \tan \theta$$

i.e., $\frac{\delta y}{\delta x}$ represents the gradient of the line.

$\therefore$ the ratio of the increase of y to the increase of x is equal to the gradient of the line and is constant for all points on the line.

It will be clear that the addition of the **constant b** to the right-hand side of the equation does not affect the gradient. In both $y = mx$ and $y = mx + b$, the gradient is m, and for any given value of m the lines are parallel (*Algebra*, § 74).

26. Meaning of a negative gradient.

The angle which a straight line makes with the x-axis is always measured in an anti-clockwise direction. When this angle is greater than a right angle, as is the case of the angle θ made by the straight line CD in Fig. 8, **its tangent is negative** (*Trigonometry*, § 69).

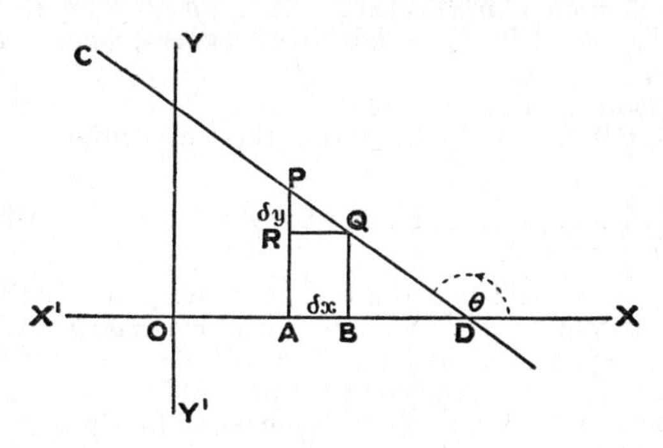

FIG. 8.

$\therefore$ **the gradient of the line is negative.**

Let P be the point (x, y), so that $OA = x$ and $PA = y$.

Let x be increased by δx to OB.

The value of the corresponding ordinate is represented by QB. Draw QR parallel to OX, *i.e.*, the ordinate PA is **decreased** by δy to QB.

Thus **while x is increased by δx, y is decreased by δy**, or, as we might express it, there is **negative** increase.

$\therefore \dfrac{\delta y}{\delta x}$ **is negative**—*i.e.*, **tan θ is negative.**

$\therefore$ **the rate of increase of y with respect to x is now negative.**

Summarising this result with the foregoing we conclude :

(1) **When y increases as x increases, the gradient is positive.**

(2) **When y decreases as x increases, the gradient is negative.**

27. Gradient of a curve.

The straight line, representing the graph of a function of the first degree, is the only graph in which the gradient is constant—*i.e.*, the same at all points on the line.

If the graph is a curve, the gradient is different at different points on the curve. It is not obvious, therefore, what is meant by the gradient of a curve, since it is continuously changing, or what is the meaning of the gradient at a point on a curve. It is necessary, therefore, to spend some little time in investigating these difficulties.

28. Graph of the motion of a body moving with uniformly increasing velocity.

In § 24 we said that if a body is moving with **uniform**—*i.e.*, **constant**—velocity, the graph which connects distance and time is a straight line. We will now consider a body moving with **uniformly increasing velocity**—*i.e.*, in equal intervals of time its velocity is increased by equal amounts. In such a case it is clear that in equal intervals of time the distances passed over are not equal. As the velocity increases, the distances passed over will also increase. The greater the velocity, the greater the distance moved. As an example we will consider the case of a falling body, in which it is clear that the velocity increases with time. The following table gives the distances passed over in successive intervals of time from rest by a body falling freely, and taking very rough approximations to the actual values.

Time (t) in s . .	0	0·4	0·8	1·2	1·6	2·0
Distance (s) in m .	0	0·8	3·2	7·2	12·8	20·0

When the corresponding values of distance and time are plotted, the graph is seen to be a smooth curve, as shown in Fig. 9. Clearly the curve slopes more and more steeply as time increases—*i.e.*, the **ratio of distance to time, or**

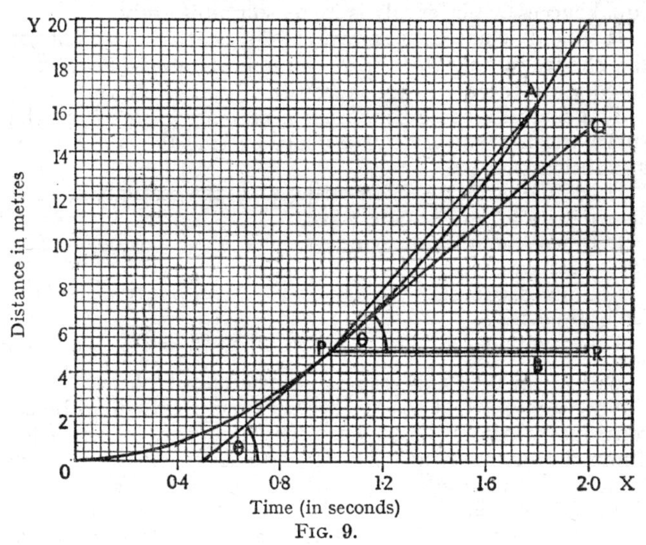

FIG. 9.

velocity, is increasing. The smooth curve indicates that this increase of velocity is uniform. Let us consider the ratio of increase of distance to increase of time over five successive intervals, as shown in the following table:

Time interval (in s) .	0 to 0·4	0·4 to 0·8	0·8 to 1·2	1·2 to 1·6	1·6 to 2·0
Distance (in m) . .	0·8	2·4	4·0	5·6	7·2
$\frac{\text{Distance}}{\text{Time}}$ . . .	2	6	10	14	18

These ratios represent the average velocities for the

corresponding intervals. They are the distances which would be passed over during the intervals, **if the body were moving with uniform velocities equal to these average velocities.** It is evident that the average velocity over equal successive intervals is increasing uniformly.

It should be noted, as shown in § 24, that the **gradients of the chords** joining the appropriate points on the curve will be equal to these average velocities.

To generalise these conclusions, take any point P on the curve and through it draw a chord cutting the curve again in another point A.

Draw the ordinate AB meeting at B the straight line PB drawn parallel to the time axis.

Let increase in **time** between the two positions be δt—*i.e.*, $PB = \delta t$.

Let increase in **distance** between the two positions be δs—*i.e.*, $AB = \delta s$.

Then **average velocity over the interval** $= \frac{\delta s}{\delta t}$.

This is equal to the gradient of the chord PA.

Now suppose that the interval of time, represented by δt, continually diminishes. Then the distance δs will also diminish, but their ratio continues to represent the average velocity during the interval and also the gradient of the chord PA, which also diminishes.

Imagine now the interval of time to become infinitely small. The interval of distance will also become infinitely small. In the limit, when A is infinitely close to P—*i.e.*, coincides—the ratio of $\frac{\delta s}{\delta t}$ approaches a finite limit, and the chord becomes a **tangent at P** (see § 20).

The limit which $\frac{\delta s}{\delta t}$ approaches will be the gradient of this tangent, and also the **velocity at P.**

Hence the term **velocity at a point is the limit of the ratio $\frac{\delta s}{\delta t}$ when these each become infinitely small.** It is also the **gradient of the curve** at the point P.

Thus the gradient of the curve at any point on the curve is equal to the gradient of the tangent to the curve drawn at that point.

In Fig. 9 draw PQ, tangent to the curve, at P.

Draw PR of unit length parallel to OX, and from R draw RS perpendicular to PR and meeting PQ in S.

$\angle QPR = \theta =$ angle made by PQ with OX.

$$\text{Gradient of } PQ = \frac{SR}{RP} = \frac{10}{1} = 10.$$

$\therefore$ velocity at the point P is 10 ms^{-1}.

i.e., the velocity at the end of one second is 10 ms^{-1}.

Students of mechanics will be able to verify that the answer should be about 9·8 ms^{-1}.

29. Gradients of the curve of $y = x^2$.

The methods employed above for obtaining the gradient at any point on a curve will now be employed to solve the problem more generally in the case of an algebraical function. The curve of $y = x^2$ has been chosen as a simple example, and one which is familiar to the student. A more general form of this function would be $y = ax^2$, but for simplicity we will take the case where $a = 1$. The methods adopted can be readily adapted for any value of a.

Fig. 10 represents the curve of $y = x^2$.

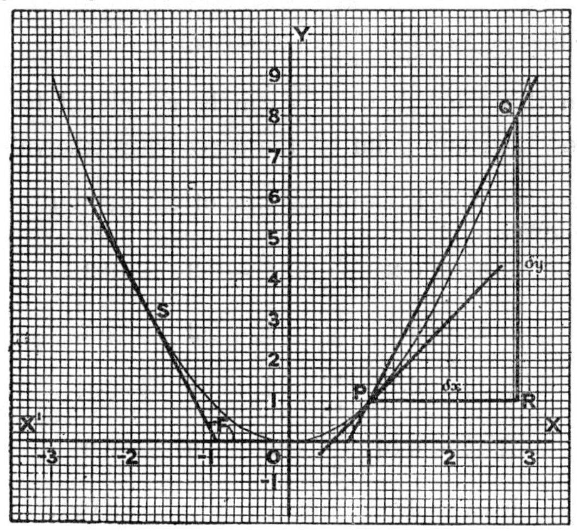

FIG. 10.

Let P be the point (1, 1).
Draw a chord PQ cutting the curve again in Q.
Draw PR parallel to OX to meet the ordinate from Q at R.
Let PR, the increase in x between P and Q, be δx.
Let QR, the corresponding increase in y, be δy.
Then gradient of the chord $PQ = \tan QPR$.

$$= \frac{\delta y}{\delta x}.$$

Also, $\frac{\delta y}{\delta x}$ is equal to the average rate of increase of y per unit increase of x between P and Q.

The algebraical expression for $\frac{\delta y}{\delta x}$ can be obtained as follows:

In the function $\quad y = x^2 \quad \ldots\ldots \quad (1)$

when x is increased by δx and y correspondingly increased by δy we get:

$$y + \delta y = (x + \delta x)^2 \quad \ldots\ldots \quad (2)$$

Subtracting (1) from (2)

$$\delta y = (x + \delta x)^2 - x^2$$

$$\therefore \quad \delta y = 2x\delta x + (\delta x)^2.$$

Dividing by δx $\quad \frac{\delta y}{\delta x} = 2x + \delta x.$

From this the value of $\frac{\delta y}{\delta x}$ can be calculated for any value of δx at any point on the curve where the value of x is known.

Thus when $x = 1$, as in the case of the point P on the curve above,

$$\text{If } \delta x = 0{\cdot}3, \quad \frac{\delta y}{\delta x} = 2 + 0{\cdot}3 = 2{\cdot}3.$$

$$\text{If } \delta x = 0{\cdot}2, \quad \frac{\delta y}{\delta x} = 2 + 0{\cdot}2 = 2{\cdot}2.$$

$$\text{If } \delta x = 0{\cdot}1, \quad \frac{\delta y}{\delta x} = 2 + 0{\cdot}1 = 2{\cdot}1.$$

$$\text{If } \delta x = 0{\cdot}01, \quad \frac{\delta y}{\delta x} = 2 + 0{\cdot}01 = 2{\cdot}01.$$

$$\text{If } \delta x = 0{\cdot}001, \quad \frac{\delta y}{\delta x} = 2 + 0{\cdot}001 = 2{\cdot}001.$$

These results exhibit the gradient of the chord PQ when δx diminishes and Q moves nearer to P. Then it is evident that the gradient of the chord approaches 2. We can therefore conclude that when Q moves to coincidence with P and the chord becomes the tangent to the curve at P,

the gradient of the tangent is 2.

We may also say that

The rate of increase of y per unit increase of x at the point P is 2.

Similar conclusions follow for any point on the curve, but the gradient of each tangent will depend on the value of x at the point; **the gradient is therefore a function of x.**

Thus at the point on the curve where $x = 3$, the gradient of the tangent will be 6.

It is evident that the conclusions reached hold for any curve whose equation is known. In general, therefore, we arrive at the following important conclusion:

The gradient at any point on a curve representing a function is equal to that of the tangent drawn to the curve at the point. It is also the rate of increase of the function for the value of x at the point.

30. Negative gradient.

In Fig. 10 let a point S be taken on the curve, corresponding to a negative value of x. Draw the tangent to the curve and produce it to meet the axis. The angle made with the axis is greater than a right angle, Consequently the **gradient is negative.** As was shown in § 26, this indicates that the rate of increase of the function is negative—*i.e.*, the function decreases. An examination of the curve shows that as x increases through negative values from $-\infty$ to 0, the function as represented by the curve is decreasing from $+\infty$ to 0 at the origin. At this point OX is tangential to the curve and the gradient of the curve is zero.

Exercise 3.

1. Draw the straight line $3x - 2y = 6$ and find its gradient. If P and Q are two points on the line such that the value of x at Q is greater by 0·8 than the value of x at P,

by how much is the value of y at Q greater than the value at P?

2. Find the gradients of the straight lines

(a) $\frac{x}{2} - \frac{y}{5} = 4.$

(b) $4x + 5y = 16.$

(c) $\frac{x}{a} + \frac{y}{b} = 1.$

3. The gradient of a straight line is 1·2. It passes through a point whose co-ordinates are (5, 10). What is the equation of the line?

4. The distance in metres passed over by a body falling from rest is given approximately by the formula $s = 4{\cdot}9t^2$. Representing an increase in time by δt and the corresponding increase in distance by δs, find by the method used in § 29 an expression for δs in terms of δt for any value of t. Hence find the value of $\frac{\delta s}{\delta t}$. Using this result find the average velocity for the following intervals:

(1) 2 s to 2·2 s (2) 2 s to 2·1 s
(3) 2 s to 2·01 s (4) 2 s to 2·001 s

From these results, deduce what the velocity at the end of 2 seconds appears to be.

5. In the curve of $y = x^2$, using the notation employed in § 29, find the value of $\frac{\delta y}{\delta x}$ as the value of x is increased from 3 to 3·1, 3·01, 3·001 and 3·0001 respectively. Deduce the gradient of the tangent to the curve at the point where $x = 3$.

6. Draw the curve of $y = x^3$ for values of x between 0 and 2.

Find an expression for δy in terms of x and δx.

Hence find an expression for $\frac{\delta y}{\delta x}$.

Taking the values of x as 2·1, 2·01, 2·001, and 2·0001, find the limit which $\frac{\delta y}{\delta x}$ approaches as the value of x approaches 2.

Hence find the gradient of the tangent to the curve at the

point where $x = 2$. Check by drawing the tangent to the curve at this point.

7. For the function $y = \frac{1}{x}$ (see Fig. 3) find an expression for δy in terms of x and δx. Hence find an expression for $\frac{\delta y}{\delta x}$.

Taking the values $x = 1{\cdot}1,\ 1{\cdot}01,\ 1{\cdot}001,\ 1{\cdot}0001$, find the limit which $\frac{\delta y}{\delta x}$ is approaching as x approaches unity. Hence find the gradient and angle of slope of the curve at the point where $x = 1$. Check your result by drawing the curve and constructing the tangent at this point.

8. Find the gradient of the tangent drawn at the point where $x = 1$ on each of the curves

(1) $y = x^2 + 2$
(2) $y = x^2 - 3$ (see *Algebra*, § 112)

9. Find the gradient of the tangent drawn at the point where $x = 2$ on each of the following curves:

(1) $y = 3x^2$
(2) $y = 2x^2 - 1$.

CHAPTER IV

DIFFERENTIAL COEFFICIENT. DIFFERENTIATION

31. Algebraical aspect of the rate of change of a function.

In this chapter we take a very important step forward in the development of our subject. It follows logically from the work of the preceding chapter. To make this clear we will briefly summarise the steps by which the subject has advanced. They are as follows:

(1) The **value of a function changes** as the variable changes upon which it depends.

(2) The **rate at which the function changes** is of great practical importance and it is necessary to be able to calculate it.

(3) The **rate of change (whether of increase or decrease) can be found geometrically** as follows:

(*a*) **When the function is of the first degree.** Such a function can be represented by a straight-line graph, and *the gradient of this straight line is equal to the rate of change of the function.*

If y is a function of x, and δx and δy are corresponding increases of x and y, **the gradient is equal to** $\frac{\delta y}{\delta x}$. This is **constant** throughout the line, *i.e.*, the **rate of change is uniform.**

(*b*) **When the function is not of the first degree** its graph will be a curve, and the rate of change of the function will differ in different parts of the curve. Its value at any point is equal to the gradient of the tangent at the point on the curve corresponding to any assigned value of x.

The geometrical method has many important applications, and is suggestive as an illustration, but in practice the gradient is not easily found by this method. For practical purposes, and for accuracy, an algebraic method is necessary.

The determination by algebraic methods in the case of the function $y = x^2$ has in effect been indicated in § 29, when, by means of arithmetic calculations, the values of $\frac{\delta y}{\delta x}$ were shown to be approaching nearer to a limit, as x approached an assigned value. For convenience the working is repeated.

$$\begin{aligned} &\text{Let} & y &= x^2. \\ &\text{Then} & y + \delta y &= (x + \delta x)^2. \\ &\text{Subtracting} & \delta y &= (x + \delta x)^2 - x^2 \\ & & &= 2x(\delta x) + (\delta x)^2. \end{aligned}$$

Dividing by δx, $\frac{\delta y}{\delta x} = 2x + \delta x$ . . . . . . (A)

We can now carry this a step further.

It has been shown geometrically that when δx approaches zero, the gradient of the chord, which represents the average rate of increase of the function over the interval represented by δx, gradually approaches the gradient of the tangent at a point corresponding to any assigned value of x.

Thus the gradient of the tangent, represented by the limit of $\frac{\delta y}{\delta x}$, is equal to the rate of increase of the function for the assigned value of x.

Since from (A) above, for any value of δx

$$\frac{\delta y}{\delta x} = 2x + \delta x$$

when $\delta x \longrightarrow 0$, $\frac{\delta y}{\delta x}$ approaches a limit and **the limit of**

$$\frac{\delta y}{\delta x} = 2x \quad . \quad . \quad . \quad . \quad . \quad . \quad \text{(B)}$$

***i.e.*, when $\delta x \longrightarrow 0$ the limit of $\frac{\delta y}{\delta x}$ represents the rate of increase of y with respect to x, for any assigned value of x.**

For example, when $x = 1$, limit of $\frac{\delta y}{\delta x} = 2$, *i.e.*, the rate of increase of y, or x^2 with respect to x is 2. (Cf. § 29.)

Similarly,

when $x = 2$, limit of $\frac{\delta y}{\delta x} = 4$. (Cf. Ex. 3, No. 8.)

when $x = 3$, limit of $\frac{\delta y}{\delta x} = 6$. (Cf. Ex. 3, No. 5.)

Using the notation of § 15, we may write (B) above as,

$$\underset{\delta x \to 0}{Lt} \frac{\delta y}{\delta x} = 2x.$$

A still more convenient notation is employed to represent this limit.

$$\underset{\delta x \to 0}{Lt} \frac{\delta y}{\delta x} \text{ is represented by } \frac{dy}{dx}$$

in which the English letter *d* is used instead of the Greek δ and the condition $\delta x \longrightarrow 0$ is understood.

Thus (B) becomes

$$\frac{dy}{dx} = 2x.$$

This limit is called the differential coefficient of the function with respect to x, the independent variable.

Thus, when $y = x^2$, $2x$ is the differential coefficient of y, or x^2, with respect to x.

A similar procedure will enable us to find the differential coefficient of any other function.

32. The Differential Coefficient.

Summarising the foregoing section we may conclude:

(1) *If* y *be a continuous function of* x, *and* δx *be any increase in the value of* x, *there will be a corresponding increase (or decrease) in the value, denoted by* δy.

(2) *The ratio* $\frac{\delta y}{\delta x}$ *represents the average rate of increase of* y *with respect to* x, *when* x *increased to* $x + \delta x$.

(3) *Since* y *is a continuous function of* x, *if* δx *becomes infinitely small, so also does* δy.

(4) *When* $\delta x \longrightarrow 0$, *the ratio* $\frac{\delta y}{\delta x}$ *in general tends to*

a finite limit, and this limit is called the differential coefficient of y with respect to x. *It is represented by the symbol* $\frac{dy}{dx}$,

i.e.,
$$\underset{\delta x \to 0}{Lt} \frac{\delta y}{\delta x} = \frac{dy}{dx}.$$

The Differential Calculus is fundamentally concerned with the variation of functions, and we can regard a **differential coefficient as a rate-measurer** in such variations. It measures the rate at which a function is changing its value compared with that of the variable upon which it depends.

Thus for the function $y = x^2$, since $\frac{dy}{dx} = 2x$, when $x = 4$, y, or x^2, is changing its value at 8 times the rate at which x is changing.

The differential coefficient $\frac{dy}{dx}$ is also called **a derivative** of y with respect to x, or the **derived function.**

Except in the case of a linear function, **the differential coefficient of y with respect to x is itself a function of x.**

Notation for the differential coefficient.

Besides the form $\frac{dy}{dx}$, the differential coefficient of y with respect to x may also be denoted by y'.

Thus if
$$y = x^2$$
$$y' = 2x.$$

In general, the differential coefficient of $y = f(x)$ may be denoted by $f'(x)$.

The same forms are used for other letters representing functions. Thus if s is a function of t, the differential coefficient of s with respect to t is written $\frac{ds}{dt}$.

33. Differentiation. Differentials.

The process of finding the differential coefficient or derivative of a function is called **Differentiation.**

The operation may be expressed by using the **operating**

symbol $\frac{d}{dx}$. Thus the differentiation of x^2 with respect to x can be written in the form $\frac{d(x^2)}{dx}$ or $\frac{d}{dx}(x^2)$.

In general, the differentiation of $f(x)$ with respect to x can be expressed by $\frac{d(f(x))}{dx}$ or $\frac{d}{dx}(f(x))$. It may also be denoted by the form D_xy or Dy when there is no doubt as to what is the independent variable.

Differentials.

The infinitely small increments of x and y which are implied in the form $\frac{dy}{dx}$ are called **differentials.** Thus $\frac{dy}{dx}$ represents the ratio of the differential of y to the differential of x.

In the example $\qquad y = x^2$

we have $\qquad \frac{dy}{dx} = 2x.$

This might be described by the statement that the ratio of the differential of y to the differential of x is equal to $2x$, or the differential of y is $2x$ times the differential of x. This could be expressed by the equation

$$dy = 2x \,.\, dx.$$

In this form $2x$ is shown as a **coefficient of the differential of** x, hence the term "differential coefficient".

The student should not at present regard $\frac{dy}{dx}$ as a fraction in which the numerator and denominator can be separated, but as a limit, as shown above.

General definition of a differential coefficient.

It will now be seen, from what has been stated above, that the general expression for the differential coefficient of any function, $f(x)$ is given by

$$\underset{\delta x \to 0}{\text{Lt}} \frac{f(x + \delta x) - f(x)}{\delta x}.$$

34. The Sign of the Differential Coefficient.

It has been shown above that the differential coefficient of a function is equal to the gradient of the tangent at a point on the curve which represents the function. It was also shown in § 26 that this gradient may be positive or negative. Consequently the differential coefficient may also be positive or negative. This will be examined further in a later chapter. For the moment the student is reminded of the conclusions stated in § 26 as to the sign of the gradient and the increase or decrease of the function. These conclusions apply also to the differential coefficient.

35. Differential coefficient of a constant.

Since a differential coefficient measures the rate of change of a variable, and a constant has no change whatever, the **differential coefficient of a constant must be zero.**

36. Differentiation of $y = mx + b$.

As the student has learnt previously, this is the general form of a function of the first degree. Its graph is a straight line (§ 25), and therefore of constant gradient. This can be shown algebraically from first principles as follows:

Let δx be an increment of x.
Let δy be the corresponding increment of y.

Substituting in $$y = mx + b$$

$$y + \delta y = m(x + \delta x) + b$$

subtracting $$\delta y = m(\delta x).$$

$$\therefore \quad \frac{\delta y}{\delta x} = m.$$

This is true for any value of δx with the corresponding value of δy, since m is a constant.

$$\therefore \quad \frac{dy}{dx} = m.$$

It will be noticed that the value of $\frac{dy}{dx}$ is independent of b. For different values of b the equation represents a series of parallel lines, having the gradient "m." See § 25.

37. Differentiation of $y = x^3$.

The following proof will provide another example of the general method which may be adopted for finding the differential coefficient of a function by first principles.

Let δx be an increment of x.
Let δy be the corresponding increment of y.
Substituting in

$$y = x^3 \quad . \quad . \quad . \quad . \quad . \quad . \quad . \quad . \quad . \quad . \quad (1)$$

$$y + \delta y = (x + \delta x)^3$$

$$= x^3 + 3x^2(\delta x) + 3x(\delta x)^2 + (\delta x)^3 \quad . \quad (2)$$

Subtracting (1) from (2)

$$\delta y = 3x^2(\delta x) + 3x(\delta x)^2 + (\delta x)^3.$$

Dividing by δx, which is not equal to zero, since it represents any increase of x

$$\frac{\delta y}{\delta x} = 3x^2 + 3x(\delta x) + (\delta x)^2.$$

Proceeding to the limiting value of $\frac{\delta y}{\delta x}$, when $\delta x \longrightarrow 0$ both $3x(\delta x)$ and $(\delta x)^2$ approach zero.

$$\underset{\delta x \longrightarrow 0}{Lt} \left(\frac{\delta y}{\delta x}\right) = 3x^2$$

i.e.

$$\frac{dy}{dx} = 3x^2.$$

38. Differentiation of $y = x^4$.

If the method of the foregoing section be applied to $y = x^4$, this would involve the expansion of $(x + \delta x)^4$, which is:

$$x^4 + 4x^3(\delta x) + 6x^2(\delta x)^2 + 4x(\delta x)^3 + (\delta x)^4.$$

After subtraction of x^4 and division by δx, there is left:

$$4x^3 + 6x^2(\delta x) + 4x(\delta x)^2 + (\delta x)^3.$$

On proceeding to the limit when $\delta x \longrightarrow 0$, every term after $4x^3$ vanishes, and we are left with $4x^3$ as the differential coefficient.

Any function of the form $y = x^n$ is dealt with in the same way, and it is evident that in the expansion of $(x + \delta x)^n$, the second term provides the differential coefficient.

For example, with $y = x^5$, $\frac{dy}{dx} = 5x^4$

,, $y = x^6$, $\frac{dy}{dx} = 6x^5$.

Generally when x *is a positive integer* it may be deduced that if

$$y = x^n$$
$$\frac{dy}{dx} = nx^{n-1}$$

A general proof of this follows.

Let $y = x^n$.

Let δx be an increment of x.

Let δy be the corresponding increment of y.

Substituting $\qquad y + \delta y = (x + \delta x)^n$.

Expanding the right-hand side by the Binomial Theorem (*Algebra*, p. 279).

$$y + \delta y = x^n + nx^{n-1}(\delta x) + \frac{n(n-1)}{\lfloor 2}x^{n-2}(\delta x)^2 + \frac{n(n-1)(n-2)}{\lfloor 3}x^{n-3}(\delta x)^3 + \dots$$

but $y = x^n$.

Subtracting

$$\delta y = nx^{n-1}(\delta x) + \frac{n(n-1)}{\lfloor 2}x^{n-2}(\delta x)^2 + \frac{n(n-1)(n-2)}{\lfloor 3}x^{n-3}(\delta x)^3 + \dots$$

Dividing throughout by δx.

$$\frac{\delta y}{\delta x} = nx^{n-1} + \frac{n(n-1)}{\lfloor 2}x^{n-2}(\delta x) + \frac{n(n-1)(n-2)}{\lfloor 3}x^{n-3}(\delta x)^2 \dots$$

Let $\delta x \longrightarrow 0$; then each term on the right-hand side after the first tends to zero.

$$\therefore \quad \underset{\delta x \to 0}{Lt}\left(\frac{\delta y}{\delta x}\right) = nx^{n-1},$$

or

$$\frac{dy}{dx} = nx^{n-1}.$$

The question now arises as to the values of n for which this result is true. Does it apply only to those cases when n is a positive integer? Evidently the validity of it depends on that of the Binomial Theorem. Does this hold when n is negative, or fractional? The question is briefly discussed in *Algebra*, p. 282. There it will be learned that, subject to certain numerical restrictions, which do not affect the above, the Theorem holds for all values of n.

The differential coefficient of $y = x^n$ can, however, be found by other methods, not involving the Binomial Theorem. If the student desires to study them, he should consult a larger treatise on the subject.

The conclusion therefore is that **for all values of n**

$$\frac{d(x^n)}{dx} = nx^{n-1}.$$

39. Differentiation of $y = ax^n$, where a is any constant. Compressing the proof given in § 38 we get the following:

$$y = ax^n$$

$$y + \delta y = a(x + \delta x)^n$$

$$= a\left\{x^n + nx^{n-1}\delta x + \frac{n(n-1)}{\lfloor 2}x^{n-2}(\delta x)^2 \ldots\right\}$$

Subtracting

$$\delta y = a\left\{nx^{n-1}(\delta x) + \frac{n(n-1)}{\lfloor 2}x^{n-2}(\delta x)^2 + \ldots\right\}$$

$$\therefore \quad \frac{\delta y}{\delta x} = a\left\{nx^{n-1} + \frac{n(n-1)}{\lfloor 2}x^{n-2}(\delta x) + \ldots\right\}$$

Let $\delta x \longrightarrow 0$.

Then $\underset{\delta x \to 0}{Lt}\left(\frac{\delta y}{\delta x}\right) = a\{nx^{n-1}\}$

The constant factor a thus is a factor of right-hand side throughout and remains as a factor of the differential coefficient.

$$\therefore \quad \frac{dy}{dx} = nax^{n-1}.$$

40. Worked examples.

Example 1. *Find from first principles the differential coefficient of*

$$y = \frac{1}{x} \quad \text{or} \quad y = x^{-1}.$$

Let δx be an increment of x.

Let δy be the corresponding increment of y.

Then $$y + \delta y = \frac{1}{x + \delta x}$$

but $$y = \frac{1}{x}.$$

Subtracting $$\delta y = \frac{1}{(x + \delta x)} - \frac{1}{x}$$
$$= \frac{x - (x + \delta x)}{x(x + \delta x)}$$
$$= \frac{-\delta x}{x(x + \delta x)}.$$

Dividing by δx, $$\frac{\delta y}{\delta x} = \frac{-1}{x^2 + x\delta x}$$

Proceeding to the limit when $\delta x \longrightarrow 0$.

$$\underset{\delta x \to 0}{Lt} \frac{\delta y}{\delta x} = -\frac{1}{x^2}$$

or $$\frac{dy}{dx} = -\frac{1}{x^2}.$$

Example 2. *Write down the differential coefficients of the following functions.*

(1) $y = x^8$; $\frac{dy}{dx} = 8x^{8-1} = 8x^7.$

(2) $y = x^{\frac{1}{2}}$; $\frac{dy}{dx} = \frac{1}{2}x^{\frac{1}{2}-1} = \frac{1}{2}x^{-\frac{1}{2}} = \frac{1}{2\sqrt{x}}.$

(3) $y = x^{-3}$; $\frac{dy}{dx} = -3x^{-3-1} = -3x^{-4} = \frac{-3}{x^4}.$

(4) $y = x^{1\cdot5}$; $\frac{dy}{dx} = 1\cdot5 \times x^{1\cdot5-1} = 1\cdot5x^{0\cdot5}.$

(5) $y = x^{-\frac{1}{3}}$; $\frac{dy}{dx} = (-\frac{1}{3}) \times (x^{-\frac{1}{3}-1}) = -\frac{1}{3}x^{-\frac{4}{3}}.$

(6) $y = x$; $\frac{dy}{dx} = x^{1-1} = x^0 = 1.$

Example 3. *Differentiate the following functions:*

(1) $y = 6x^4$; $\frac{dy}{dx} = 6 \times 4 \times x^{4-1} = 24x^3.$

(2) $y = 4\sqrt[3]{x}$ or $y = 4x^{\frac{1}{3}}$

$$\frac{dy}{dx} = 4 \times \tfrac{1}{3} \times x^{\frac{1}{3}-1} = \tfrac{4}{3}x^{-\frac{2}{3}}$$

$$= \tfrac{4}{3} \div \sqrt[3]{x^2} = \frac{4}{3\sqrt[3]{x^2}}.$$

(3) $y = px^{2q}$; $\dfrac{dy}{dx} = p \times 2q \times x^{2q-1}$.

$$= 2pqx^{2q-1}.$$

(4) $s = 16t^2$; $\dfrac{ds}{dt} = 2 \times 16 \times t^{2-1} = 32t$. (Cf. § 28.)

Example 4. *Find the gradient of the tangent to the curve* $y = \dfrac{1}{x}$ *at the point where* $x = 1$.

The gradient is given by the value of the differential coefficient at the point.

Now $\dfrac{d}{dx}\left(\dfrac{1}{x}\right) = -\dfrac{1}{x^2}$. (Example 1.)

when $x = 1$

$$\frac{dy}{dx} = -1 \quad \text{or} \quad \tan 135°.$$

(Cf. Ex. 3, No. 7.)

Exercise 4.

1. Write down the differential coefficients of the following functions with respect to x:

$$x^7;\ 5x;\ \frac{x}{3};\ 0{\cdot}06x;\ \tfrac{1}{4}x^5;\ 15x^4;\ \frac{2x^6}{3};\ 1{\cdot}5x^3;\ (4x)^2.$$

2. Differentiate with respect to x:

$$bx^4;\ \frac{ax^6}{b};\ ax^p;\ x^{2a};\ 2x^{2b+1};\ 4\pi x^2.$$

3. Differentiate with respect to x:

$$6x + 4;\ 0{\cdot}54x - 6;\ -3x + 2;\ px + q.$$

4. Of what functions of x are the following the differential coefficients:

$$x;\ 3x;\ x^2;\ \tfrac{1}{4}x^2;\ x^5;\ x^n;\ x^{2a};\ \tfrac{2}{3}x^3;\ 4ax^2?$$

5. If $v = u + at$, where u and a are constants, find $\frac{dv}{dt}$.

6. If $s = \frac{1}{2} at^2$, where a is a constant, find $\frac{ds}{dt}$ when $a = 20$

7. If $A = \pi r^2$, find $\frac{dA}{dr}$.

8. If $V = \frac{4}{3}\pi r^3$, find $\frac{dV}{dr}$.

9. Differentiate the following functions of x:

$$5\sqrt{x}; \ \frac{5}{x}; \ \frac{5}{\sqrt{x}}; \ \sqrt[3]{x^2}; \ \sqrt[4]{2x^3}.$$

10. Differentiate with respect to x:

$$x^{0 \cdot 4}; \ 8x^{0 \cdot 2}; \ \frac{8}{x^{0 \cdot 2}}; \ 6x^{-4}; \ x^{-p}.$$

11. Differentiate with respect to x:

$$6x^{3 \cdot 2}; \ 2x^{-1 \cdot 5}; \ 29x^{0 \cdot 7}; \ \frac{6}{\sqrt[5]{x^3}}.$$

12. If $p = \frac{20}{v^2}$, find $\frac{dp}{dv}$.

13. Find the gradient of the curve of $y = \frac{1}{4}x^2$ at the point on the curve where $x = 3$. For what value of x is the gradient of the curve equal to zero?

14. Find the gradient of the curve of $y = 2x^3$, at the point where $x = 2$.

15. Find the gradients of the curve of $y = \frac{2}{x}$ at the points where $x = 10, 2, 1, \frac{1}{2}$.

16. Find from first principles the differential coefficient of $y = \frac{1}{x^2}$.

17. At what point on the curve of x^2 is the gradient of the curve equal to 2?

18. At what point on the curve of $y = x^3$ does the tangent to the curve make an angle of 45° with the x-axis?

19. At what point on the curve of $y = \sqrt{x}$ is the gradient equal to 2?

20. It is required to draw a tangent to the curve $y = 0 \cdot 5x^2$ which shall be parallel to the straight line $2x - 4y = 3$. At what point on the curve must it be drawn?

CHAPTER V

SOME RULES FOR DIFFERENTIATION

SUCCESSIVE DIFFERENTIATION

41. Differentiation of a Sum.

THE functions which were differentiated in the preceding chapter were expressions of **one term** only, with the exception of $y = mx + b$ (§ 36). This was found from first principles.

We now proceed to consider in general the differentiation of a function which is itself the sum of two or more functions of the same variable, such as $y = 5x^3 + 14x^2 - 7x$. The proof given below is a general one for the **sum of any number of functions of the same variable.**

Let u and v be functions of x.

Let y be their sum, so that

$$y = u + v.$$

Let x receive the increment δx.

Then u, v and y, being functions of x, will receive corresponding increments.

Let δu, δv and δy be these increments, so that

$$\begin{array}{lll} u & \text{becomes} & u + \delta u \\ v & \text{,,} & v + \delta v \\ y & \text{,,} & y + \delta y. \end{array}$$

$\therefore$ From $\quad y = u + v$

we have $\quad y + \delta y = (u + \delta u) + (v + \delta v)$.

Subtracting $\quad \delta y = \delta u + \delta v$.

Dividing by δx $\quad \dfrac{\delta y}{\delta x} = \dfrac{\delta u}{\delta x} + \dfrac{\delta v}{\delta x}.$

This is true for all values of δx and the corresponding increments δu, δv, δy.

Also their limits are equal. (Th. I, Limits, § 21.)

$\therefore$ If $\quad \delta x \longrightarrow 0$

$$\underset{\delta x \to 0}{Lt} \frac{\delta y}{\delta x} = \underset{\delta x \to 0}{Lt} \left\{\frac{\delta u}{\delta x} + \frac{\delta v}{\delta x}\right\}$$

$$= \underset{\delta x \to 0}{Lt} \left(\frac{\delta u}{\delta x}\right) + \underset{\delta x \to 0}{Lt} \left(\frac{\delta v}{\delta x}\right). \quad \text{(Th. 2, § 21.)}$$

Replacing these forms by the corresponding symbols for differential coefficients

$$\frac{dy}{dx} = \frac{du}{dx} + \frac{dv}{dx}$$

clearly the theorem will hold for any number of functions. Hence the **Rule for differentiation of a sum.**

The differential coefficient of the sum of a number of functions is equal to the sum of the differential coefficients of these functions.

42. Worked examples.

Example 1. *Differentiate with respect to x.*

$$y = 3x^3 + 7x^2 - 9x + 20.$$

Using the above rule

$$\frac{dy}{dx} = 9x^2 + 14x - 9.$$

Example 2. *Find the gradient at that point on the curve of $y = x^2 - 4x + 3$ where $x = 3$.*

What is the point of zero gradient on this curve?

If

$$y = x^2 - 4x + 3.$$

$$\frac{dy}{dx} = 2x - 4.$$

$\therefore$ When

$$x = 3, \ \frac{dy}{dx} = (2 \times 3) - 4.$$

$$= 2.$$

When the gradient is zero

$$2x - 4 = 0.$$

$$\therefore \quad x = 2.$$

Example 3. *If $s = 80t - 16t^2$, find $\frac{ds}{dt}$. When $\frac{ds}{dt} = 16$, find t.*

$$s = 80t - 16t^2$$

$$\therefore \quad \frac{ds}{dt} = 80 - 32t.$$

Then

$$\frac{ds}{dt}, \text{ i.e. } 80 - 32t = 16$$

$$32t = 64$$

$$t = 2.$$

Exercise 5.

Differentiate the following functions of x:

1. $6x^2 + 5x$.
2. $3x^3 + x - 1$.
3. $4x^4 + 3x^2 - x$.
4. $\frac{1}{2}x^2 + \frac{1}{7}x + \frac{1}{4}$.
5. $\frac{5}{x} + 4x$.
6. $7 + \frac{4}{x} - \frac{2}{x^2}$.
7. $x(5 - x + 3x^2)$.
8. $8\sqrt{x} + \sqrt{10}$.

Find $\frac{ds}{dt}$ when

9. $s = ut + \frac{1}{2}at^2$.
10. $s = 5t + 16t^2$.
11. $s = 3t^2 - 4t + 7$.
12. Find $\frac{dy}{dx}$ when $y = ax^3 + bx^2 + cx + d$.
13. Differentiate with respect to x, $\left(x + \frac{1}{x}\right)^2$.
14. Differentiate with respect to x, $\sqrt{x} + \frac{1}{\sqrt{x}}$.
15. Differentiate with respect to x, $(1 + x)^3$.
16. If $y = x^{2n} - nx^2 + 5n$, find $\frac{dy}{dx}$.
17. Find $\frac{dy}{dx}$ when $y = \sqrt{x} + \sqrt[3]{x} + \frac{2}{x}$.
18. Find the gradient at that point on the curve of $y = 2x^2 - 3x + 1$ where $x = 1{\cdot}5$. For what value of x will the curve have zero gradient?
19. For what values of x will the curve of $y = x(x^2 - 12)$ have zero gradient?
20. What are the gradients of the curve

$$y = x^3 - 6x^2 + 11x - 6$$

when x has the values 1, 2, 3?

21. What are the points of zero gradient on the curve of $y = x + \frac{1}{x}$?

43. Differentiation of a Product.

The differential coefficient of some products such as $(x + 2)^3$ or $3x(x + 2)$ can be found by multiplying out and using the rule for the differentiation of a sum. In most cases, however, that cannot be done, as, for example, $x^2\sqrt{1 - x}$ and $x^3 \sin x$.

The differential coefficient of a product is **not** equal to the product of the differential coefficients of the factors, as will be apparent on testing such an example as $3x(x + 2)$.

A general rule for use in all cases is found as follows:

Let u and v be functions of x.

Let $y = u \times v$,

consequently y is also a function of x.

Let δx be an increment of x.

Let δu, δv and δy be the corresponding increments of u, v and y.

Substituting the new values of u, v and y in

$$y = u \times v \quad . \quad . \quad . \quad . \quad . \quad . \quad . \quad (1)$$

$$y + \delta y = (u + \delta u)(v + \delta v)$$

or $$y + \delta y = uv + u(\delta v) + v \,.\, (\delta u) + (\delta u)(\delta v)$$

Subtracting (1) $\delta y = u(\delta v) + v(\delta u) + (\delta u)(\delta v)$.

Dividing by δx, $\dfrac{\delta y}{\delta x} = u \,.\, \dfrac{\delta v}{\delta x} + v \,.\, \dfrac{\delta u}{\delta x} + \delta u \,.\, \dfrac{\delta v}{\delta x}$.

Let $\delta x \longrightarrow 0$.

Then δu, δv, δy all approach zero.

$\therefore$ by Limits Th. 2.

$$\underset{\delta x \to 0}{Lt} \frac{\delta y}{\delta x} = \underset{\delta x \to 0}{Lt} \left(u \,.\, \frac{\delta v}{\delta x}\right) + \underset{\delta x \to 0}{Lt} \left(v \,.\, \frac{\delta u}{\delta x}\right) + \underset{\delta x \to 0}{Lt} \left(\delta u \,.\, \frac{\delta v}{\delta x}\right).$$

In the limit, since $\delta u \longrightarrow 0$, the last term—viz. $\delta u \times \dfrac{\delta v}{\delta x}$—also approaches zero.

$\therefore$ with the usual notation

$$\frac{dy}{dx} = u\frac{dv}{dx} + v\frac{du}{dx}.$$

This important rule may be expressed as follows:

(1) Differentiate each factor in turn and multiply by the other factor.

(2) The sum of the products is $\dfrac{dy}{dx}$.

This rule may be extended to more than two factors.

Thus if $$y = uvw$$

where u, v, w are factors of x

Then

$$\frac{dy}{dx} = \left(\frac{du}{dx} \times vw\right) + \left(\frac{dv}{dx} \times uw\right) + \left(\frac{dw}{dx} \times uv\right).$$

44. Worked examples.

Example 1. *Differentiate* $(x^2 - 5x + 2)(2x^2 + 7)$.

Let

$$y = (x^2 - 5x + 2)(2x^2 + 7).$$

Then

$$\frac{dy}{dx} = \left\{\frac{d(x^2 - 5x + 2)}{dx} \times (2x^2 + 7)\right\} + \left\{\frac{d(2x^2 + 7)}{dx} \times (x^2 - 5x + 2\right\}$$

$$= (2x - 5)(2x^2 + 7) + 4x(x^2 - 5x + 2).$$

This result can be simplified, if necessary.

Example 2. *Differentiate* $(x^2 - 1)(2x + 1)(x^3 + 2x^2 + 1)$.

$$\frac{dy}{dx} = \left\{\frac{d(x^2 - 1)}{dx} \times (2x + 1)(x^3 + 2x^2 + 1)\right\} + \left\{\frac{d(2x + 1)}{dx} \times (x^2 - 1)(x^3 + 2x^2 + 1)\right\} + \left\{\frac{d(x^3 + 2x^2 + 1)}{dx} \times (x^2 - 1)(2x + 1)\right\}$$

$$= 2x(2x + 1)(x^3 + 2x^2 + 1) + 2(x^2 - 1)(x^3 + 2x^2 + 1) + (3x^2 + 4x)(x^2 - 1)(2x + 1).$$

This can be further simplified.

Exercise 6.

Differentiate the following by means of the rule for products.

1. $(3x + 1)(2x + 1)$.
2. $(x^2 + 1)(\frac{1}{2}x + 1)$.
3. $(3x - 5)(x^2 + 2x)$.
4. $(x^2 + 3)(2x^2 - 1)$.
5. $(x^2 + 4x)(3x^2 - x)$.
6. $(x^2 + x + 1)(x - 1)$.
7. $(x^2 - x + 1)(x + 1)$.
8. $(x^2 + 4x + 5)(x^2 - 2)$.
9. $(x^2 - 5)(x^2 + 5)$.
10. $(x^2 - x + 1)(x^2 + x - 1)$.
11. $(x - 2)(x^2 + 2x + 4)$.
12. $(2x^2 - 3)(3x^2 + x - 1)$.
13. $(x - 1)(x + 1)(x^2 + 1)$.
14. $(x + 1)(2x + 1)(3x + 2)$.
15. $(ax^2 + bx + c)(px + q)$.
16. $\sqrt{x}(2x - 1)(x^2 + x + 1)$.
17. $2x^{\frac{3}{2}}(\sqrt{x} + 2)(\sqrt{x} - 1)$.

45. Differentiation of a quotient.

In § 40 the differential coefficient of a simple example of a quotient, viz. $\frac{1}{x}$, was found by first principles. This method, however, is apt to become very tedious in more complicated examples. The general rule which is explained below is that which is usually employed.

Let u and v be functions of x.

Let
$$y = \frac{u}{v}.$$

y is then a function of x.

Using the notation employed in the preceding section for increments of these

$$y + \delta y = \frac{u + \delta u}{v + \delta v}$$

subtracting
$$\delta y = \frac{u + \delta u}{v + \delta v} - \frac{u}{v}$$
$$= \frac{v(u + \delta u) - u(v + \delta v)}{v(v + \delta v)}$$
$$= \frac{v\delta u - u\delta v}{v(v + \delta v)}.$$

Dividing by δx,
$$\frac{\delta y}{\delta x} = \frac{v \cdot \frac{\delta u}{\delta x} - u \cdot \frac{\delta v}{\delta x}}{v(v + \delta v)}.$$

Let $\delta x \longrightarrow 0$; in consequence δu, δv, and δy tend to zero. Then

$$\underset{\delta x \to 0}{Lt}\left(\frac{\delta y}{\delta x}\right) = \frac{\underset{\delta x \to 0}{Lt}\left(v \cdot \frac{\delta u}{\delta x}\right) - \underset{\delta x \to 0}{Lt}\left(u \cdot \frac{\delta v}{\delta x}\right)}{\underset{\delta x \to 0}{Lt}\ v(v + \delta v)} \quad \text{(Th. 4, Limits).}$$

The limits in the numerator can be expressed by

$$v \cdot \frac{du}{dx} - u \cdot \frac{dv}{dx}$$

and the limit of the denominator is v^2, since $\delta v \longrightarrow 0$.

$$\therefore \quad \frac{dy}{dx} = \frac{v \cdot \frac{du}{dx} - u \cdot \frac{dv}{dx}}{v^2}.$$

This can be written:

$$\frac{dy}{dx} = \frac{(\text{den.} \times \text{d.c. of num.}) - (\text{num.} \times \text{d.c. of den.})}{(\text{den.})^2}.$$

46. Worked examples.

Example 1. *Differentiate* $\dfrac{3x}{x - 1}$.

Using
$$\frac{dy}{dx} = \frac{v \cdot \dfrac{du}{dx} - u \cdot \dfrac{dv}{dx}}{v^2}$$

where $u = 3x$ and $v = x - 1$.

$$\frac{dy}{dx} = \frac{\{(x - 1)(3)\} - \{3x(1)\}}{(x - 1)^2}$$
$$= \frac{3x - 3 - 3x}{(x - 1)^2}$$
$$= \frac{-3}{(x - 1)^2}.$$

Example 2. $y = \dfrac{x^3 + 1}{x^3 - 1}$.

Using the formula quoted above

$$\frac{dy}{dx} = \frac{\{(x^3 - 1) \times (3x^2)\} - \{(x^3 + 1) \times (3x^2)\}}{(x^3 - 1)^2}$$
$$= \frac{3x^5 - 3x^2 - 3x^5 - 3x^2}{(x^3 - 1)^2}$$
$$= \frac{-6x^2}{(x^3 - 1)^2}.$$

Example 3. *Differentiate* $\dfrac{x(x + 1)}{x^2 - 3x + 2}$.

We have
$$y = \frac{x^2 + x}{x^2 - 3x + 2}.$$

Using the above rule

$$\frac{dy}{dx} = \frac{\{(x^2 - 3x + 2)(2x + 1)\} - \{(x^2 + x)(2x - 3)\}}{(x^2 - 3x + 2)^2}$$

$$= \frac{(2x^3 - 5x^2 + x + 2) - (2x^3 - x^2 - 3x)}{(x^2 - 3x + 2)^2}$$

$$= \frac{-4x^2 + 4x + 2}{(x^2 - 3x + 2)^2}.$$

Exercise 7.

Differentiate the following functions of x.

1. $\frac{3}{2x - 1}$.
2. $\frac{1}{1 - 3x^2}$.
3. $\frac{x}{x + 2}$.
4. $\frac{x + 1}{x + 2}$.
5. $\frac{3x - 1}{2x + 3}$.
6. $\frac{x + b}{x - b}$.
7. $\frac{x - b}{x + b}$.
8. $\frac{x^2}{x - 4}$.
9. $\frac{x^2}{x^2 - 4}$.
10. $\frac{\sqrt{x}}{x + 1}$.
11. $\frac{x + 1}{\sqrt{x}}$.
12. $\frac{\sqrt{x} + 1}{\sqrt{x} - 1}$.
13. $\frac{x^3 - 1}{x^3 + 1}$.
14. $\frac{x^2 + x + 1}{x^2 - x + 1}$.
15. $\frac{2x^2 - x + 1}{3x^2 + x - 1}$.
16. $\frac{1 + x + x^2}{x}$.
17. $\frac{2x^4}{a^2 - x^2}$.
18. $\frac{2x - 3}{2 - 3x}$.
19. $\frac{x(x - 1)}{x - 2}$.
20. $\frac{x^{\frac{1}{2}} + 2}{x^{\frac{3}{2}}}$.

47. Function of a function.

To understand the meaning of "*function of a function*" consider the trigonometrical function $\sin^2 x$, i.e. $(\sin x)^2$. This function, being the square of $\sin x$, is a function of $\sin x$, just as x^2 is a function of x, or u^2 is a function of u.

But $\sin x$ is itself a function of x.

$\therefore$ $\sin^2 x$ is a function of $\sin x$, which is a function of x, *i.e.*, $\sin^2 x$ is a function, of a function of x.

Similarly $\sqrt{x^2 + 4x}$ is a function of $x^2 + 4x$, just as $\sqrt{x}$ is a function of x.

But $x^2 + 4x$ is itself a function of x.

$\therefore$ $\sqrt{x^2 + 4x}$ is a function of $x^2 + 4x$, which is a function of x. The idea of a "**function of a function**" may be extended. For example, we have seen that $\sin^2 x$ is a function of a function of x. But $\sin^2(\sqrt{x})$ is a function of $\sin\sqrt{x}$, which is a function of $\sqrt{x}$, which in its turn is a function of x. The idea of "a function of a function" often puzzles the beginner, and there is a tendency to overlook it and to omit application of the rule for differentiating it which we shall discover later. For example, it may be overlooked that such a familiar function as $\sin 2x$ is a function of a function, since $2x$ is a function of x.

We cannot proceed further with the example above of $\sin^2 x$, since the rules for differentiating trigonometrical functions are dealt with in a subsequent chapter.

An algebraic function, say $y = (x^2 - 5)^4$, will be used as an example in discovering the rule for differentiating a function of a function.

Now $(x^2 - 5)^4$ is a function—the fourth power—of $x^2 - 5$, which is itself a function of x.

If $\qquad u = (x^2 - 5)$

we can write $\qquad y = u^4$.

Differentiating y with respect to u, according to rule

$$\frac{dy}{du} = 4u^3.$$

But we require $\frac{dy}{dx}$; therefore the following method is adopted to find it.

Let δx be an increment of x.

,, δu be the corresponding increment of u.

,, δy ,, ,, ,, y.

These being finite increments, it is obvious that by the law of fractions

$$\frac{\delta y}{\delta x} = \frac{\delta y}{\delta u} \times \frac{\delta u}{\delta x}.$$

Let $\delta x \longrightarrow 0$; in consequence δu and δy will proceed to zero. Then each of the ratios $\frac{\delta y}{\delta x}, \frac{\delta y}{\delta u}, \frac{\delta u}{\delta x}$ approaches a limit.

$$\therefore \quad \underset{\delta x \to 0}{Lt} \frac{\delta y}{\delta x} = \underset{\delta x \to 0}{Lt} \left\{ \frac{\delta y}{\delta u} \times \frac{\delta u}{\delta x} \right\}$$

$$= \underset{\delta x \to 0}{Lt} \left(\frac{\delta y}{\delta u} \right) \times \underset{\delta x \to 0}{Lt} \left(\frac{\delta u}{\delta x} \right)$$

by the third law of limits, § 21.

$$\therefore \quad \frac{dy}{dx} = \frac{dy}{du} \times \frac{du}{dx}.$$

Applying this result to the above we have:

$$\frac{dy}{du} = 4u^3$$

and as $u = x^2 - 5$

$$\frac{du}{dx} = 2x$$

since

$$\frac{dy}{dx} = \frac{dy}{du} \times \frac{du}{dx}.$$

$$\therefore \quad \frac{dy}{dx} = 4u^3 \times 2x$$

i.e., $\frac{dy}{dx} = 4(x^2 - 5)^3 \times 2x = 8x(x^2 - 5)^3.$

Worked Examples.

Example 1. *Differentiate* $y = \sqrt{1 - x^2}$

$$\sqrt{1 - x^2} = (1 - x^2)^{\frac{1}{2}}.$$

Let $u = 1 - x^2$; then $\frac{du}{dx} = -2x$ and $y = u^{\frac{1}{2}}$.

$$\therefore \quad \frac{dy}{du} = \tfrac{1}{2}u^{-\frac{1}{2}}$$

$$= \tfrac{1}{2}(1 - x^2)^{-\frac{1}{2}}.$$

Since

$$\frac{dy}{dx} = \frac{dy}{du} \times \frac{du}{dx}$$

$\therefore$ substituting

$$\frac{dy}{dx} = \tfrac{1}{2}(1 - x^2)^{-\frac{1}{2}} \times (-2x)$$

$$= -x(1 - x^2)^{-\frac{1}{2}}$$

$$= \frac{-x}{\sqrt{1 - x^2}}.$$

Example 2. *Differentiate* $y = (x^2 - 3x + 5)^3$.

Let $u = x^2 - 3x + 5$; then $\frac{du}{dx} = 2x - 3$ and $y = u^3$.

$$\therefore \quad \frac{dy}{du} = 3u^2$$
$$= 3(x^2 - 3x + 5)^2.$$

Substituting in $\frac{dy}{dx} = \frac{dy}{du} \times \frac{du}{dx}$

$$\frac{dy}{dx} = 3(x^2 - 3x + 5)^2 \times (2x - 3)$$
$$= 3(2x - 3)(x^2 - 3x + 5)^2.$$

After some practice the student will probably find that usually he will be able to dispense with the use of " u " and can write down the result. The above example is a convenient one for trying this procedure.

Example 3. *Differentiate* $y = (3x^2 - 5x + 4)^{\frac{3}{2}}$.

Working this without introducing u, the solution can be written down in two stages, as follows :

$$\frac{dy}{dx} = \tfrac{3}{2}(3x^2 - 5x + 4)^{\frac{3}{2}-1} \times \text{d.c. of } (3x^2 - 5x + 4)$$
$$= \tfrac{3}{2}(3x^2 - 5x + 4)^{\frac{1}{2}} \times (6x - 5).$$

Example 4. *Differentiate* $y = (x^2 + 5)\sqrt[3]{x^2 + 1}$.

This being a product of two functions, we employ the rule

$$\frac{d(uv)}{dx} = u \cdot \frac{dv}{dx} + v \cdot \frac{du}{dx}.$$

Hence

$$\frac{dy}{dx} = (x^2 + 5)\{\text{d.c. of } \sqrt[3]{x^2 + 1}\}$$
$$+ \{\sqrt[3]{x^2 + 1} \times \text{d.c. of } (x^2 + 5)\} \quad \text{(A)}$$

Of these $\sqrt[3]{x^2 + 1}$ is a function of a function.

It is better to work this separately and substitute afterwards:

$$\frac{d\{(x^2 + 1)^{\frac{1}{3}}\}}{dx} = \tfrac{1}{3}(x^2 + 1)^{\frac{1}{3}-1} \times \{(\text{d.c.}) \text{ of } (x^2 + 1)\}$$
$$= \tfrac{1}{3}(x^2 + 1)^{-\frac{2}{3}} \times 2x$$
$$= \frac{2x}{3(x^2 + 1)^{\frac{2}{3}}}.$$

Substituting in (A)

$$\frac{dy}{dx} = (x^2 + 5)\left\{\frac{2x}{3(x^2 + 1)^{\frac{2}{3}}}\right\} + \{\sqrt[3]{x^2 + 1} \times 2x\}$$

$$= \frac{2x(x^2 + 5)}{3\sqrt[3]{(x^2 + 1)^2}} + 2x\sqrt[3]{x^2 + 1}.$$

This might be further simplified.

Example 5. *Differentiate* $\dfrac{\sqrt{1 + 3x}}{4x}$.

Employing the formula for a quotient, viz.:

$$\frac{d}{dx}\left(\frac{u}{v}\right) = \frac{v \cdot \dfrac{du}{dx} - u \cdot \dfrac{dv}{dx}}{v^2},$$

and substituting

$$\frac{dy}{dx} = \frac{4x \times \{\text{d.c. of} \sqrt{1 + 3x}\} - \{\sqrt{1 + 3x} \times \text{d.c. of } 4x\}}{(4x)^2} \quad \text{(A)}$$

Of these $\sqrt{1 + 3x}$ or $(1 + 3x)^{\frac{1}{2}}$ is a function of a function. Applying the method above

$$\frac{d}{dx}(1 + 3x)^{\frac{1}{2}} = \tfrac{1}{2}(1 + 3x)^{-\frac{1}{2}} \times 3 = \frac{3}{2\sqrt{1 + 3x}}.$$

Substituting in (A)

$$\frac{dy}{dx} = \frac{4x \times \dfrac{3}{2\sqrt{1 + 3x}} - 4\sqrt{1 + 3x}}{16x^2}$$

$$= \frac{\dfrac{6x}{\sqrt{1 + 3x}} - 4\sqrt{1 + 3x}}{16x^2}$$

$$= \frac{6x - 4(1 + 3x)}{16x^2\sqrt{1 + 3x}}$$

$$= \frac{6x - 4 - 12x}{16x^2\sqrt{1 + 3x}} = \frac{-4 - 6x}{16x^2\sqrt{1 + 3x}}.$$

$$= -\frac{2 + 3x}{8x^2\sqrt{1 + 3x}}.$$

48. Differentiation of implicit functions.

It was pointed out in § 11 that it frequently happens, when y is a function of x, that the relation between x and y is not explicitly stated, but the two variables occur in the form of an equation from which y can be obtained in terms of x, though sometimes this is not possible. Even when y can be found in terms of x, it is in such a form that differentiation may be tedious or difficult. This is apparent from the examples of Implicit functions given in § 11.

In such cases the method adopted is to differentiate term by term throughout the equation, remembering that in differentiating functions of y we are differentiating a function of a function.

Example 1. *Find* $\frac{dy}{dx}$ *from the following equation.*

$$x^2 - y^2 + 3x = 5y.$$

Differentiating $2x - 2y \cdot \frac{dy}{dx} + 3 = 5\frac{dy}{dx}$.

The differential coefficient $\frac{dy}{dx}$ remains, as we have not yet determined it. It will be seen, however, that the equation can be solved for $\frac{dy}{dx}$.

Thus collecting terms involving it

$$2y \cdot \frac{dy}{dx} + 5\frac{dy}{dx} = 2x + 3$$

or

$$\frac{dy}{dx}(2y + 5) = 2x + 3$$

$$\therefore \quad \frac{dy}{dx} = \frac{2x + 3}{2y + 5}.$$

It will be observed that the solution gives $\frac{dy}{dx}$ in terms of the two variables x and y. When corresponding values of x and y are known, the numerical value of $\frac{dy}{dx}$ can be determined. An example of this follows.

Example 2. *Find the slope of the tangent to the curve $x^2 + xy + y^2 = 4$ at the point $(2, -2)$.*

Differentiating $x^2 + xy + y^2 = 4$ as shown above, and remembering that xy is a product

$$2x + y + x \cdot \frac{dy}{dx} + 2y\frac{dy}{dx} = 0.$$

$$\therefore \quad \frac{dy}{dx}(x + 2y) = -(2x + y)$$

and

$$\frac{dy}{dx} = -\frac{2x + y}{x + 2y}.$$

$\therefore$ when

$$x = 2,\ y = -2$$

$$\frac{dy}{dx} = -\frac{4 - 2}{2 - 4}$$

$$= 1.$$

$\therefore$ the gradient of the tangent to the curve at this point is 1 and the **angle of slope** is 45°.

Exercise 8.

Differentiate the following:

1. $(2x + 5)^2$; $(1 - 5x)^4$; $(3x + 7)^{\frac{1}{3}}$.
2. $\dfrac{1}{1 - 2x}$; $(1 - 2x)^2$; $\sqrt{1 - 2x}$.
3. $(x^2 - 4)^5$; $(1 - x^2)^{\frac{3}{2}}$; $\sqrt{3x^2 - 7}$.
4. $\dfrac{1}{1 - 2x^2}$; $\sqrt{1 - 2x^2}$; $x\sqrt{1 - x^2}$.
5. $\dfrac{1}{4 - x}$; $\dfrac{1}{\sqrt{4 - x}}$; $\dfrac{1}{(4 - x)^2}$.
6. $\dfrac{1}{x^2 - 1}$; $\dfrac{1}{\sqrt{x^2 - 1}}$; $\dfrac{x}{\sqrt{1 + x^2}}$.
7. $\dfrac{1}{\sqrt{1 - x^2}}$; $\sqrt{\left(\dfrac{x}{1 - x}\right)}$; $\sqrt{\left(\dfrac{1 - x}{1 + x}\right)}$.
8. $x\sqrt{\left(\dfrac{1 - x}{1 + x}\right)}$; $\sqrt[3]{x^2 + 1}$.
9. $\sqrt{a^2 + x^2}$; $\dfrac{1}{\sqrt{a^2 + x^2}}$.
10. $\sqrt{1 - x + x^2}$; $(1 - 2x^2)^n$.

11. $\dfrac{x^2}{\sqrt{a^2 - x^2}}$; $\left(x + \dfrac{1}{x}\right)^2$.

12. $\dfrac{1}{\sqrt{1 + x^3}}$; $\dfrac{\sqrt{1 + 2x}}{x}$.

13. $\dfrac{x}{\sqrt{1 + x^2}}$; $\dfrac{\sqrt{1 + x^2}}{x}$.

14. $\dfrac{1}{\sqrt{2x^2 - 3x + 4}}$; $x^2\sqrt{1 - x}$.

15. $\dfrac{\sqrt{1 - x^2}}{1 - x}$; $x\sqrt{2x + 3}$.

Find $\dfrac{dy}{dx}$ from the following implicit functions:

16. $3x^2 + 7xy + 9y^2 = 6$.
17. $(x^2 + y^2)^2 - (x^2 - y^2) = 0$.
18. $x^3 + y^3 = 3xy$.
19. $x^n + y^n = a^n$.
20. Find the gradient, at the point (1, 1), of the tangent to the curve $x^2 + y^2 - 3x + 4y - 3 = 0$.

49. Successive differentiation.

It was pointed out in § 32 that the differential coefficient of a function of x, unless it be a linear function, is itself a function of x.

For example, if
$$y = 3x^4$$
$$\frac{dy}{dx} = 12x^3.$$

Since $12x^3$ is a function of x, it can be differentiated with respect to x and
$$\frac{d}{dx}(12x^3) = 36x^2.$$

This expression is called the second differential coefficient of the original function, and the operation can be indicated by $\dfrac{d}{dx}\left(\dfrac{dy}{dx}\right)$.

The symbol $\dfrac{d^2y}{dx^2}$ is employed to represent the second differential coefficient. In this symbol the figure "2" in the numerator and denominator x is not an index, but signifies that y, the original function, has been twice differentiated and each time with respect to x.

Thus, $\frac{d^2y}{dx^2}$ measures the rate at which $\frac{dy}{dx}$ is changing with respect to x, just as $\frac{dy}{dx}$ measures the rate at which y is changing with respect to x.

The second differential coefficient is also a function of x, unless $\frac{dy}{dx}$ is a linear function or a constant. Consequently $\frac{d^2y}{dx^2}$ can also be differentiated with respect, and the result is the **third differential coefficient** of y with respect to x.

It is represented by $\frac{d^3y}{dx^3}$.

Thus in the above example in which

$$\frac{d^2y}{dx^2} = 36x^2.$$

$$\frac{d^3y}{dx^3} = 72x.$$

Thus it is possible to have a succession of differential coefficients. This process of successive differentiation can be continued indefinitely or until one of the differential coefficients is a constant. This can be illustrated by the example of $y = x^n$, as follows:

Successive differential coefficients of x^n.

$$y = x^n$$

$$\frac{dy}{dx} = nx^{n-1}$$

$$\frac{d^2y}{dx^2} = n(n-1)x^{n-2}$$

$$\frac{d^3y}{dx^3} = n(n-1)(n-2)x^{n-3}.$$

If n is a positive integer this process can be continued until ultimately $n - n$ is reached as the index of x and the differential coefficient becomes $n(n-1)(n-2)\ldots 3, 2, 1$ —*i.e.*, **factorial** n or $\lfloor n$. The next and subsequent differential coefficients are zero. If n is not a positive integer the process can be continued indefinitely. The following example will serve as an illustration.

Find the successive differential coefficients of

$$y = x^3 - 7x^2 + 6x + 3.$$

Then
$$\frac{dy}{dx} = 3x^2 - 14x + 6$$
$$\frac{d^2y}{dx^2} = 6x - 14$$
$$\frac{d^3y}{dx^3} = 6$$
$$\frac{d^4y}{dx^4} = 0.$$

50. Alternative notation for differential coefficients.

The successive differential coefficients are also called the **derivatives** or derived functions of the original function. They may conveniently be denoted by the following alternative symbols:

(1) When the **functional** notation is employed:

If			
$f(x)$ or $\phi(x)$	denotes a function of x.		
$f'(x)$ or $\phi'(x)$	denotes the 1st diff. coefficient.		
$f''(x)$ or $\phi''(x)$	,,	2nd	,,
$f'''(x)$ or $\phi'''(x)$	,,	3rd	,,
$f^{IV}(x)$ or $\phi^{IV}(x)$	,,	4th	,, etc.

(2) Or y may be retained with a suffix or an accent:

Thus if			
y	denotes the function.		
y_1	,,	1st diff. coefficient.	
y_2	,,	2nd	,,
y_3	,,	3rd	,, , etc.

or sometimes the terms y', y'', y''' . . . are used.

51. Derived curves.

If has been shown above that successive differentiation of a function of x produces a set of derivatives each of which is also a function of x. These derivatives can be represented by their graphs. Consequently the derived functions give rise to a series of **derived curves,** between which definite relations exist.

Consider the function

$$y = x^2 - 4x + 3$$

then $$y_1 = 2x - 4$$

and $$y_2 = 2.$$

Fig. 11 shows (1) the graph of $y = x^2 - 4x + 3$, (2) the graph of $y_1 = 2x - 4$, and (3) $y_2 = 2$, the graphs of the two derived functions. The following connections between the curve of the original function and of its two derivatives will be obvious.

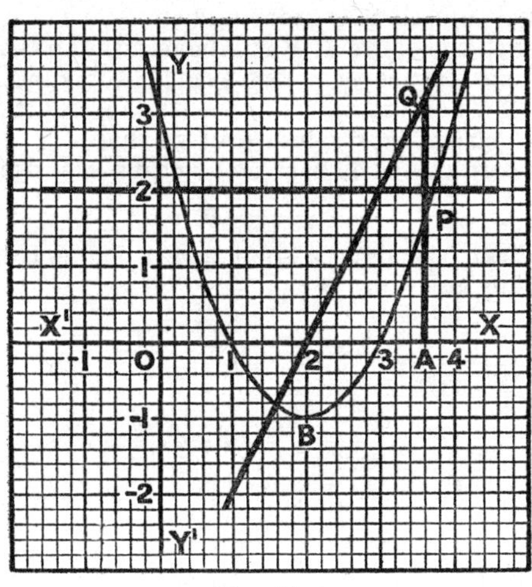

FIG. 11.

(1) Since y_1, the first derived function, gives the rate of increase of y with respect to x, **its value for any assigned value of x equals the gradient at the corresponding point on the curve.**

Take any point A on OX where $x = 3{\cdot}6$. Drawing the ordinate at A, P is the corresponding point on the curve, and Q the point on the straight line $y_1 = 2x - 4$, the first derived function.

Then, **the value of the ordinate QA is equal to the gradient of the curve at P.** This value is seen to be $3{\cdot}2$ units. By calculation, substituting $x = 3{\cdot}6$ in

$y_1 = 2x - 4$, the differential coefficient, this is equal to $(2 \times 3\cdot6) - 4 = 3\cdot2$.

(2) The **graph of the second derived function,** viz. $y_2 = 2$, being parallel to OX and having a constant value for any ordinate, shows that the gradient of $y_1 = 2x - 4$ is constant, viz. 2.

(3) **At the lowest point B on the curve of** $y = x^2 - 4x + 3$ the value of the ordinate at the corresponding point on the first derived curve, $y_1 = 2x - 4$ is zero; it cuts OX at this point. Thus **the gradient of the original function is zero, when** $x = 2$. A tangent drawn to the curve at B will be parallel to OX.

(4) **For values of x less than 2, the function $x^2 - 4x + 3$ is decreasing while the derived function values are negative. For values greater than 2, $x^2 - 4x + 3$ is increasing and the derived function $2x - 4$ is positive.**

Exercise 9.

Write down the first, second and third derivatives of the following functions of x:

1. $x^2(x - 1)$. 2. x^{2b}.
3. $5x^4 - 3x^3 + 2x^2 - x + 1$.
4. $10x^5 - 4x^3 + 5x - 2$.
5. $\dfrac{1}{x}$. 6. $\sqrt{x}$.
7. $\sqrt{2x + 1}$. 8. $\dfrac{1}{x^2}$.

---

9. Find the **nth** differential coefficient of $\dfrac{1}{a^2 - x^2}$.

$$\left[\text{Hint, } \frac{1}{a^2 - x^2} = \frac{1}{2a}\left(\frac{1}{a + x} + \frac{1}{a - x}\right)\right].$$

10. If $f(x) = 6x^2 - 5x + 3$, find $f'(0)$. For what value of x is $f'(x) = 0$? To what point on the curve of $f(x)$ does this correspond?

11. If $f(x) = x^3 - 5x^2 + 7$, find $f'(1)$ and $f''(2)$. For what values of x does $f'(x)$ vanish?

12. Find the values of x for which the curve of $f(x) = \frac{1}{3}x^3 - \frac{5}{2}x^2 + 6x + 1$ has zero gradient. For what value of x is the gradient of $f'(x)$ equal to zero? To what value of $f'(x)$ does this correspond?

CHAPTER VI

MAXIMA AND MINIMA VALUES. POINTS OF INFLEXION

52. Sign of the differential coefficient.

The sign of the differential coefficient, to which brief references were made in §§ 34 and 51, must now be examined in more detail.

If y is a continuous function of x, and if x receives an increment δx, then y will be increased or decreased by a finite quantity δy.

If y **is increased,** then δy must be positive, and δx being always positive, the rate of change as expressed by the limit of $\frac{\delta y}{\delta x}$—*i.e.*, $\frac{dy}{dx}$—**must be positive.**

If, however, y **is decreased,** δy must be regarded as negative. Hence the rate of change as expressed by the limit of $\frac{\delta y}{\delta x}$ must also be negative—*i.e.*, $\frac{dy}{dx}$ **must be negative.**

More concisely:

(1) **If y increases as x increases, $\frac{dy}{dx}$ is positive.**

(2) **If y decreases as x increases, $\frac{dy}{dx}$ is negative.**

Similar conclusions were reached in connection with the gradient of a curve at a point. Since algebraical functions can be represented graphically, the form of the curve, as shown below, will indicate whether the function is increasing or decreasing, and consequently whether the differential coefficient is positive or negative.

A. **Functions which are increasing** are shown by portions of their curves in Figs. 12(a) and 12(b), where P is a point on the curve and Q the point corresponding to an increase of δx in the value of x.

(1) Curves may be **concave upwards and rising,** as in Fig. 12(a). Examples are $y = x^2$ (for positive values of x), $y = 10^x$, $y = \tan x$ $\left(\text{between } 0 \text{ and } \frac{\pi}{2}\right)$.

(2) Or they may be **concave downwards and rising,** as in Fig. 12(*b*). Examples are, $y = \sqrt{x}$, $y = \log x$, $y = \sin x$ $\left(\text{between 0 and } \frac{\pi}{2}\right)$.

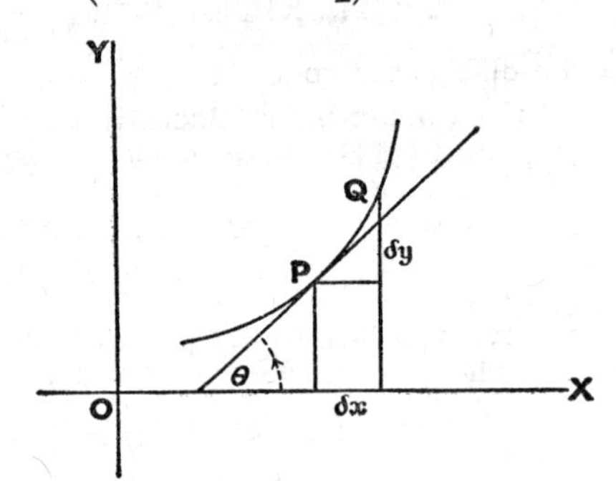

Fig. 12(*a*).

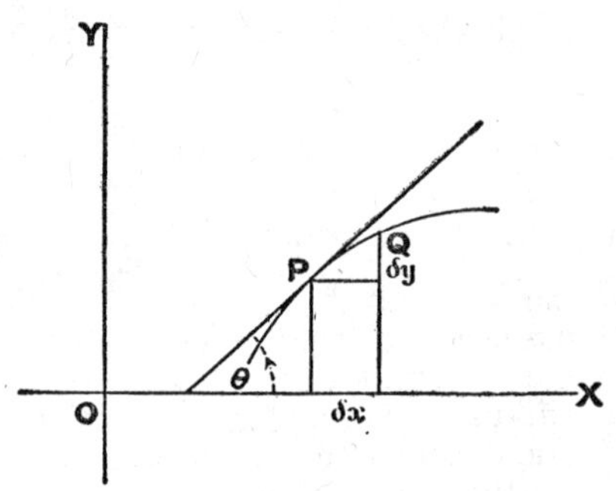

Fig. 12(*b*).

In both kinds the curve rises upwards to the right as x increases.

As is evident from the figures, as x is increased by δx, y is increased by δy,

Thus $\frac{\delta y}{\delta x}$ **and its limit are positive.**

Geometrically it is evident that in each case the tangent

to the curve at P makes an *acute* angle with OX. Hence the gradient, given by $\tan\theta$, is positive. It is also evident that in Fig. 12(*a*) $\frac{dy}{dx}$ is increasing, in Fig. 12(*b*) decreasing.

B. Functions which are decreasing can be similarly represented by portions of their curves in Figs. 13(*a*) and 13(*b*).

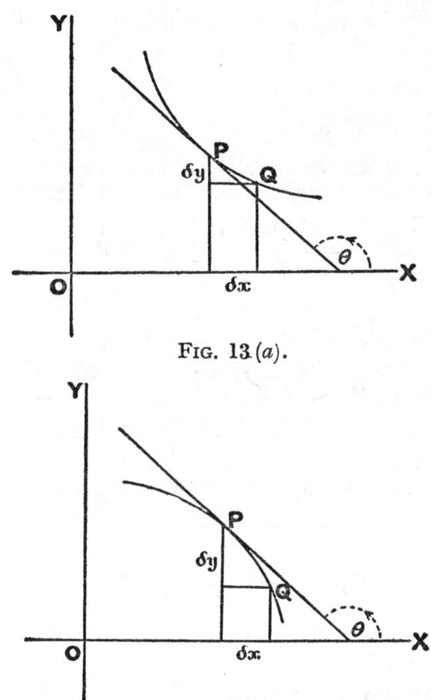

Fig. 13(*a*).

Fig. 13(*b*).

Using the same letters and notation as in Figs. 12(*a*) and 12(*b*), it is evident that in each case, as x at the point P receives an increment δx, the new value of the function at x is less. Hence δy must be regarded as negative and $\frac{\delta y}{\delta x}$, with its limit $\frac{dy}{dx}$, are negative.

As before, there are two types:

(1) The **curve concave upwards falling,** as in Fig. 13(a). Examples are:

$y = x^2$ (for negative values of x), $y = \frac{1}{x}$, $y = \cot x$ $\left(\text{between } 0 \text{ and } \frac{\pi}{2}\right)$, etc.

(2) The **curve concave downwards falling,** as in Fig. 13(b). Examples are:

$y = \sin x$ $\left(\text{between } \frac{\pi}{2} \text{ and } \pi\right)$, $y = -x^2$ (positive values of x), etc.

The tangents drawn to both of these curves make obtuse angles with OX. Consequently $\tan\theta$, the gradient of the line, is negative.

It is also clear that $\frac{dy}{dx}$ **is itself increasing in 13(a) but decreasing in Fig. 13(b).**

53. Stationary values.

Two of the cases illustrated above—viz., Figs. 12(a) and 13(a)—occur in the graph of $y = x^2 - 4x + 3$ which was shown in Fig. 11. This is repeated in Fig. 14, and we will examine it further.

Since

$$y = x^2 - 4x + 3$$

$$\frac{dy}{dx} = 2x - 4.$$

This latter is represented in Fig. 14 by the straight line AB.

The following changes in the curve and function can be seen from the graph:

(1) **While x increases from $-\infty$ to $+2$, y is decreasing.** Values of $\frac{dy}{dx}$—represented by the line AB—are **negative** (see Fig. 13(a)).

(2) **While x increases from $+2$ to $+\infty$, y is increasing** (see Fig. 12(a)). Hence values of $\frac{dy}{dx}$ are **positive.**

(3) **At C the curve ceases to decrease and begins**

to increase. Thus when $x = 2$, the value of y is momentarily not changing, but is stationary. There is therefore no rate of change, and $\frac{dy}{dx}$ is zero. The straight line AB thus cuts OX at this point.

Hence when $x = 2$, the function is said to have a stationary value, and C is called a stationary point on the curve.

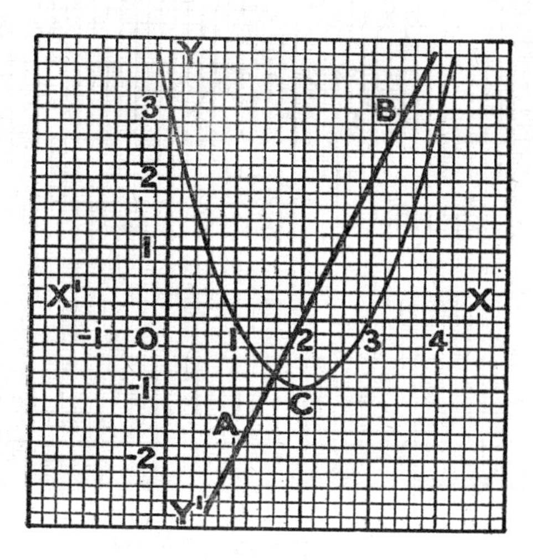

FIG. 14.

These important conclusions may be summarised as follows:

(1) *If* $x < +2$, *y is decreasing and* $\frac{dy}{dx}$ *is negative.*

(2) *If* $x > +2$, *y is increasing and* $\frac{dy}{dx}$ *is positive.*

(3) *When* $x = 2$, *at C y is momentarily neither increasing nor decreasing*—i.e., *the function has a stationary value and* $\frac{dy}{dx} = 0$.

Next we will consider the function:

$$y = 3 + 2x - x^2.$$

$$\therefore \quad \frac{dy}{dx} = 2 - 2x.$$

The graphs of these are shown in Fig. 15, in which the

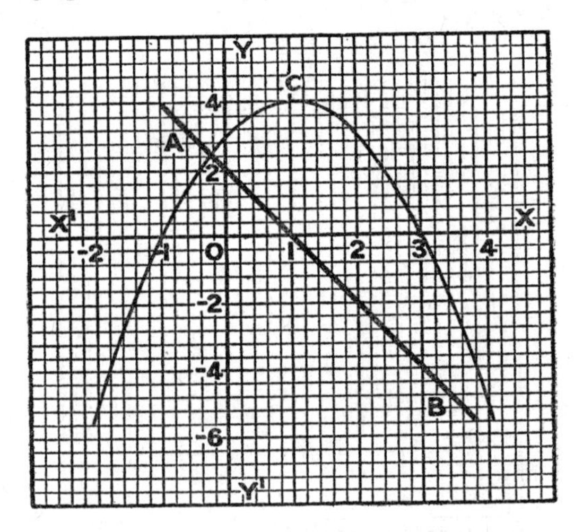

FIG. 15.

straight line AB represents the derived function $2 - 2x$. Examining these, as was done above, we see:

(1) *When $x < +1$, y is increasing and $\frac{dy}{dx}$ is positive.*

(2) *When $x > +1$, y is decreasing and $\frac{dy}{dx}$ is negative.*

(3) *When $x = 1$ (at C) y has ceased to increase and begins to decrease.*

$\therefore$ *the value of the function at C is stationary and the curve has a stationary point.*

54. Turning points.

Comparing the stationary points in Figs. 14 and 15 of the curves of

$$y = x^2 - 4x + 3$$

and

$$y = 3 + 2x - x^2$$

we note the following important differences.

In $y = x^2 - 4x + 3$, at the stationary point,

(1) The **curve is changing from concave upwards falling to concave upwards rising** (Figs. 13(*a*) and 12(*a*). The slope, θ, changes from an **obtuse** angle, through zero, to an **acute** angle.

(2) The values of the function are **decreasing before and increasing after.**

(3) Consequently $\dfrac{dy}{dx}$ is **negative before** and **positive after.**

In $y = 3 + 2x - x^2$

(1) The **curve is changing from concave downwards rising, to concave downwards falling;** but θ is changing from an **acute** angle before the point to an **obtuse** after (cf. Figs. 12(*b*) and 13(*b*)).

(2) The values of the function are **increasing before** and **decreasing after.**

(3) Consequently $\dfrac{dy}{dx}$ is **positive before** and **negative after.**

Thus at both points:

(1) The function decreases before and increases after, or vice versa.

(2) $\dfrac{dy}{dx} = 0$ and **changes sign.**

Such points on a curve are called "**turning points**". We shall see later that not all stationary points are turning points.

It should be noted that for both stationary and turning points an essential condition is that $\dfrac{dy}{dx} = 0$. It is the behaviour of the function before and after, and consequently that of the differential coefficient, which determines the difference.

55. Worked examples.

Example 1. *For what value of* x *is there a turning point on the curve of* $y = 2x^2 - 6x + 9$?

If
$$y = 2x^2 - 6x + 9$$
$$\frac{dy}{dx} = 4x - 6.$$

For a stationary point $\frac{dy}{dx} = 0$.

$$\therefore \quad 4x - 6 = 0$$
and
$$x = 1{\cdot}5.$$

For values of $x < 1{\cdot}5$, $\frac{dy}{dx}$ is negative.

$\therefore$ function is **decreasing**.

For values of $x > 1{\cdot}5$, $\frac{dy}{dx}$ is positive.

$\therefore$ function is **increasing**.

As the function is decreasing before the stationary point and increasing after,

$\therefore$ there is a **turning point when** $x = 1{\cdot}5$.

Example 2. *Examine* $y = 1 - 2x - x^2$ *for turning points.*

If
$$y = 1 - 2x - x^2$$
$$\frac{dy}{dx} = -2 - 2x.$$

For stationary values $\frac{dy}{dx} = 0$.

$$\therefore \quad -2x - 2 = 0 \quad \text{and} \quad x = -1.$$

$\therefore$ there is a stationary point where $x = -1$.

If $x < -1$, $\frac{dy}{dx}$ is positive; $\therefore$ y is increasing.

If $x > -1$, $\frac{dy}{dx}$ is negative; $\therefore$ y is decreasing.

$\therefore$ **y is increasing before and decreasing after the stationary point.**

$\therefore$ **the stationary point is also a turning point when $x = -1$.**

Note.—The student is recommended to draw the curves of the above two functions.

56. Maximum and minimum values.

There is a very important difference between the turning points of the curves of the functions examined in § 53, viz.:

$$y = x^2 - 4x + 3$$

and

$$y = 3 + 2x - x^2,$$

as the student will have observed by an examination of Figs. 14 and 15.

(1) In $y = x^2 - 4x + 3$ (Fig. 14) the turning point C, is the **lowest** point on the curve—*i.e.*, at that point y has its **least** value. If points are taken on the curve close to and on either side of C, the value of the function at each of them is **greater** than at C, the turning point.

Such a point is called a minimum point, and the function is said to have a minimum value for the corresponding value of x.

It should be observed that values of the function decrease to the minimum point and increase after it.

(2) In $y = 3 + 2x - x^2$ (Fig. 15) the turning point, C, is the **highest** point on the curve—*i.e.*, at that point y has its **greatest** value. If, again, points are taken on the curve, close to and on either side of C, the value of the function at each of them is **less** than at C.

Such a point is called a maximum point, and the function is said to have a maximum value for the corresponding value of x.

Values of the function increase to the maximum value and decrease after it.

The values of the function at the maximum and minimum points, while greater or less than values at points close to them on the curve, are not necessarily the greatest and least values respectively which some functions may have. This will be apparent in a function such as that which is examined in the next section. Examples of both maximum and minimum values may be found in the same graph.

57. The curve of $y = (x - 1)(x - 2)(x - 3)$.

This function will vanish when $x - 1 = 0$, $x - 2 = 0$, and $x - 3 = 0$—*i.e.*, when $x = 1$, $x = 2$ and $x = 3$.

Consequently the curve will cut the x axis for these values of x. If the function is a continuous one—*i.e.*, small changes in x always produce correspondingly small changes in y—then, **between two consecutive values for which the curve cuts the axis there must be a turning point.**

Consequently for the curve of the above function there must be **two turning points.**

(1) Between the points $x = 1$ and $x = 2$.
(2) Between the points $x = 2$ and $x = 3$.

We note further by examination of the function:

(1) If $x < 1$, y is always negative.
(2) If $x > 1$ and < 2 y is positive.
(3) If $x > 2$ and < 3 y is negative.
(4) If $x > 3$, y is always positive.

These last two sets of results lead us to the conclusion

(1) That there is a maximum point (positive) between $x = 1$ and $x = 2$.

(2) That there is a minimum point (negative) between $x = 2$ and $x = 3$.

Making the usual table of corresponding values of x and y and making use of the above conclusions, the curve can be drawn as shown in Fig. 16. It would need, however, much tedious calculation to obtain with any high degree of accuracy either the value of the maximum or minimum points, or the corresponding values of x.

We therefore proceed to the algebraical treatment of the problem. Multiplying out the function, we have:

$$y = x^3 - 6x^2 + 11x - 6.$$

$$\therefore \quad \frac{dy}{dx} = 3x^2 - 12x + 11.$$

It is a necessary condition for turning points that

$$\frac{dy}{dx} = 0.$$

$$\therefore \quad 3x^2 - 12x + 11 = 0.$$

Solving the equation, the two roots are $x = 1{\cdot}42$ and $x = 2{\cdot}58$ (both approx.).

For these values of x, therefore (marked P and Q on Fig. 16), there are turning points on the curve.

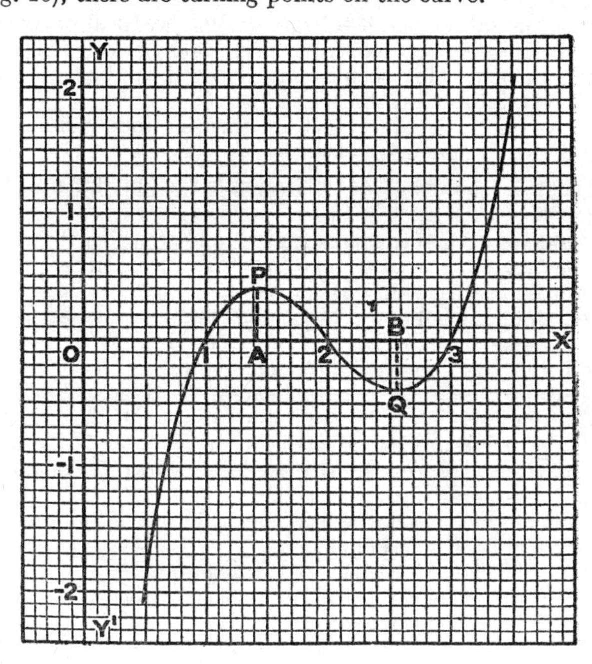

FIG. 16.

Substituting the values in the function, we get for the values of the turning points:

$$y = +\,0{\cdot}385 \text{ (}P\text{ on Fig. 16)}$$

and

$$y = -\,0{\cdot}385 \text{ (}Q\text{ on Fig. 16).}$$

The conclusion therefore is that:

(1) y has a **maximum** value of $0{\cdot}385$ when $x = 1{\cdot}42$.

(2) y has a **minimum** value of $-\,0{\cdot}385$ when $x = 2{\cdot}58$.

58. To distinguish between maximum and minimum values.

In the preceding example it was possible to decide which was a maximum and which a minimum value by reference to the curve of the function. This method, though valuable as an illustration, is not satisfactory for practical purposes. Accordingly, we proceed to examine algebraical methods, which are general in their application and can be employed with certainty and ease.

Three methods can be used; they all follow from the conclusions previously reached.

Test I. Examination of changes in the function near the turning points.

A maximum point was defined as one at which the value of the function is **greater** than for values of x, a little greater or a little less than that at the turning point.

A minimum point was similarly defined as one at which the value of the function is **less** than for values of x slightly greater or less than at the turning point.

Test I consists in the application of these definitions. Values of x slightly greater and less than that at the turning point are substituted in the function. From a comparison of the results we can decide which of the above definitions is satisfied.

This might be expressed in general terms as follows:

Let $f(x)$ be a function of x.

Let a be the value of x at a turning point.

Then $f(a)$ is the value of the function at the point.

Let h be a small number.

Then $f(a + h)$ is a value of the function slightly greater than at the turning point and $f(a - h)$ is a value of the function slightly less.

Then for a maximum $f(a)$ is greater than both $f(a + h)$ and $f(a - h)$.

Test II. Changes in the value of the differential coefficient before and after the turning point.

(1) **Maximum point.** We have seen above that:

The function is **increasing** before and **decreasing after.**

$\therefore \dfrac{dy}{dx}$ **must be positive before** and **negative after.**

To discover this, substitute in the differential coefficient values of x a little greater and a little less than the value at the point.

If it is changing sign from positive to negative through the zero value the point is a maximum.

(2) **Minimum point.** Similarly, since $\frac{dy}{dx}$ must be negative before and positive after, if on substitution as before it is changing sign from negative to positive, the point is a minimum.

Test III. Sign of the second differential coefficient.

This method is based upon the fact that $\frac{d^2y}{dx^2}$ is the differential coefficient of $\frac{dy}{dx}$, and indicates, therefore, the variations of that function.

(1) **Maximum point.**

(*a*) The function is increasing before and decreasing after.

(*b*) $\therefore$ $\frac{dy}{dx}$ is positive before and negative after.

(*c*) $\therefore$ at a maximum point $\frac{dy}{dx}$ is decreasing.

(*d*) $\therefore$ $\frac{d^2y}{dx^2}$ **must be negative.**

(2) **Minimum point.**

(*a*) The function is decreasing before and increasing after.

(*b*) $\therefore$ $\frac{dy}{dx}$ is negative before and positive after.

(*c*) $\therefore$ at a minimum point $\frac{dy}{dx}$ is increasing.

$\therefore$ $\frac{d^2y}{dx^2}$ **must be positive.**

59. Graphical illustrations.

All these conclusions can be exemplified by further consideration of the curve of

$$y = (x - 1)(x - 2)(x - 3)$$

or

$$y = x^3 - 6x^2 + 11x - 6$$

which was examined for turning points in § 57.

Since
$$f(x) = x^3 - 6x^2 + 11x - 6$$
$$f'(x) = 3x^2 - 12x + 11$$
$$f''(x) = 6x - 12.$$

All three curves are shown in Fig. 17.

Testing for turning points
$$3x^2 - 12x + 11 = 0$$
whence $x = 1{\cdot}42$ and $2{\cdot}58$ (as above).

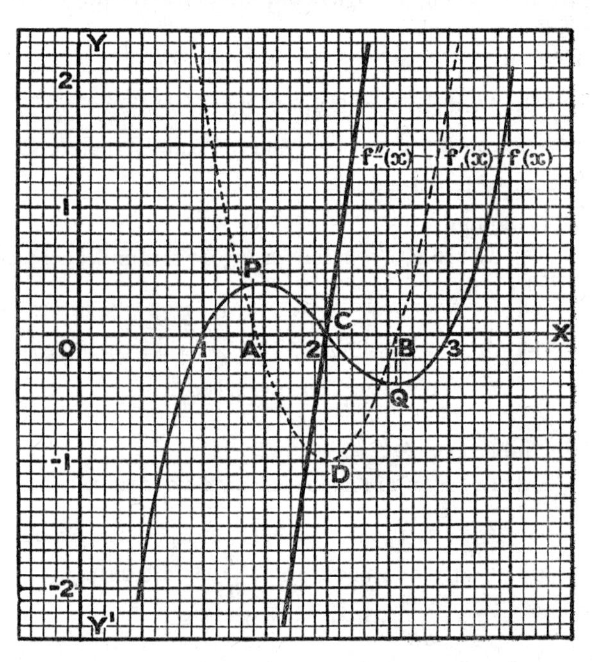

FIG. 17.

Corresponding to these values of x, marked A and B in Fig. 17, are the turning points P and Q, and it was found in § 57 that at P, $f(x) = 0{\cdot}385$ and at Q, $f(x) = -0{\cdot}385$.

Test III above may be employed to distinguish algebraically which is the maximum and which the minimum.

Accordingly we substitute in $\frac{d^2y}{dx^2}$ or $f''(x)$ those values of x which produce turning points.

From above $\frac{d^2y}{dx^2} = 6x - 12$

(1) When $x = 1{\cdot}42$, $6x - 12 = 8{\cdot}52 - 12$
$= -3{\cdot}48$,

i.e. $\frac{d^2y}{dx^2}$ is negative

$\therefore$ P must be a maximum point.

(2) When $x = 2{\cdot}58$, $6x - 12 = 15{\cdot}48 - 12$
$= +3{\cdot}48$,

i.e., $\frac{d^2y}{dx^2}$ is positive.

$\therefore$ Q must be a minimum point.

Turning to Fig. 17, we will now examine these turning points, comparing for the maximum and minimum values the corresponding curves, viz., $f(x)$, $f'(x)$ or $\frac{dy}{dx}$, $f''(x)$ or $\frac{d^2y}{dx^2}$.

A. At the maximum point P.

(1) $f'(x) = 0$, the essential condition of a turning point.
(2) $f(x)$ is increasing before P, decreasing after.
(3) $f'(x)$ is positive before P, negative after.
(4) $\therefore$ $f'(x)$ is decreasing.
(5) $\therefore$ $f''(x)$ or $\frac{d^2y}{dx^2}$ is negative.

B. At the minimum point Q.

(1) $f'(x) = 0$, the essential condition.
(2) $f(x)$ is decreasing before Q, increasing after.
(3) $f'(x)$ is negative before Q, positive after.
(4) $\therefore$ $f'(x)$ or $\frac{dy}{dx}$ is increasing.
(5) $\therefore$ $f''(x)$ or $\frac{d^2y}{dx^2}$ is positive.

All of these conclusions are illustrated in Fig. 17.

Of the three methods given above for the discrimination between maximum and minimum values of a function:

Test I is a sound one fundamentally, though the calculations are apt to be tedious.

Test II is also sound, but often laborious.

Test III is generally the easiest and most useful, but there is an exception which will be discussed later.

60. Worked examples.

Example 1. *Find the maximum or minimum value when* $y = 2x^2 - 6x + 3$.

$$\frac{dy}{dx} = 4x - 6.$$

For a maximum or minimum

$$\frac{dy}{dx} = 0.$$

$$\therefore \quad 4x - 6 = 0$$

$$x = 1{\cdot}5.$$

There is a turning point on the curve when $x = 1{\cdot}5$. To distinguish between maximum and minimum:

(*a*) Considering the expression for $\frac{dy}{dx}$, viz. $4x - 6$:

(1) If $x < 1{\cdot}5$, $\frac{dy}{dx}$ is negative.

(2) If $x > 1{\cdot}5$ $\frac{dy}{dx}$ is positive.

$\therefore$ $\frac{dy}{dx}$ is increasing as x increases.

$\therefore$ by Test II y is a minimum when $x = 1{\cdot}5$.

(*b*) $\frac{d^2y}{dx^2} = 4.$

This is always positive.

$\therefore$ by Test III y is a minimum when $x = 1{\cdot}5$.

Example 2. *Examine* $y = 5 - x - x^2$ *for turning points and distinguish between maximum and minimum.*

Since
$$y = 5 - x - x^2$$
$$\frac{dy}{dx} = -1 - 2x.$$

For a turning point $\quad -1 - 2x = 0,$

whence
$$x = -\tfrac{1}{2}.$$

(*a*) If $x < -\frac{1}{2}$, $-1 - 2x$ is positive.
If $x > -\frac{1}{2}$, $-1 - 2x$ is negative.

$\therefore \frac{dy}{dx}$ is decreasing as x increases.

$\therefore$ by Test II y is a maximum when $x = -\frac{1}{2}$.

(*b*) Also $\frac{d^2y}{dx^2} = -2$.

This is always negative.

$\therefore$ by Test III the turning point is a maximum.

Example 3. *Find the turning points on the curve of* $y = x^3 - 6x^2 + 9x - 2$, *and distinguish between maximum and minimum.*

$$y = x^3 - 6x^2 + 9x - 2.$$

$$\therefore \quad \frac{dy}{dx} = 3x^2 - 12x + 9.$$

and $$\frac{d^2y}{dx^2} = 6x - 12.$$

For turning points,

$$\frac{dy}{dx} = 0.$$

$$\therefore \quad 3x^2 - 12x + 9 = 0$$

or $$x^2 - 4x + 3 = 0.$$

$$\therefore \quad x = 3 \text{ or } 1.$$

$\therefore$ There are turning points when $x = 1$ and $x = 3$.

To distinguish between maximum and minimum, we use Test III and examine $\frac{d^2y}{dx^2}$.

From above $\frac{d^2y}{dx^2} = 6x - 12$.

If $x = 1$, $\frac{d^2y}{dx^2} = -6$. $\therefore$ a maximum point.

If $x = 3$, $\frac{d^2y}{dx^2} = +6$. $\therefore$ a minimum point.

$\therefore$ the curve has a maximum point when $x = 1$ and a minimum point when $x = 3$.

The values can be found by substituting these values for x in the function $x^3 - 6x^2 + 9x - 2$.

They are maximum value $+ 2$.
minimum value $- 2$.

Example 4. *When a body is projected vertically upwards with a velocity of* 7 ms^{-1}, *the height* (s) *reached by the body after* t *seconds is given in metres by the formula* $s = 7t - 4{\cdot}9t^2$. *Find the greatest height to which the body will rise, and the time taken.*

s is a function of t: $s = 7t - 4{\cdot}9t^2$.

Differentiating with respect to t gives

$$\frac{ds}{dt} = 7 - 9{\cdot}8t.$$

But when s is greatest

$$\frac{ds}{dt} = 0.$$

$$\therefore \quad 7 - 9{\cdot}8t = 0,$$

whence
$$t = \frac{1}{1{\cdot}4}$$

Also
$$\frac{d^2s}{dt^2} = -9{\cdot}8,$$

which is always negative, $\therefore$ the value of s when $t = \frac{1}{1{\cdot}4}$ is a maximum.

Substituting $t = \frac{1}{1{\cdot}4}$ in $s = 7t - 4{\cdot}9t^2$

we get $s = 10$ m.

Example 5. *The cost, £*C, *per mile of an electric cable is given by* $C = \frac{120}{x} + 600x$, *where* x *is its cross-section in* cm^2. *Find the cross-section for which the cost is least, and the least cost per mile.*

$$C = \frac{120}{x} + 600x.$$

$$\therefore \quad \frac{dC}{dx} = -\frac{120}{x^2} + 600.$$

For a maximum or minimum value of C, $\frac{dC}{dx} = 0$.

$$\therefore \quad -\frac{120}{x^2} + 600 = 0$$

$$\therefore \quad x^2 = \frac{120}{600}$$

$$x = \pm\sqrt{0{\cdot}2} = \pm\, 0{\cdot}447 \text{ cm}^2 \text{ (approx.)}.$$

The negative root has no meaning in this connection, and is disregarded.

To discover whether this value of x corresponds to a maximum or minimum, we use Test III.

Then $$\frac{d^2y}{dx^2} = \frac{240}{x^3}.$$

When $x = 0{\cdot}447$ this is positive.

$\therefore$ the cost is a **minimum** for this cross-section.

Substituting for x in $\frac{120}{x} + 600x$, we get the minimum cost.

Thus $$C = \frac{120}{0{\cdot}447} + 600 \times 0{\cdot}447.$$

$$= £537 \text{ (approx.)}.$$

Example 6. *A cylindrical gasometer is to be constructed so that its volume is V m³. Find the relation between the radius of the base and the height of the gasometer so that the cost of construction of the metal part, not including the base, shall be the least possible. Find also the radius of the base, r, in terms of V.*

Let h be the height of the gasometer.

Let A be the area of surface, excluding the base.

The cost will be least when A is least.

Using the formulae for a cylinder, without base,

$$A = \pi r^2 + 2\pi rh \quad . \quad . \quad . \quad . \quad . \quad (1)$$

and $$V = \pi r^2 h \quad . \quad . \quad . \quad . \quad . \quad . \quad (2)$$

These equations contain two independent variables, r and h. We accordingly eliminate one of them, h, between the two equations and obtain A in terms of r and V, which is a constant.

From (2) $$h = \frac{V}{\pi r^2}.$$

Substituting in (1)

$$A = \pi r^2 + \left(2\pi r \times \frac{V}{\pi r^2}\right).$$
$$= \pi r^2 + \frac{2V}{r}.$$

A is a function of r; $\therefore$ differentiate A with respect to r.

Then $$\frac{dA}{dr} = 2\pi r - \frac{2V}{r^2}.$$

Since A is to be a minimum, $\frac{dA}{dr}$ must equal zero.

$$\therefore \quad 2\pi r - \frac{2V}{r^2} = 0.$$
$$\therefore \quad \frac{V}{r^2} = \pi r$$

and $$V = \pi r^3.$$
$$r = \sqrt[3]{\frac{V}{\pi}}.$$

Also since $$V = \pi r^2 h$$
$$\pi r^3 = \pi r^2 h.$$
$$\therefore \quad h = r.$$

Note.—The student should not make the mistake of attempting to differentiate A with respect to r in Equation (1) as it stands. Care should be taken to distinguish between constants and variables in such equations. In addition to containing two variables, this equation does not contain the constant V. It is therefore necessary to eliminate h and obtain A in terms of r and V.

61. Points of inflexion.

When studying how to discriminate between maximum and minimum values of a function, one of the tests applied (Test III) was that of the sign of $\frac{d^2y}{dx^2}$; viz., that for a maximum it is negative, and for a minimum, positive. To complete this test it is necessary further to consider what happens when $\frac{d^2y}{dx^2} = 0$.

The following brief investigation will also include consideration of a case in which $\frac{dy}{dx} = 0$ but the function is neither a maximum nor minimum.

We will first illustrate these points by considering the case of $y = x^3$.

Then $\frac{dy}{dx} = 3x^2$

and $\frac{d^2y}{dx^2} = 6x.$

The curves of the function and its first two derivatives are shown in Fig. 18.

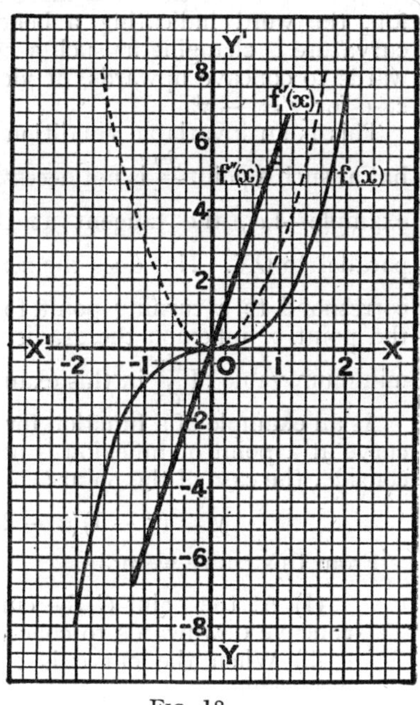

Fig. 18.

It will be seen that the curve of $y = x^3$ passes through the origin, and at that point its curvature changes from **concave downwards and rising** (Fig. 12*b*) **to concave upwards and rising** (Fig. 12*a*). Thus it is rising in each part—*i.e.*, it is **increasing throughout,** except at the origin, when the curve is momentarily stationary. At that point, therefore, there is a stationary value, the gradient is zero, and the tangent to the curve is the axis of x. It does not therefore fulfil the condition for a turning point—viz., increasing before and decreasing after, or vice versa.

The curve of its differential coefficient—*i.e.*, of $y' = 3x^2$—is shown in the parabola, which is dotted. This curve is always positive, which was to be expected from the fact that the function $y = x^3$ is always increasing. Its value is zero at the origin. This indicates that the gradient of $y = x^3$ is zero at that point, which is a minimum point for $y = 3x^2$. It shows further that $y = x^3$ has a minimum gradient at the point.

Such a point as this on a curve is called a **Point of Inflexion,** the word indicating a bending in the curve. The curvature is changing at such a point from concave downwards to concave upwards, or vice versa, as would be the case for $y = -x^3$.

This is an invariable condition for a point of inflexion, but at such a point $\frac{dy}{dx}$ is not necessarily zero, as in the above example—*i.e.*, the tangent at the point is not always parallel to OX. Nor does the zero value of $\frac{dy}{dx}$ necessarily correspond to a turning point for the function. But for the point of inflexion the gradient is a minimum and the minimum value of $\frac{dy}{dx}$ in this example is zero.

As an example of a function for which the tangent to the curve at a point of inflexion is not parallel to OX, we can consider the case of the point C on the curve of

$$y = (x - 1)(x - 2)(x - 3). \qquad \text{(Fig. 17.)}$$

From this curve we note:

(1) At the point C the curvature is changing from concave downwards to concave upwards.

(2) When the curve is concave downwards $\frac{dy}{dx}$ is decreasing.

$$\therefore \quad \frac{d^2y}{dx^2} \text{ is negative} \quad (\S\, 52).$$

When the curve is concave upwards $\frac{dy}{dx}$ is increasing.

$$\frac{d^2y}{dx^2} \text{ is positive} \quad (\S\, 52).$$

(3) At the point of change—*i.e.*, at the point of inflexion—$\frac{d^2y}{dx^2}$ is zero.

(4) Consequently $\frac{d^2y}{dx^2}$ is changing sign at the point of inflexion.

(5) At the point of inflexion, C, $\frac{dy}{dx}$, is a minimum for the corresponding value of x.

(6) This value of $\frac{dy}{dx}$—viz., -1—gives the gradient of the curve at the point of inflexion. It is therefore the gradient of the tangent at the point. If θ be the slope of the tangent, then $\tan\theta = -1$, and $\theta = 135°$.

Summing up, it may be stated that **at a point of inflexion** on a curve:

(1) *The curvature changes from concave upwards to concave downwards, or vice versa.*

(2) *Consequently* $\frac{dy}{dx}$ *will be increasing before and decreasing after, or vice versa.*

(3) *Therefore* $\frac{d^2y}{dx^2}$ *will be positive before and negative after, or vice versa.*

(4) $\frac{dy}{dx}$ *will also be a maximum or minimum.*

$$\therefore \quad \frac{d^2y}{dx^2} = 0.$$

Thus $\frac{d^2y}{dx^2}$ *changes sign through the point of inflexion.*

62. The tests for discrimination between maximum and minimum values of a function may be summarised as follows:

	Maximum.	Minimum.	Point of Inflexion.
$y = f(x)$	(1) Increasing before. (2) Decreasing after.	(1) Decreasing before (2) Increasing after.	Changing from concave up to concave down or vice-versa.
$\frac{dy}{dy}$.	(1) Positive before (2) Negative after. (3) Equal to 0 at the point. ∴ decreasing.	(1) Negative before (2) Positive after. (3) Equal to 0 at the point. ∴ increasing.	A maximum or minimum.
$\frac{d^2y}{dy^2}$.	Negative.	Positive.	Zero and changing sign.

Exercise 10.

1. Draw the curve of $y = x^2 - 2x$. Find $\frac{dy}{dx}$ and obtain its value when $x = -1, 0, 2, 3$, checking the values from the graph. For what value of x is there a turning point on the curve? Is this a maximum or minimum point? What is the sign of $\frac{d^2y}{dx^2}$?

2. Draw the curve of $y = 3x - x^2$. Find $\frac{dy}{dx}$ and calculate its value when $x = 0, 1, 2, 3$. For what value of x is $\frac{dy}{dx}$ zero? What is the sign of $\frac{d^2y}{dx^2}$ for the same value of x? Is the function a maximum or a minimum for this value?

3. Find the turning points for the following function and ascertain whether the function is a maximum or minimum in each case:

(1) $4x^2 - 2x$. (2) $x - 1{\cdot}5x^2$.
(3) $x^2 + 4x + 2$. (4) $2x^2 + x - 1$.

4. Find the maximum and minimum values of the following functions and state the corresponding values of x:

(1) $x^3 - 12x$. (2) $2x^3 - 9x^2 + 12x$.
(3) $x^3 - 6x^2 + 12$. (4) $4x^3 + 9x^2 - 12x + 13$.
(5) $2 - 9x + 6x^2 - x^3$.

5. Find the maximum and minimum values of $(x + 1)(x - 2)^2$ and the corresponding values of x.

6. Find the maximum and minimum values of $4x + \frac{1}{x}$.

7. Divide 10 into two parts such that their product is a maximum.

8. A particle is projected at an angle θ with initial velocity u. Its horizontal displacement (x) and height (y) are connected by the equation:

$$y = x \tan \theta - \frac{gx^2}{2u^2 \cos^2 \theta}$$

What is the maximum height to which the particle rises, and how far has it travelled horizontally? (g is the acceleration due to gravity, a constant.)

9. A closed cylindrical drum is to be manufactured to contain 40 m³. If the minimum amount of metal is to be

used, what is the ratio of the height of the drum to the diameter of its base?

10. An open tank is to be made of sheet iron; it must have a square base and sides perpendicular to the base. Its capacity is to be 8m^3. Find the side of the square base and the depth, so that the least amount of sheet iron may be used.

11. If $\frac{ds}{dt} = 4{\cdot}8 - 3{\cdot}2t$ and $s = 5$ when $t = 0{\cdot}5$, express s as a function of t and find its maximum value.

12. If $H = pV$ and $p = 3 - \frac{1}{2}V$, find the maximum value of H.

13. A rectangular sheet of tin, 30 cm $\times$ 24 cm, has four equal squares cut out at the corners, and the sides are then turned up to form a rectangular box. What must be the length of the side of each square cut away, so that the volume of the box may be as great as possible?

14. The strength of a rectangular beam of given length is proportional to bd^3 where b represents the breadth and d the depth. If the cross-section of a beam has a perimeter of 4 m, find the breadth and depth of the strongest beam.

15. Find the values of x corresponding to (1) a maximum value, (2) a minimum value, (3) a point of inflexion on the curve of $y = 2x^3 + 3x^2 - 36x + 10$.

16. Find the maximum and minimum values of the curve of the function $y = x(x^2 - 1)$. Find also the gradient of the curve at the point of inflexion.

17. Find the value of x at the point of inflexion of the curve of $y = 3x^3 - 4x + 5$.

18. The distance s travelled by a body propelled vertically upward in time t is given approximately by the formula

$$s = 120t - 4{\cdot}9t^2.$$

Find the greatest height which the body will reach and the time taken.

19. The bending moment (M) of a beam, supported at one end, at a distance x from one end is given by the formula

$$M = \tfrac{1}{2}wlx - \tfrac{1}{2}wx^2,$$

where l is the length and w is the uniform-load per unit length. Find the point on the beam at which the bending moment is a maximum.

CHAPTER VII

DIFFERENTIATION OF THE TRIGONOMETRIC FUNCTIONS

63. The circular measure of angles.

When considering the differentiation of the Trigonometric or Circular functions it must be remembered that the angle whose function is being examined is assumed to be measured in circular measure. Thus, when finding the differential coefficient of $\sin \theta$—*i.e.*, the rate of increase of $\sin \theta$ with respect to θ—it is clearly necessary that θ should be measured in *absolute* units, and not in arbitrarily chosen units such as degrees. Unless it is specially indicated to the contrary, in all further work in this volume angles will be regarded as measured in radians, often expressed in the convenient form of fractions or multiples of π radians.

Students who are at all hazy about circular measure should revise it before proceeding further. (*Trigonometry*, Chap. X.)

64. Differentiation of sin *x*.

Let $$y = \sin x.$$

Let δx be an increment of x.

Let δy be the corresponding increment of y.

Then $$y + \delta y = \sin (x + \delta x)$$

but $$y = \sin x.$$

$\therefore$ Subtracting $$\delta y = \sin (x + \delta x) - \sin x.$$

Dividing by δx, $$\frac{\delta y}{\delta x} = \frac{\sin (x + \delta x) - \sin x}{\delta x} \qquad \text{(A)}$$

Our next step is to find the value of the limit of the right-hand side as $\delta x \longrightarrow 0$. This requires some manipulation.

We first change the numerator from a sum to a product by employing the trigonometric formula

$$\sin P - \sin Q = 2 \cos \frac{P + Q}{2} \sin \frac{P - Q}{2}$$

(*Trigonometry*, § 87.)

where $(x + \delta x)$ takes the place of P, and x the place of Q.

Transforming the numerator in (A) we get:

$$\frac{\delta y}{\delta x} = \frac{2 \cos \frac{\{(x + \delta x) + x\}}{2} \sin \frac{\{(x + \delta x) - x\}}{2}}{\delta x}$$

$$= \frac{2 \cos \left(x + \frac{\delta x}{2}\right) \sin \frac{\delta x}{2}}{\delta x}$$

or, re-arranging

$$\frac{\delta y}{\delta x} = 2 \cos \left(x + \frac{\delta x}{2}\right) \times \frac{\sin \frac{\delta x}{2}}{\delta x}.$$

Transferring the numerical factor, 2, on the right side to the denominator, we have:

$$\frac{\delta y}{\delta x} = \cos \left(x + \frac{\delta x}{2}\right) \times \frac{\sin \frac{\delta x}{2}}{\frac{\delta x}{2}} \quad . \quad . \quad . \quad (B)$$

In this form the second factor takes the form of $\frac{\sin \theta}{\theta}$, the limit of which, when $\theta \longrightarrow 0$, was found in § 19. From this we know that, proceeding to the limit in B,

$$\underset{\delta x \rightarrow 0}{Lt} \frac{\sin \frac{\delta x}{2}}{\frac{\delta x}{2}} = 1.$$

Taking limits, therefore, we have:

$$\underset{\delta x \rightarrow 0}{Lt} \frac{\delta y}{\delta x} = \underset{\delta x \rightarrow 0}{Lt} \left\{ \cos \left(x + \frac{\delta x}{2}\right) \times \frac{\sin \frac{\delta x}{2}}{\frac{\delta x}{2}} \right\}$$

$$\therefore \quad \frac{dy}{dx} = \cos x \quad \left(\text{since } \frac{\delta x}{2} \longrightarrow 0\right)$$

Geometric proofs of the above, as well as of those which follow, are of interest, and will be found in larger books on the subject.

65. Differentiation of cos x.

Employing the notation and method used with sin x we obtain:—

$$\delta y = \cos(x + \delta x) - \cos x.$$

Using the formula

$$\cos P - \cos Q = -2 \sin \frac{P+Q}{2} \sin \frac{P-Q}{2}$$

(*Trigonometry*, § 87.)

$$\delta y = -2 \sin\left(x + \frac{\delta x}{2}\right) \sin \frac{\delta x}{2}.$$

Dividing by δx

$$\frac{\delta y}{\delta x} = -2 \sin\left(x + \frac{\delta x}{2}\right) \times \frac{\sin \frac{\delta x}{2}}{\delta x}$$

$$= -\sin\left(x + \frac{\delta x}{2}\right) \times \frac{\sin \frac{\delta x}{2}}{\frac{\delta x}{2}} \qquad \text{(as in § 64)}$$

$$\therefore \quad \underset{\delta x \to 0}{Lt} \frac{\delta y}{\delta x} = - \underset{\delta x \to 0}{Lt} \left\{ \sin\left(x + \frac{\delta x}{2}\right) \times \frac{\sin \frac{\delta x}{2}}{\frac{\delta x}{2}} \right\}$$

whence $\quad \dfrac{dy}{dx} = -\sin x.$

66. Differentiation of tan x.

This can be found most easily by making use of the differential coefficients of sin x and cos x as obtained above.

Since $\tan x = \dfrac{\sin x}{\cos x}$

$$\therefore \quad \frac{dy}{dx} = \frac{(\cos x \times \cos x) - \{\sin x \times (-\sin x)\}}{\cos^2 x}$$

(quotient rule)

$$= \frac{\cos^2 x + \sin^2 x}{\cos^2 x}$$

$$= \frac{1}{\cos^2 x} \qquad (\textit{Trig.}, \text{ § 65})$$

$$\therefore \quad \frac{dy}{dx} = \sec^2 x.$$

A proof from first principles can readily be obtained by the method employed above for $\sin x$ and $\cos x$, using the appropriate trigonometrical formula.

67. Differentiation of sec x, cosec x, cot x.

The differential coefficients of these functions can be found from first principles as above, but they are more easily obtained by expressing them as reciprocals of **cos x**, **sin x** and **tan x**, and using the rule for the differentiation of a quotient.

(*a*) $y = \operatorname{cosec} x$.

Then $$y = \frac{1}{\sin x}.$$

$$\therefore \quad \frac{dy}{dx} = \frac{0 - \cos x}{\sin^2 x} = \frac{-\cos x}{\sin^2 x} \quad \text{(quotient rule).}$$

This is more useful in the following form:

$$\frac{-\cos x}{\sin^2 x} = -\frac{\cos x}{\sin x} \times \frac{1}{\sin x}$$

$$\therefore \quad \frac{dy}{dx} = -\operatorname{cosec} x \cot x.$$

(*b*) $y = \sec x$.

$$y = \sec x = \frac{1}{\cos x}$$

$$\therefore \quad \frac{dy}{dx} = \frac{-(-\sin x)}{\cos^2 x} \quad \text{(quotient rule)}$$

$$= \frac{\sin x}{\cos^2 x}$$

$$= \frac{1}{\cos x} \times \frac{\sin x}{\cos x}$$

$$\therefore \quad \frac{dy}{dx} = \sec x \tan x.$$

(c) $y = \cot x$.

$$y = \frac{1}{\tan x}$$

$$\therefore \frac{dy}{dx} = \frac{-\sec^2 x}{\tan^2 x} \qquad \text{(quotient rule)}$$

$$= -\frac{1}{\cos^2 x} \times \frac{\cos^2 x}{\sin^2 x}$$

$$= -\frac{1}{\sin^2 x}$$

$$\therefore \frac{dy}{dx} = -\operatorname{cosec}^2 x.$$

68. Summary.

The above results are summarised as follows for convenience:

Function.	$\frac{dy}{dx}$.
$\sin x$	$\cos x$
$\cos x$	$-\sin x$
$\tan x$	$\sec^2 x$
$\operatorname{cosec} x$	$-\operatorname{cosec} x \cot x$
$\sec x$	$\sec x \tan x$
$\cot x$	$-\operatorname{cosec}^2 x$

69. Differentiation of modified forms.

The differentiation of the trigonometric functions frequently requires the application of the rule for "*a function of a function*". A very common form involves a multiple of x—for example, ax. This is a function of a function, and its differential coefficient is a. Hence this must appear as a factor of the differential coefficient.

Thus if $\quad y = \sin ax, \quad \frac{dy}{dx} = a \cos ax$

$$y = \cos ax, \ \frac{dy}{dx} = -a \sin ax$$

$$y = \tan ax, \ \frac{dy}{dx} = a \sec^2 ax$$

and similarly for their reciprocals.

Thus
$$\frac{d}{dx}\sin 2x = 2\cos 2x$$

$$\frac{d}{dx}\cos\frac{x}{2} = -\frac{1}{2}\sin\frac{x}{2}$$

$$\frac{d}{dx}\tan\frac{ax}{b} = \frac{a}{b}\sec^2\frac{ax}{b}.$$

Slightly more complicated forms are such as the following:

If
$$y = \sin(ax + b), \ \frac{dy}{dx} = a\cos(ax + b)$$

$$y = \sin(\pi + nx), \ \frac{dy}{dx} = n\cos(\pi + nx)$$

$$y = \tan(1 - x), \ \frac{dy}{dx} = -\sec^2(1 - x)$$

$$y = \sin\left(\frac{1}{x^2}\right), \quad \frac{dy}{dx} = -\frac{2}{x^3}\cos\left(\frac{1}{x^2}\right).$$

70. Worked examples.

Example 1. *Differentiate* $y = \sin^2 x$.

i.e.
$$y = (\sin x)^2$$

$$\therefore \quad \frac{dy}{dx} = 2\sin x \times \frac{d(\sin x)}{dx}$$

$$= 2\sin x\cos x$$

$$= \sin 2x.$$

Example 2. *Differentiate* $y = \sin\sqrt{x}$.

$$y = \sin x^{\frac{1}{2}}$$

$$\therefore \quad \frac{dy}{dx} = \cos(x^{\frac{1}{2}}) \times (\tfrac{1}{2}x^{-\frac{1}{2}})$$

$$= \tfrac{1}{2}\cos x^{\frac{1}{2}} \times \frac{1}{\sqrt{x}}.$$

$$= \frac{\cos\sqrt{x}}{2\sqrt{x}}.$$

Example 3. *Differentiate* $y = \sqrt{\sin x}$;

i.e. $= (\sin x)^{\frac{1}{2}}$

$$\therefore \quad \frac{dy}{dx} = \tfrac{1}{2}(\sin x)^{-\frac{1}{2}} \times \frac{d(\sin x)}{dx}$$

$$= \frac{\cos x}{2(\sin x)^{\frac{1}{2}}} = \frac{\cos x}{2\sqrt{\sin x}}.$$

Example 4. *Differentiate* $y = \sin^2 (x^2)$. (See § 47.)

$$y = (\sin x^2)^2.$$

$$\therefore \quad \frac{dy}{dx} = 2 \sin x^2 \times \cos x^2 \times 2x$$

$$= 4x \sin x^2 \cos x^2.$$

Exercise 11.

Differentiate the following:

1. $3 \sin x$.
2. $\sin 3x$.
3. $\cos \frac{x}{2}$.
4. $\tan \frac{x}{3}$.
5. $\sec 0{\cdot}6x$.
6. $\operatorname{cosec} \frac{x}{6}$.
7. $\sin 2x + \cos 2x$.
8. $\sin 3x - \cos 3x$.
9. $\sec x + \tan x$.
10. $\sin 4x + \cos 5x$.
11. $\cos \frac{1}{2}\theta + \sin \frac{1}{4}\theta$.
12. $\sin \left(2x + \frac{\pi}{2}\right)$.
13. $\cos (3\pi - x)$.
14. $\operatorname{cosec} (a - \frac{1}{2}x)$.
15. $\sin^3 x$.
16. $\sin (x^3)$.
17. $\cos^3 (2x)$.
18. $\sec (x^2)$.
19. $\tan \sqrt{1 - x}$.
20. $a \sin nx + b \cos nx$.
21. $a(1 - \cos x)$.
22. $2 \tan \frac{x}{2}$.
23. $\cos \left(2x + \frac{\pi}{2}\right)$.
24. $\tan 2x - \tan^2 x$.
25. $x^2 + 3 \sin \frac{1}{2}x$.
26. $\cos \frac{a}{x}$.
27. $x \sin x$.
28. $\frac{x}{\sin x}$.
29. $x \tan x$.
30. $\frac{x}{\tan x}$.
31. $\frac{\tan x}{x}$.
32. $\sin 2x + \sin (2x)^2$.
33. $\cos^3 (x^2)$.
34. $x^2 \tan x$.

35. $\cot(5x + 1)$.
36. $\cot^2 3x$.
37. $\sqrt{\cos x}$.
38. $\sin 2x \cos 2x$.
39. $\sin^2 x + \cos^2 x$.
40. $\sin^2 x - \cos^2 x$.
41. $\dfrac{1}{1 + \cos x}$.
42. $\dfrac{1 - \cos x}{1 + \cos x}$.
43. $\dfrac{\sqrt{x}}{\sin x}$.
44. $x^2 \cos 2x$.
45. $\dfrac{x^2}{\cos 2x}$.
46. $\dfrac{\tan x - 1}{\sec x}$.
47. $x\sqrt{\sin x}$.
48. $\dfrac{\sin^2 x}{1 + \sin x}$.
49. $\dfrac{1}{1 - \tan x}$.
50. $\sec^2 x \operatorname{cosec} x$.

71. Successive derivatives.

Let $$y = \sin x.$$

Then
$$\frac{dy}{dx} = \cos x$$
$$\frac{d^2y}{dx^2} = -\sin x$$
$$\frac{d^3y}{dx^3} = -\cos x$$
$$\frac{d^4y}{dy^4} = \sin x.$$

Clearly these derivatives will repeat in sets of four, identical with the first four above.

From *Trigonometry* we know that $\cos x = \sin\left(x + \frac{\pi}{2}\right)$.
$\therefore$ the above may be written:

$$\frac{dy}{dx} = \cos x = \sin\left(x + \frac{\pi}{2}\right)$$
$$\frac{d^2y}{dx^2} = \frac{d}{dx}\left\{\sin\left(x + \frac{\pi}{2}\right)\right\} = \cos\left(x + \frac{\pi}{2}\right) = \sin\left(x + \frac{2\pi}{2}\right)$$
$$\frac{d^3y}{dx^3} = \frac{d}{dx}\left\{\sin\left(x + \frac{2\pi}{2}\right)\right\} = \cos\left(x + \frac{2\pi}{2}\right) = \sin\left(x + \frac{3\pi}{2}\right).$$

They may be continued indefinitely, $\frac{\pi}{2}$ being added in each successive derivative, the sine form being retained.

Thus it may be deduced that

$$\frac{d^n y}{dx^n} = \sin\left(x + \frac{n\pi}{2}\right).$$

Successive derivatives of $\cos x$ may be similarly obtained. Those of $\tan x$, $\sec x$, $\operatorname{cosec} x$, and $\cot x$ become complicated after a few steps in differentiation, and cannot be expressed by a general formula.

72. Maximum and minimum values of trigonometric functions.

Note.—Unless the student is familiar with functions of an angle of any magnitude, he should revise *Trigonometry*, §§ 130–136.

(1) $y = \sin x$, $y = \cos x$.

When

$$y = \sin x$$

$$\frac{dy}{dx} = \cos x$$

$$\frac{d^2y}{dx^2} = -\sin x.$$

The graph of $\sin x$ is represented by the thickest curve in Fig. 19. The broken curve is that of $\frac{dy}{dx}$, and the thin one $\frac{d^2y}{dx^2}$.

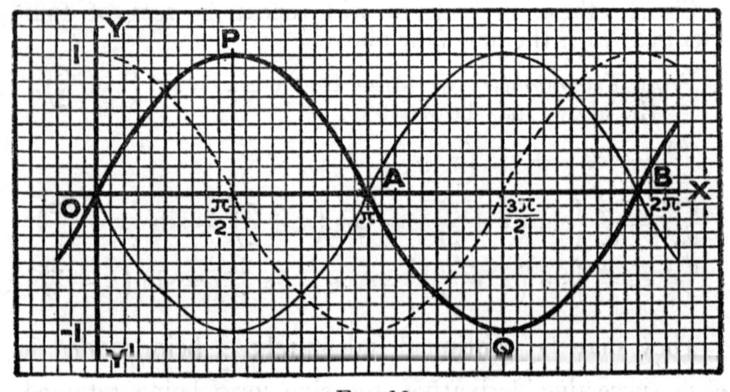

FIG. 19.

A periodic function. Since $\sin x = \sin(x + 2\pi)$, the portion of the curve between $x = 0$ and $x = 2\pi$ will be repeated for intervals of 2π as x increases. There will be similar sections for negative angles.

Thus the section of the curve between 0 and 2π will be repeated an infinite number of times between $-\infty$ and $+\infty$, the whole forming one **continuous** curve.

$\sin x$ is an example of what is termed a **periodic function,** and the number 2π is called the **period** of the function.

The following **characteristics of the curve of $\sin x$** illustrate much of the work of the preceding chapter.

(*a*) **Types of curvature.** The curve between 0 and 2π provides examples of the four types of curvature illustrated in Figs. 12(*a*) and (*b*) and Fig. 13(*a*) and (*b*), while that of $\frac{dy}{dx}$ illustrates the connection between these forms of curvature and the sign of the differential coefficient (see § 52).

(*b*) **Turning points.** The curve between 0 and 2π shows that between these two values of x there are **two turning points,** at P and Q, the values being $+1$ and -1.

At P, when $x = \frac{\pi}{2}$, $\frac{dy}{dx} = 0$, and $\frac{d^2y}{dx^2}$ **is negative.**

$\therefore$ P is a **maximum point.**

At Q, when $x = \frac{3\pi}{2}$, $\frac{dy}{dx} = 0$ and $\frac{d^2y}{dx^2}$ **is positive.**

$\therefore$ Q is a **minimum point.**

This is true for any section of 2π as x increases. Consequently throughout the curve from $-\infty$ to $+\infty$ there is an **infinite sequence of turning points, alternately maximum and minimum.**

(*c*) **Points of Inflexion.** There are two points of inflexion on this section of the curve at A and B. At A the curve changes from concave down to concave up, $\frac{dy}{dx}$ is a minimum, viz., -1, $\frac{d^2y}{dx^2} = 0$ and is changing from negative to positive.

Hence A is a point of **minimum gradient.** Its

gradient is given by the value of $\frac{dy}{dx}$ at the point, viz., -1. As this is the tangent of the angle of slope, the curve crosses the axis at an angle of $\frac{3\pi}{4}$.

At B this is reversed. The curve changes from concave up to concave down, $\frac{dy}{dx}$ is a maximum, and $\frac{d^2y}{dx^2} = 0$, and is changing sign from positive to negative. B is therefore a point of **maximum** gradient. This is equal to $+1$, and the curve cuts the axis at $\frac{\pi}{4}$. There is also a point of inflexion at the origin.

It will be seen that Fig. 19 illustrates graphically the whole of the summary in § 62.

The **graph of cos** x is that of $\sin x$, moved $\frac{\pi}{2}$ to the left along OX. The curve of $\frac{dy}{dx}$ shows its shape and position in Fig. 19. Consequently with angles, where they occur, diminished by $\frac{\pi}{2}$, the above remarks respecting **sin** x are applicable to $\cos x$.

(2) $y = \tan x$, $y = \cot x$.

When:

$$y = \tan x$$
$$\frac{dy}{dx} = \sec^2 x$$
$$\frac{d^2y}{dx^2} = 2 \sec^2 x \tan x$$

When:

$$y = \cot x$$
$$\frac{dy}{dx} = -\operatorname{cosec}^2 x$$
$$\frac{d^2y}{dx^2} = 2 \operatorname{cosec}^2 x \cot x$$

The graphs of $\tan x$ and of its differential coefficient $\sec^2 x$ are represented in Fig. 20, the latter curve being dotted.

The following characteristics of the curve of $y = \tan x$ may be noted:

(*a*) **The curve is discontinuous.** When $x \longrightarrow \frac{\pi}{2}$, $\tan x \longrightarrow +\infty$. On passing through $\frac{\pi}{2}$, an infinitely small increment of x results in the angle being in the second quadrant. Its tangent is therefore negative, while still numerically infinitely great. With this infinitely small increase in x tan θ changes from $+\infty$ to $-\infty$. The curve of the function is therefore discontinuous. Similar changes occur when $x = \frac{3\pi}{2}$, $\frac{5\pi}{2}$, etc. This can be observed in Fig. 20.

(*b*) **The curve of tan x is consequently periodic and the period is π.**

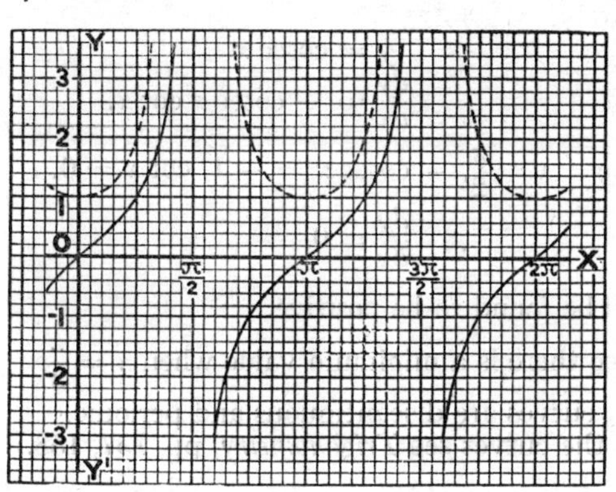

FIG. 20.

(*c*) **The function is always increasing,** and this is indicated by the fact that $\frac{dy}{dx}$, viz., $\sec^2 x$, is always positive.

(*d*) **There is a point of inflexion when $x = \pi$.** The curve is changing from concave down to concave up, the differential coefficient, $\sec^2 x$, is a minimum, and its value is $+1$. Consequently the curve crosses OX at an angle of $\frac{\pi}{4}$. Similar points occur for $x = 0$ and any integral multiple of π.

Since $\cot x = \frac{1}{\tan x}$ its curve is the inversion of that of tan x. It is always decreasing ($-\operatorname{cosec}^2 x$ is always negative); it is periodic and has points of inflexion when $x = \frac{\pi}{2}, \frac{3\pi}{2}$ . . . The student should draw it as an exercise.

(3) $y = \text{cosec } x$, $y = \sec x$.

Turning points on these curves may be deduced from those of their reciprocals. When $\sin x$ is a maximum, $\text{cosec } x$ is a minimum; consequently the curves are periodic, and maximum, and minimum values occur alternately.

If
$$y = \text{cosec } x$$
$$\frac{dy}{dx} = -\text{cosec } x \cot x.$$

When $x = \frac{\pi}{2}$, $-\text{cosec } x = -1$, $\cot x = 0$.

$$\therefore \quad \frac{dy}{dx} = 0.$$

$\frac{d^2y}{dx^2}$ will be found to be positive.

Hence there is a minimum value when $x = \frac{\pi}{2}$.

Both curves are discontinuous and periodic.
(For the curves see *Trigonometry*, pp. 157, 158.)

73. Worked example.

Find the turning points on the curve of $y = \sin x + \cos x$.

If
$$y = \sin x + \cos x$$
$$\frac{dy}{dx} = \cos x - \sin x.$$

For turning points
$$\frac{dy}{dx} = 0.$$

Putting
$$\cos x - \sin x = 0$$
$$\sin x = \cos x$$
and
$$\tan x = 1.$$
$$\therefore \quad x = \frac{\pi}{4}.$$

But this is the smallest of a series of angles whose tangent is $+ 1$.

All these angles are included in the general formula

$$n\pi + \frac{\pi}{4} \qquad (\textit{Trig.}, \S\ 136.)$$

$\therefore$ the angles for which there are turning points in the above function are

$$\frac{\pi}{4}, \frac{5\pi}{4}, \frac{9\pi}{4} \ldots$$

Also $\qquad \frac{d^2y}{dx^2} = -\sin x - \cos x.$

This is negative when

$$x = \frac{\pi}{4}, \frac{9\pi}{4} \ldots$$

and positive when

$$x = \frac{5\pi}{4} \ldots$$

$\therefore$ the curve is periodic, and maximum and minimum values occur alternately:

Maximum when $\qquad x = \frac{\pi}{4}, \frac{9\pi}{4} \ldots$

Minimum when $\qquad x = \frac{5\pi}{4}, \frac{13\pi}{4} \ldots$

Max. value $\qquad = \sin\frac{\pi}{4} + \cos\frac{\pi}{4}$

$$= \frac{1}{\sqrt{2}} + \frac{1}{\sqrt{2}} = \sqrt{2}.$$

Similarly minimum value $= -\sqrt{2}$.

The curve is represented in Fig. 21. P is the maximum point and Q the minimum. A is obviously a point of inflexion.

The curve can be drawn by first drawing the curves of $\sin x$ and $\cos x$, and then adding the ordinates of the two curves for various values of x.

The curve is a simple example of what are termed **Harmonic Curves**, or wave diagrams, which are of importance in Electrical Engineering (see *Trigonometry*, § 139).

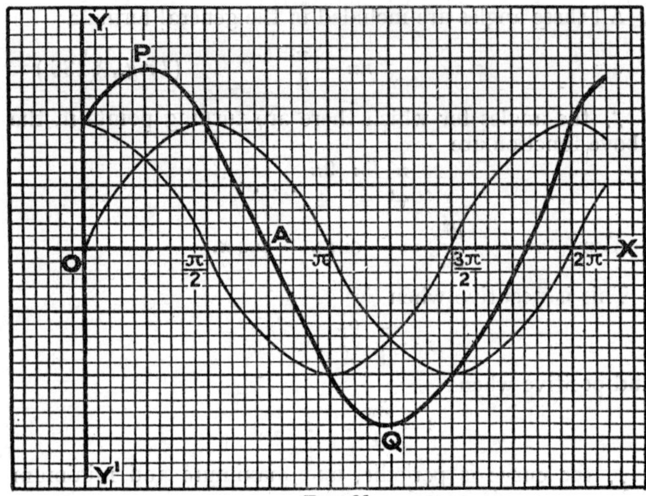

FIG. 21.

Exercise 12.

For what values of x, not greater than π are there maximum or minimum values of the following?

1. $\sin 2x - x$.
2. $\sin^2 x \cos^2 x$.
3. $\sin x + \sin x \cos x$.
4. $\dfrac{\sin x}{1 + \tan x}$.
5. $2 \sin x + \cos x$.
6. $\sin x + \cos 2x$.
7. $2 \sin x - \sin 2x$.
8. $\sin x \sin 2x$.
9. What is the smallest value of x for which $2 \sin x + 3 \cos x$ is a maximum.
10. Find the smallest value of x for which $\tan^2 x - 2 \tan x$ is a maximum or minimum.
11. Show that the maximum value of $a \sin \theta + b \cos \theta$ is $\sqrt{a^2 + b^2}$ and the minimum value $-\sqrt{a^2 + b^2}$.

(*Trig.*, § 139.)

74. Inverse circular functions.

When we write $y = \sin x$ the sine is expressed as a function of the angle denoted by x. When x varies, the sine varies in consequence—*i.e.*, the angle is the independent variable and the sine the dependent variable.

But we may require to reverse this relation, *i.e.*, **to express the angle as a function of the sine.** Thus we express the fact that when the sine is varied, the angle varies in consequence. **The sine now becomes the independent variable and the angle the dependent variable.** This relation, as the student knows from Trigonometry, is expressed by the form

$$y = \sin^{-1} x$$

which means, **y is the angle of which x is the sine.** From this meaning we can write down the direct function relation, viz.:

$$x = \sin y.$$

It must be noted that the -1 is not an index, but a part of the symbol $\sin^{-1}$, which expresses the inverse function.

All the other circular functions can similarly be expressed as inverse functions.

75. Differentiation of $\sin^{-1} x$ and $\cos^{-1} x$.

Let $\qquad y = \sin^{-1} x$

Then, as shown $\qquad x = \sin y \quad . \quad . \quad . \quad . \quad . \quad$ (A)

Differentiating x with respect to y.

$$\frac{dx}{dy} = \cos y.$$

$$\therefore \quad \frac{dy}{dx} = \frac{1}{\frac{dx}{dy}} = \frac{1}{\cos y}.$$

From the relation $\sin^2 y + \cos^2 y = 1$ we have

$$\cos y = \sqrt{1 - \sin^2 y} = \sqrt{1 - x^2} \quad \text{(from A)}$$

Hence $\qquad \dfrac{dy}{dx} = \dfrac{1}{\sqrt{1 - x^2}}.$

Similarly if $$y = \cos^{-1} x$$

$$\frac{dy}{dx} = -\frac{1}{\sqrt{1 - x^2}}.$$

The following points should be noted about these functions and their differential coefficients. They can be examined more easily by means of the graph of the function $y = \sin^{-1} x$ (Fig. 22).

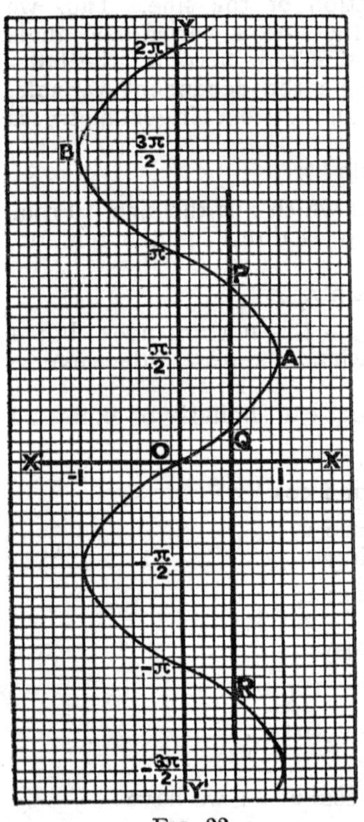

FIG. 22.

(1) The function is a many valued function—*i.e.*, for any assigned value of x there is an infinite number of values of y. $y = \sin x$ is a single valued function.

(2) Since $\sin y$ lies between $+1$ and -1, the function $\sin^{-1} x$ exists between these values of x only.

(3) Since there is an infinite number of angles having a given sine, so for any value of x between $+1$ and -1 there is an infinite number of points on the curve. For example, if $x = \frac{1}{2}$, the values of y at P, Q and R represent three of the angles whose sine is $\frac{1}{2}$, and that at Q is the smallest positive angle.

(4) The differential coefficient of $\sin^{-1} x$, viz., $\frac{1}{\sqrt{1 - x^2}}$, may be positive or negative. Referring to Fig. 22, it will be seen that at all such points as Q, where the

gradient of the curve is the tangent of an acute angle, the d.c. will be positive, while at such points as P and R, where the angle of slope is obtuse, the d.c. will be negative.

(5) Since x lies between $+1$ and -1, $\frac{1}{\sqrt{1-x^2}}$ cannot vanish. Therefore there are no maximum or minimum points on the curve. If $x = \pm 1$, $\frac{1}{\sqrt{1-x^2}}$ becomes infinite. Consequently, at such points as A and B the tangent to the curve is perpendicular to the x-axis.

76. Differentiation of $\tan^{-1} x$ and $\cot^{-1} x$.

Let $\quad y = \tan^{-1} x$

Then $\quad x = \tan y$

Differentiating with respect to y

$$\frac{dx}{dy} = \sec^2 y$$

and

$$\frac{dy}{dx} = \frac{1}{\sec^2 y}$$

$$= \frac{1}{1 + \tan^2 y}$$

$$= \frac{1}{1 + x^2}$$

$$\therefore \quad \frac{dy}{dx} = \frac{1}{1 + x^2}.$$

Similarly, if $y = \cot^{-1} x$ we can show that

$$\frac{dy}{dx} = -\frac{1}{1 + x^2}.$$

In this case there is no ambiguity of sign.

The following points are illustrated by the graph of $\tan^{-1} x$ in Fig. 23.

(1) $\frac{dy}{dx}$ is always positive; $\therefore y$ is always increasing.

(2) $\frac{dy}{dx}$ does not vanish for any value of x; $\therefore$ there is no turning point.

(3) Points of inflexion occur when $y = 0, \pi, 2\pi, -\pi$, etc.

The gradient is positive.

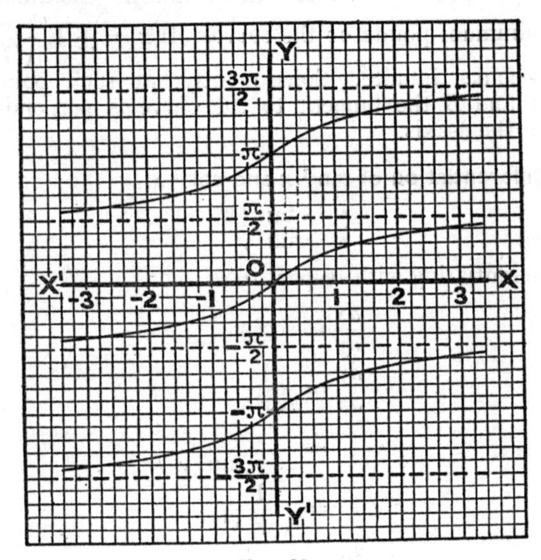

Fig. 23.

The graph of $y = \cot^{-1} x$ is the reverse of this curve.

$\frac{dy}{dx}$ is always negative, $\therefore$ the function is always decreasing.

There are no turning points, but a series of points of inflexion at which the gradient is negative.

The drawing of the curve is left as an exercise to the student.

77. Differentiation of $y = \sec^{-1} x$ and $y = \operatorname{cosec}^{-1} x$.

Let $\qquad y = \sec^{-1} x.$

Then $\qquad x = \sec y.$

$$\therefore \quad \frac{dx}{dy} = \sec y \tan y$$

and $$\frac{dy}{dx} = \frac{1}{\sec y \tan y}$$

but $$\tan y = \sqrt{\sec^2 y - 1} = \sqrt{x^2 - 1}.$$

$$\therefore \quad \frac{dy}{dx} = \frac{1}{x\sqrt{x^2 - 1}}.$$

Similarly if $y = \operatorname{cosec}^{-1} x$

$$\frac{dy}{dx} = -\frac{1}{x\sqrt{x^2 - 1}}.$$

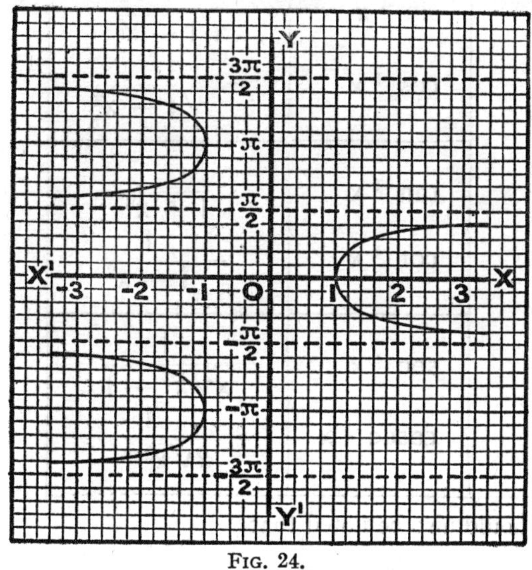

FIG. 24.

Fig. 24 represents part of the curve of $\sec^{-1} x$. It is a many-valued discontinuous curve with no part of it between $x = +1$ and $x = -1$.

$\frac{dy}{dx}$—*i.e.*, $\frac{1}{x\sqrt{x^2 - 1}}$ cannot vanish for any finite value of x. There are therefore no turning points, but when $x = \pm 1$, $\frac{dy}{dx}$ becomes infinite, as was the case in the curve of $\sin^{-1} x$ (Fig. 22). The curve of $\text{cosec}^{-1} x$ is similar.

78. Summary of formulae.

The differential coefficients of the inverse functions are collected together below for reference.

Function.	$\frac{dy}{dx}$.
$\sin^{-1} x$	$\frac{1}{\sqrt{1 - x^2}}$
$\cos^{-1} x$	$-\frac{1}{\sqrt{1 - x^2}}$
$\tan^{-1} x$	$\frac{1}{1 + x^2}$
$\cot^{-1} x$	$-\frac{1}{1 + x^2}$
$\sec^{-1} x$	$\frac{1}{x\sqrt{x^2 - 1}}$
$\text{cosec}^{-1} x$	$-\frac{1}{x\sqrt{x^2 - 1}}$

It should also be noted that

$$\sin^{-1} \frac{x}{a} = \frac{1}{\sqrt{a^2 - x^2}}$$

$$\tan^{-1} \frac{x}{a} = \frac{a}{a^2 + x^2}$$

$$\sec^{-1} \frac{x}{a} = \frac{a}{x\sqrt{x^2 - a^2}}.$$

Similarly for the other three functions.

79. Worked examples.

Example 1. *Differentiate* $\sin^{-1} x^2$.

Using the rule for a "function of a function"

$$\frac{dy}{dx} = \frac{1}{\sqrt{1-(x^2)^2}} \times \frac{d}{dx}(x^2)$$

$$= \frac{2x}{\sqrt{1-x^4}}.$$

Example 2. *Differentiate* $\tan^{-1} \frac{1}{x^2}$.

$$\frac{dy}{dx} = \frac{1}{1+\left(\frac{1}{x^2}\right)^2} \times \frac{d}{dx}\left(\frac{1}{x^2}\right)$$

$$= \frac{1}{1+\frac{1}{x^4}} \times \frac{-2}{x^3}$$

$$= \frac{x^4}{x^4+1} \times \frac{-2}{x^3}$$

$$= \frac{-2x}{x^4+1}.$$

Example 3. *Differentiate* $x^2 \sin^{-1}(1-x)$.

Using the rule for differentiation of a product

$$\frac{dy}{dx} = 2x \sin^{-1}(1-x) + x^2 \times \frac{1}{\sqrt{1-(1-x)^2}} \times \frac{d}{dx}(1-x)$$

$$= 2x \sin^{-1}(1-x) + \frac{x^2}{\sqrt{1-(1-2x+x^2)}} \times (-1)$$

$$= 2x \sin^{-1}(1-x) - \frac{x^2}{\sqrt{2x-x^2}}.$$

Exercise 13.

Differentiate the following functions:

1. (*a*) $\sin^{-1} 4x$; (*b*) $\sin^{-1} \frac{x}{2}$.
2. (*a*) $b \cos^{-1}\left(\frac{x}{a}\right)$; (*b*) $\cos^{-1} \frac{x}{3}$.

3. (a) $\tan^{-1}\frac{x}{a}$; (b) $\tan^{-1}(a - x)$.
4. (a) $\cos^{-1} 2x^2$; (b) $\sin^{-1}\sqrt{x}$.
5. (a) $x \sin^{-1} x$; (b) $\sin^{-1}\frac{1}{x}$.
6. (a) $\sin^{-1}(3x - 1)$; (b) $\operatorname{cosec}^{-1}\frac{x}{2}$.
7. (a) $\tan^{-1}(x + 1)$; (b) $(x^2 + 1)\tan^{-1} x$.
8. (a) $\tan^{-1}\sqrt{1 - x}$; (b) $\sin^{-1}\sqrt{1 - x^2}$.
9. (a) $\sec^{-1} 5x$; (b) $\sec^{-1} x^2$.
10. (a) $\sin^{-1}(\sin x)$; (b) $\sin^{-1}\sqrt{\sin x}$.
11. (a) $2\sec^{-1} ax$; (b) $\tan^{-1}\sqrt{x}$.
12. (a) $\tan^{-1}\frac{2x}{1 - x^2}$; (b) $\tan^{-1}\frac{\sqrt{1 - x^2}}{x}$.
13. (a) $\sec^{-1}\frac{a}{\sqrt{a^2 - x^2}}$; (b) $\sec^{-1}\frac{x^2 + 1}{x^2 - 1}$.
14. (a) $\sin^{-1}\frac{x}{\sqrt{1 + x^2}}$; (b) $\operatorname{cosec}^{-1}\frac{1}{2x^2 - 1}$.
15. (a) $x \tan^{-1} x$; (b) $\tan x \sin^{-1} x$.

CHAPTER VIII

EXPONENTIAL AND LOGARITHMIC FUNCTIONS

80. Compound Interest Law of Growth.

THE student will be familiar with two methods of payment of interest on money, termed Simple and Compound Interest (*Algebra*, § 207). In each the interest bears a fixed ratio to the magnitude of the sum of money involved. But while with Simple Interest the principal remains the same from year to year, with Compound Interest it is added to the principal at the end of each year, over a period, and the interest for the succeeding year is calculated on the sum of principal and interest.

Let P = the Principal.
Let r = the rate per cent. per annum.
Interest added at end of 1st year $= P \times \frac{r}{100}$.

$$\therefore \text{ Amount at end of 1st year} = P + \frac{Pr}{100}$$

$$= P\left(1 + \frac{r}{100}\right).$$

This is the principal for the new year.
∴ by the same working as for the 1st year

$$\text{Amount at end of 2nd year} = P\left(1 + \frac{r}{100}\right)^2$$

$$\text{,, \quad ,, \quad 3rd ,,} = P\left(1 + \frac{r}{100}\right)^3$$

$$\text{,, \quad ,, \quad } t\text{th ,,} = P\left(1 + \frac{r}{100}\right)^t.$$

Suppose the interest is added at the end of **each half year** instead of at the end of each year, then:

$$\text{Amount at end of 1st half-year} = P\left(1 + \frac{r}{2 \times 100}\right)$$

$$\text{,, \quad ,, \quad 1st year} = P\left(1 + \frac{r}{2 \times 100}\right)^2$$

$$\text{,, \quad ,, \quad 2nd ,,} = P\left(1 + \frac{r}{2 \times 100}\right)^4$$

$$\therefore \text{ Amount at end of } t \text{ years} = P\left(1 + \frac{r}{2 \times 100}\right)^{2t}$$

If the interest is added 4 times a year:

Amount at end of 1st year $= P\left(1 + \frac{r}{4 \times 100}\right)^4$

" " t " $= P\left(1 + \frac{r}{4 \times 100}\right)^{4t}$

Similarly, if the interest is added monthly, *i.e.*, 12 times a year:

Amount at end of t year $= P\left(1 + \frac{r}{12 \times 100}\right)^{12t}$.

If the interest is added m times a year:

Amount at end of t years $= P\left(1 + \frac{r}{100m}\right)^{mt}$.

In this result let

$$\frac{r}{100m} = \frac{1}{n}.$$

Then $$m = \frac{nr}{100}$$

$\therefore$ the amount after t years $= P\left(1 + \frac{1}{n}\right)^{\frac{nrt}{100}}$

$$= P\left\{\left(1 + \frac{1}{n}\right)^n\right\}^{\frac{rt}{100}}$$

Now suppose that n becomes indefinitely large, *i.e.*, the interest is added on at indefinitely small intervals, so that the growth of the principal may be regarded as continuous.

Then the amount reached will be the limit of

$$P\left\{\left(1 + \frac{1}{n}\right)^n\right\}^{\frac{rt}{100}}$$

when n becomes infinitely large.

To find this we require to find the limit of $\left(1 + \frac{1}{n}\right)^n$ as $n \longrightarrow \infty$, *i.e.*,

$$\text{Amount} = P\left\{\underset{n \to \infty}{Lt}\left(1 + \frac{1}{n}\right)^n\right\}^{\frac{rt}{100}}$$

It becomes necessary, therefore, to find the value of

$$\underset{n \to \infty}{Lt}\left(1 + \frac{1}{n}\right)^n.$$

81. The value of $\underset{n \to \infty}{Lt} \left(1 + \frac{1}{n}\right)^n$.

Expanding $\left(1 + \frac{1}{n}\right)^n$ by means of the Binomial Theorem

$$\left(1 + \frac{1}{n}\right)^n = 1 + n \cdot \frac{1}{n} + \frac{n(n-1)}{\lfloor 2} \cdot \frac{1}{n^2} + \frac{n(n-1)(n-2)}{\lfloor 3} \cdot \frac{1}{n^3} + \ldots$$

Simplifying by dividing the factors in the numerators by $n, n^2, n^3 \ldots$

Then

$$\left(1 + \frac{1}{n}\right)^n = 1 + 1 + \frac{1 - \frac{1}{n}}{\lfloor 2} + \frac{\left(1 - \frac{1}{n}\right)\left(1 - \frac{2}{n}\right)}{\lfloor 3} + \ldots + \frac{\left(1 - \frac{1}{n}\right)\left(1 - \frac{2}{n}\right) \ldots \left(1 - \frac{r-1}{n}\right)}{\lfloor r} + \ldots$$

But the limit of $\left(1 + \frac{1}{n}\right)^n$ is equal to the sum of the limits. (Th. limits No. 2.)

Also

$$\underset{n \to \infty}{Lt} \frac{1 - \frac{1}{n}}{\lfloor 2} = \frac{1}{\lfloor 2} \ldots, \quad \underset{n \to \infty}{Lt} \frac{\left(1 - \frac{1}{n}\right)\left(1 - \frac{2}{n}\right)}{\lfloor 3} = \frac{1}{\lfloor 3}$$

and $\underset{n \to \infty}{Lt} \dfrac{\left(1 - \frac{1}{n}\right)\left(1 - \frac{2}{n}\right) \ldots \left(1 - \frac{r-1}{n}\right)}{\lfloor r} = \dfrac{1}{\lfloor r}$, etc.

$$\therefore \quad \underset{n \to \infty}{Lt} \left(1 + \frac{1}{n}\right)^n = 1 + \frac{1}{1} + \frac{1}{\lfloor 2} + \frac{1}{\lfloor 3} + \frac{1}{\lfloor 4} + \ldots$$

The limit is thus represented by an infinite series. It can be proved that as the number of terms is increased without limit, the sum of all the terms approaches a finite limit, *i.e.*, the series is **convergent** (see § 18). Its value has been calculated to hundreds of places of decimals, and can be found arithmetically as follows to any required degree of accuracy. Each term can be found from the preceding by simple division of the preceding by the new factor in the denominator. Thus:

1st term	= 1·000000
2nd ,,	= 1·000000
3rd ,,	= 0·500000 (dividing 2nd by 2)
4th ,,	= 0·166667
5th ,,	= 0·041667
6th ,,	= 0·008333
7th ,,	= 0·001389
8th ,,	= 0·000198
9th ,,	= 0·000025
10th ,,	= 0·000003
Sum of 10 terms	= 2·718282

Thus its value to 6 significant figures is 2·71828.

This constant is always denoted by the letter e.

$$\text{or} \qquad e = \underset{n \to \infty}{Lt} \left(1 + \frac{1}{n}\right)^n$$

$$= 1 + \frac{1}{1} + \frac{1}{\underline{|2}} + \frac{1}{\underline{|3}} + \frac{1}{\underline{|4}} + \ldots$$

We have seen above that the amount (A) at C.I. after t years when the interest is continuously added is

$$A = P\left\{\left(1 + \frac{1}{n}\right)^n\right\}^{\frac{rt}{100}}$$

when n becomes infinitely large.

Replacing $\left(1 + \frac{1}{n}\right)^n$ by its limit when $n \longrightarrow \infty$, we get:

$$A = Pe^{\frac{rt}{100}}.$$

Let $$\frac{rt}{100} = x.$$

Then we can write:

$$A = Pe^x$$

e^x is called an **exponential function** because the index or exponent is the variable part of the function, whether it be t as above or x in general.

82. The Compound Interest Law.

The fundamental principle employed in arriving at the above result is that the growth of the principal is continuous in time and does not take place by sudden increases at regular intervals. In practice, compound interest is added at definite intervals of time, but the phenomenon of continuous growth is a natural law of organic growth and change. In many physical, chemical, electrical and engineering processes the mathematical expressions of them involve functions in which **the variation is proportional to the functions themselves.** In such cases the exponential function will be involved, and as the fundamental principle is that which entered into the Compound Interest investigations above, this law of growth was called by Lord Kelvin the **Compound Interest Law.**

83. The Exponential Series.

We shall next proceed to show that the function, e^x, can be expressed in a series involving ascending powers of x, a result which might have been anticipated, since a series was used to arrive at the limit of $\left(1 + \frac{1}{n}\right)^n$ when n became infinite.

Since $$e = \underset{n \to \infty}{Lt} \left(1 + \frac{1}{n}\right)^n.$$

Then $$e^x = \left\{\left(1 + \frac{1}{n}\right)^n\right\}^x$$

$$= \left(1 + \frac{1}{n}\right)^{nx}.$$

Expanding thus by the Binomial Theorem

$$\left(1+\frac{1}{n}\right)^{nx} = 1 + nx\cdot\frac{1}{n} + \frac{nx(nx-1)}{\underline{|2}}\cdot\frac{1}{n^2} + \frac{nx(nx-1)(nx-2)}{\underline{|3}}\cdot\frac{1}{n^3} + \dots$$

$$= 1 + x + \frac{x\left(x-\frac{1}{n}\right)}{\underline{|2}} + \frac{x\left(x-\frac{1}{n}\right)\left(x-\frac{2}{n}\right)}{\underline{|3}} + \dots$$

$$\therefore \quad \underset{n\to\infty}{Lt}\left(1+\frac{1}{n}\right)^{nx} = 1 + x + \frac{x^2}{\underline{|2}} + \frac{x^3}{\underline{|3}} + \dots$$

i.e.,

$$e^x = 1 + x + \frac{x^2}{\underline{|2}} + \frac{x^3}{\underline{|3}} + \dots$$

This series can be shown to be **convergent.**

Replacing x by $-x$ we get

$$e^{-x} = 1 - x + \frac{x^2}{\underline{|2}} - \frac{x^3}{\underline{|3}} + \dots$$

Similarly:

$$e^{ax} = 1 + ax + \frac{a^2x^2}{\underline{|2}} + \frac{a^3x^3}{\underline{|3}} + \dots$$

$$e^{-ax} = 1 - ax + \frac{a^2x^2}{\underline{|2}} - \frac{a^3x^3}{\underline{|3}} + \dots$$

84. Differentiation of e^x.

This can be performed by assuming the series for e^x as above and differentiating it term by term.

Since $\quad e^x = 1 + x + \dfrac{x^2}{\underline{|2}} + \dfrac{x^3}{\underline{|3}} + \dfrac{x^4}{\underline{|4}} + \dots$

$$\frac{d}{dx}(e^x) = 0 + 1 + \frac{2x}{\underline{|2}} + \frac{3x^2}{\underline{|3}} + \frac{4x^3}{\underline{|4}} + \dots$$

$$= 1 + x + \frac{x^2}{\underline{|2}} + \frac{x^3}{\underline{|3}} + \dots$$

But this is the series for e^x

$\therefore$ If
$$y = e^x$$
$$\frac{dy}{dx} = e^x.$$

This property, viz. that the differential coefficient of e^x is equal to itself, is possessed by no other function of x. It was to be expected, since we have seen that fundamentally e^x is a function such that its rate of change is proportional to itself.

Similarly, if
$$y = e^{-x}, \quad \frac{dy}{dx} = -e^{-x}$$
$$y = e^{ax}, \quad \frac{dy}{dx} = ae^{ax}$$
$$y = e^{-ax}, \quad \frac{dy}{dx} = -ae^{-ax}.$$

The differentiation of e^x can also be readily performed by using first principles.

85. The exponential curve.

(1) If
$$y = e^x$$
$$\frac{dy}{dx} = e^x$$
$$\frac{d^2y}{dx^2} = e^x.$$

Since $\frac{dy}{dx}$ is always positive, the curve of the function e^x must be positive and **always increasing.** $\therefore$ it has **no turning points.**

Since $\frac{d^2y}{dx^2} = e^x$, this does not vanish for any finite value of x.

$\therefore$ there is **no point of inflexion.**

(2) If
$$y = e^{-x}$$
$$\frac{dy}{dx} = -e^{-x}$$
$$\frac{d^2y}{dx^2} = e^{-x}.$$

Applying the same reasoning as above, $\frac{dy}{dx}$ is always negative. $\therefore$ curve is **always decreasing**. There are no turning points and no point of inflexion.

The two curves are shown in Fig. 25. In drawing them, values of the two functions will be found in the tables on pp. 379, 380.

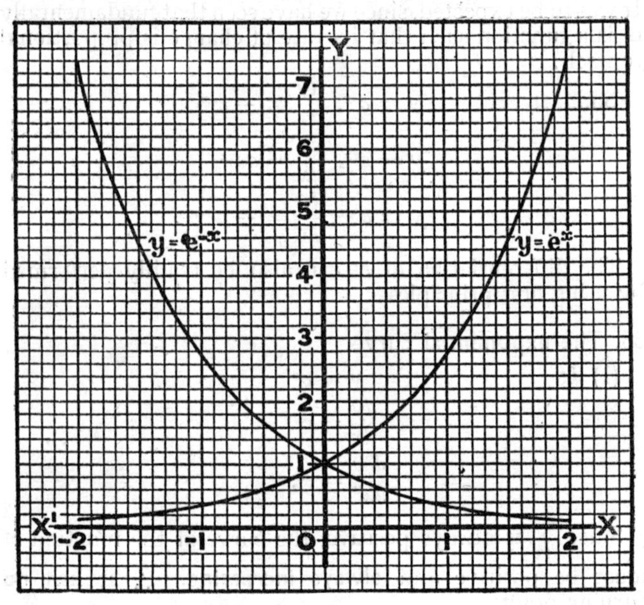

Fig. 25.

The **curve of e^x** illustrates the continuous increase of a function according to the Compound Interest law.

The **curve of e^{-x}** shows a law of decrease common in chemical and physical processes, representing a "**dying away**" law, the decrement being proportional to the magnitude of that which is diminishing at any instant. The loss of temperature in a cooling body is an example.

86. Napierian, Hyperbolic, or Natural Logarithms.

In § 81 we arrived at the formula

$$A = Pe^{\frac{rt}{100}}.$$

This may be written:

$$\frac{A}{P} = e^{\frac{rt}{100}}.$$

Let $$\frac{rt}{100} = x.$$

Then we can write:

$$\frac{A}{P} = e^x.$$

In this form it is seen that x represents **the logarithm of $\frac{A}{P}$ to base e.** In many similar examples **e** arises naturally as the base of a system of logarithms. So it came about that when logarithms were first given to the world by Lord Napier in 1614, the base of his system involved **e.** Hence such logarithms are called **Napierian logarithms.** They are also called **Hyperbolic** logs, from their association with the hyperbola, and sometimes **natural** logarithms. The introduction of 10 as a base was subsequently made by a mathematician named Briggs, who saw how valuable they would be in calculations. A short table of Napierian logarithms is given on pp. 377, 378.

In subsequent work in this book, unless it is stated to the contrary, the logs employed will be those to base **e.**

87. Differentiation of $\log_e x$.

The differential coefficient of $\log_e x$ can be readily obtained by the method of first principles, the work involving the limit of $\left(1 + \frac{1}{n}\right)^n$ as n proceeds to infinity. Or the differentiation of e^x may be demonstrated as follows:

Let $$y = \log_e x$$

Then $$x = e^y.$$

$$\therefore \quad \frac{dx}{dy} = e^y$$

and
$$\frac{dy}{dx} = \frac{1}{e^y} = \frac{1}{x}.$$

$$\therefore \quad \frac{d}{dx}(\log_e x) = \frac{1}{x}.$$

If the logarithm involved a different base, say a, then it can be changed to base e by the usual method. (*Algebra*, § 153.)

Thus if
$$y = \log_a x$$
then
$$y = \log_e x \times \log_a e$$
and
$$\frac{dy}{dx} = \frac{1}{x} \times \log_a e.$$

As a special case, if

$$y = \log_{10} x$$
$$\frac{dy}{dx} = \frac{1}{x} \times \log_{10} e$$
$$= \frac{1}{x} \times 0{\cdot}4343.$$

88. Differentiation of the general exponential functions.

e^x is a special case of a^x where a is any positive number.

Let
$$y = a^x$$
Then
$$\log_e y = x \log_e a$$
or
$$x = \log_e y \times \frac{1}{\log_e a}$$
$$\therefore \quad \frac{dx}{dy} = \frac{1}{y} \times \frac{1}{\log_e a}$$
and
$$\frac{dy}{dx} = y \times \log_e a$$
$$\therefore \quad \frac{dy}{dx} = a^x \times \log_e a.$$

As a special case, if

$$y = 10^x$$
$$\frac{dy}{dx} = 10^x \times \log_e 10.$$

89. Summary of formulae.

Function.	Diff. Coeff.
e^x e^{-x} a^z	e^x $-e^{-x}$ $a^x \times \log_e a$
$\log_e x$	$\frac{1}{x}$

90. Worked examples.

Example 1. *Differentiate* $y = e^{3x^2}$.

Employing the rule for the function of a function

If
$$y = e^{3x^2}$$
$$\frac{dy}{dx} = e^{3x^2} \times \frac{d}{dx}(3x^2)$$
$$= 6x \times e^{3x^2}.$$

Example 2. *Differentiate* $y = \log x^2$.

$$\frac{dy}{dx} = \frac{1}{x^2} \times \frac{d}{dx}(x^2)$$
$$= \frac{1}{x^2} \times 2x$$
$$= \frac{2}{x}.$$

Or it can be obtained by noting that $\log x^2 = 2 \log x$.

Example 3. *Differentiate* $\log \frac{x^2}{\sqrt{x^2 - 1}}$.

This may be written

$$y = \log x^2 - \log (x^2 - 1)^{\frac{1}{2}}$$

$$\therefore \quad \frac{dy}{dx} = \left(\frac{1}{x^2} \times 2x\right) - \frac{1}{(x^2-1)^{\frac{1}{2}}} \times \frac{d}{dx}(x^2-1)^{\frac{1}{2}}$$

$$= \frac{2}{x} - \frac{1}{(x^2-1)^{\frac{1}{2}}} \times (\tfrac{1}{2}(x^2-1)^{-\frac{1}{2}} \times 2x)$$

$$= \frac{2}{x} - \frac{x}{x^2-1}$$

$$= \frac{x^2-2}{x(x^2-1)}.$$

Example 4. *Differentiate* $y = e^{-ax} \sin (bx + c)$.

This is important in many electrical and physical problems, such as, for example, the "dying away" of the swing of a pendulum in a resisting medium.

Let $\quad y = e^{-ax} \sin (bx + c)$.

Then $\dfrac{dy}{dx} = \{e^{-ax} \times b \cos (bx + c)\} - \{ae^{-ax} \sin (bx + c)\}$

$$= e^{-ax}\{b \cos (bx + c) - a \sin (bx + c)\}.$$

Exercise 14.

Differentiate the following functions:

1. (*a*) e^{5x}; (*b*) $e^{\frac{x}{2}}$; (*c*) $e^{\sqrt{x}}$.
2. (*a*) e^{-2x}; (*b*) $e^{-\frac{5x}{2}}$; (*c*) $e^{(5-2x)}$.
3. (*a*) e^{-px}; (*b*) $e^{\frac{x}{a}}$; (*c*) $e^{(ax+b)}$.
4. (*a*) $\dfrac{e^x + e^{-x}}{2}$; (*b*) $\dfrac{e^x - e^{-x}}{2}$; (*c*) e^{x^2}.
5. (*a*) xe^x; (*b*) xe^{-x}; (*c*) x^2e^{-x}.
6. (*a*) $(x + 4)e^x$; (*b*) $e^x \sin x$; (*c*) $10e^x$.
7. (*a*) 2^x; (*b*) 10^{2x}; (*c*) $e^{\sin x}$.
8. (*a*) x^na^x; (*b*) a^{2x+1}; (*c*) $e^{\cos x}$.
9. (*a*) a^{bx^2}; (*b*) $(a + b)^x$; (*c*) $e^{\tan x}$.
10. (*a*) $\log \dfrac{x}{a}$; (*b*) $\log (ax^2 + bx + c)$.
11. (*a*) $\log x^2$; (*b*) $\log (x^3 + 3)$.
12. (*a*) $x \log x$; (*b*) $\log (px + q)$.

13. (*a*) $\log(\sin x)$; (*b*) $\log(\cos x)$.

14. (*a*) $\log\dfrac{a+x}{a-x}$; (*b*) $\log(e^{x}+e^{-x})$.

15. (*a*) $\log\{x+\sqrt{x^{2}+1}\}$; (*b*) $\sqrt{x}-\log(1+\sqrt{x})$.

16. (*a*) $\log\tan\dfrac{x}{2}$; (*b*) $\log\sqrt{x^{2}+1}$; (*c*) $\dfrac{e^{x}}{\sqrt{x}}$.

17. (*a*) $x^{2}e^{4x}$; (*b*) $ae^{-kx}\sin kx$.

18. (*a*) $\dfrac{e^{ax}}{x^{\frac{1}{2}}}$; (*b*) $\log(\sqrt{\sin x})$.

19. (*a*) x^{x}; (*b*) $\log\{\sqrt{x-1}+\sqrt{x+1}\}$.

20. (*a*) $\log\dfrac{e^{x}}{1+e^{x}}$; (*b*) $\sin x\times\log\sin x$.

21. (*a*) $\log\dfrac{\sqrt{a}+\sqrt{x}}{\sqrt{a}-\sqrt{x}}$; (*b*) $e^{ax}\sin^{2}x$.

22. (*a*) $a^{3x^{2}}$; (*b*) $\dfrac{e^{x}-e^{-x}}{e^{x}+e^{-x}}$.

23. (*a*) $\sin^{-1}\log x$; (*b*) $\cos^{-1}e^{-x}$.

24. (*a*) $e^{ax}\cos(bx+c)$; (*b*) $e^{-ax}\cos 3x$; (*c*) $e^{-\frac{1}{2}x}\sin\left(\pi x+\dfrac{\pi}{2}\right)$.

25. (*a*) $\log\dfrac{x}{a-\sqrt{a^{2}-x^{2}}}$; (*b*) $\sin^{-1}\dfrac{e^{x}-e^{-x}}{e^{x}+e^{-x}}$.

26. Find the 2nd, 3rd, 4th and nth derivatives of (*a*) $y=e^{ax}$; (*b*) $y=e^{-ax}$; (*c*) $y=\log x$.

CHAPTER IX

HYPERBOLIC FUNCTIONS

91. Definitions of Hyperbolic Functions.

In Fig. 25 there were shown the graphs of the exponential function e^x and e^{-x}. These two curves are reproduced in Fig. 26, together with two other curves marked A and B.

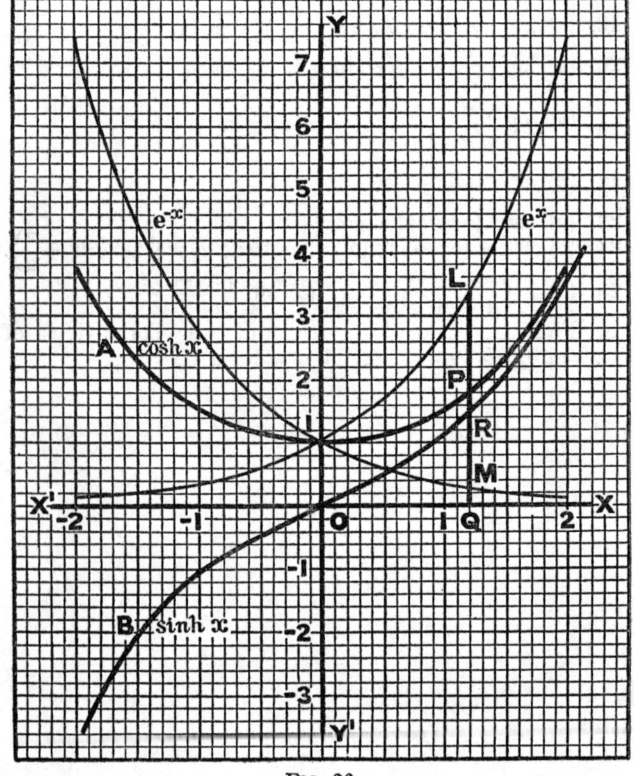

FIG. 26.

(1) In curve A the ordinate of any point on it is one half of the sum of the corresponding ordinates of e^x and e^{-x}. For example, at the point P, its ordinate PQ is half the sum of LQ and MQ.

∴ for every point on the curve

$$y = \frac{e^x + e^{-x}}{2}.$$

(2) On curve B, the ordinate of any point is one half of the difference of the ordinates of the other two curves.

Thus $\quad RQ = \frac{1}{2}(LQ - MQ)$

i.e., for any point $\quad y = \frac{1}{2}(e^x - e^{-x})$.

The two curves therefore represent two functions of x, and their equations are given by

$$y = \tfrac{1}{2}(e^x + e^{-x})$$

and

$$y = \tfrac{1}{2}(e^x - e^{-x}).$$

It is found that these two functions have properties which in many respects are analogous to those of $y = \cos x$ and $y = \sin x$. It can be shown that they bear a similar relation to the hyperbola that the trigonometric or circular functions do to the circle. Hence the function $y = \frac{1}{2}(e^x + e^{-x})$ is called the hyperbolic cosine, and $y = \frac{1}{2}(e^x - e^{-x})$ is called the hyperbolic sine.

These are abbreviated to $\cosh x$ and $\sinh x$, the added h indicating the hyperbolic cos, etc. The names are usually pronounced " cosh " and " shine," respectively.

They are defined by the equations stated above, viz.:—

$$\cosh x = \tfrac{1}{2}(e^x + e^{-x})$$
$$\sinh x = \tfrac{1}{2}(e^x - e^{-x}).$$

From these definitions, also

$$\cosh x + \sinh x = e^x$$
$$\cosh x - \sinh x = e^{-x}$$

There are four other hyperbolic functions corresponding to the other circular functions, viz.:

$$\tanh x = \frac{\sinh x}{\cosh x} = \frac{e^x - e^{-x}}{e^x + e^{-x}} = \frac{e^{2x} - 1}{e^{2x} + 1}$$

$$\text{cosech } x = \frac{1}{\sinh x} = \frac{2}{e^x - e^{-x}}$$

$$\text{sech } x = \frac{1}{\cosh x} = \frac{2}{e^x + e^{-x}}$$

$$\coth x = \frac{1}{\tanh x} = \frac{e^x + e^{-x}}{e^x - e^{-x}}.$$

These functions can be expressed in exponential form by derivation from their reciprocals.

The names of these are pronounced " than," " coshec," " shec " and " coth."

The curve of cosh x, marked A in Fig. 26, is an important one. It is called the **catenary**, and is the curve formed by a uniform flexible chain which hangs freely with its ends fixed.

These functions can be expressed in the form of series which are derived from the series for e^x, found in § 83.

Thus $$e^x = 1 + x + \frac{x^2}{\lfloor 2} + \frac{x^3}{\lfloor 3} + \ldots$$

and $$e^{-x} = 1 - x + \frac{x^2}{\lfloor 2} - \frac{x^3}{\lfloor 3} + \ldots$$

Hence by addition and subtraction:

$$\cosh x = 1 + \frac{x^2}{\lfloor 2} + \frac{x^4}{\lfloor 4} + \ldots$$

$$\sinh x = x + \frac{x^3}{\lfloor 3} + \frac{x^5}{\lfloor 5} + \ldots$$

92. Formulae connected with hyperbolic functions.

There is a close correspondence between formulae expressing relations between hyperbolic functions and similar relations between circular functions. Consider the two following examples:

(1) $\cosh^2 x - \sinh^2 x$

$$= \left(\frac{e^x + e^{-x}}{2}\right)^2 - \left(\frac{e^x - e^{-x}}{2}\right)^2$$

$$= \tfrac{1}{4}\{(e^{2x} + e^{-2x} + 2) - (e^{2x} + e^{-2x} - 2)\}$$

$$= 1.$$

$$\therefore \quad \cosh^2 x - \sinh^2 x = 1.$$

This should be compared with the trigonometrical result

$$\cos^2 x + \sin^2 x = 1.$$

(2) $\cosh^2 x + \sinh^2 x$

$$= \left(\frac{e^x + e^{-x}}{2}\right)^2 - \left(\frac{e^x - e^{-x}}{2}\right)^2$$

$$= \frac{2e^{2x} + 2e^{-2x}}{4}$$

$$= \frac{e^{2x} + e^{-2x}}{2}$$

$$= \cosh 2x$$

i.e., $\qquad \mathbf{\cosh^2 x + \sinh^2 x = \cosh 2x.}$

This is analogous to $\cos^2 x - \sin^2 x = \cos 2x$.

Similarly, any formula for circular functions has its counterpart in hyperbolic functions. It will be noticed that in the above two cases there is a difference in the signs used, and this applies only to $\sinh^2 x$. This has led to the formulation of Osborne's rule, by which formulae for hyperbolic functions can be at once written down from the corresponding formulae for circular functions.

Osborne's Rule.

In any formula connecting circular functions of general angles, the corresponding formula connecting hyperbolic functions can be obtained by replacing each circular function by the corresponding hyperbolic function, if the sign of every product or implied product of two sines is changed.

For example $\qquad \sec^2 x = 1 + \tan^2 x$

becomes $\qquad \operatorname{sech}^2 x = 1 - \tanh^2 x$

since $\qquad \tanh^2 x = \dfrac{\sinh x \times \sinh x}{\cosh x \times \cosh x}.$

93.

The more important of these corresponding formulae are summarised for convenience.

Hyperbolic Functions.	Circular Functions.
$\cosh^2 x - \sinh^2 x = 1$	$\cos^2 x + \sin^2 x = 1$
$\sinh 2x = 2 \sinh x \cosh x$	$\sin 2x = 2 \sin x \cos x$
$\cosh 2x = \cosh^2 x + \sinh^2 x$	$\cos 2x = \cos^2 x - \sin^2 x$
$\text{sech}^2 x = 1 - \tanh^2 x$	$\sec^2 x = 1 + \tan^2 x$
$\text{cosech}^2 x = \coth^2 x - 1$	$\text{cosec}^2 x = \cot^2 x + 1$
$\sinh (x \pm y) = \sinh x \cosh y \pm \cosh x \sinh y$	$\sin (x \pm y) = \sin x \cos y \pm \cos x \sin y$
$\cosh (x \pm y) = \cosh x \cosh y \pm \sinh x \sinh y$	$\cos (x \pm y) = \cos x \cos y \mp \sin x \sin y$

The following striking connections between the two sets of functions are given for the information of the student. For a full treatment any book on advanced trigonometry should be consulted.

$$\cosh x = \tfrac{1}{2}(e^x + e^{-x});\ \cos x = \tfrac{1}{2}(e^{ix} + e^{-ix})$$
$$\sinh x = \tfrac{1}{2}(e^x - e^{-x});\ \sin x = \tfrac{1}{2i}(e^{ix} - e^{-ix})$$
$$\sinh x = \frac{1}{i} \sin ix$$
$$\cosh x = \cos ix$$

where $i = \sqrt{-1}$. (See *Algebra*, Appendix, p. 284.)

94. Differential coefficients of hyperbolic functions.

(1) sinh x.

Let
$$y = \sinh x = \frac{e^x - e^{-x}}{2}$$
Then
$$\frac{dy}{dx} = \frac{e^x + e^{-x}}{2} = \cosh x.$$

(2) cosh x.

Let
$$y = \cosh x = \frac{e^x + e^{-x}}{2}$$
Then
$$\frac{dy}{dx} = \frac{e^x - e^{-x}}{2} = \sinh x.$$

(3) tanh x.

The differential coefficient may be found from the exponential definition, or we may use the above result.

Let $y = \tanh x = \frac{\sinh x}{\cosh x}$.

Then $\frac{dy}{dx} = \frac{\cosh x \,.\, \cosh x - \sinh x \,.\, \sinh x}{\cosh^2 x}$

(Quotient rule.)

$$= \frac{\cosh^2 x - \sinh^2 x}{\cosh^2 x}$$

$$= \frac{1}{\cosh^2 x} \qquad (\S\ 92)$$

$$= \textbf{sech}^2\, x.$$

Similarly, it may be shown that, if

$$y = \text{cosech}\ x, \quad \frac{dy}{dx} = -\ \text{cosech}\ x \coth x$$

$$y = \text{sech}\ x, \quad \frac{dy}{dx} = -\ \text{sech}\ x \tanh x$$

$$y = \coth x, \quad \frac{dy}{dx} = -\ \text{cosech}^2\ x.$$

These results should be compared with the differential coefficients of the corresponding circular functions.

95. Curves of the hyperbolic functions.

The curves of $\cosh x$ and $\sinh x$ in Fig. 26 should be examined again with the assistance of their differential coefficients.

(1) $y = \cosh x$; $\frac{dy}{dx} = \sinh x$, $\frac{d^2y}{dx^2} = \cosh x$.

$\frac{dy}{dx}$ vanishes only when $x = 0$. There is therefore a turning point on the curve (curve A). Also, since $\sinh x$ is negative before this point and positive after, while $\frac{d^2x}{dx^2}$ is positive, the point is a **minimum**. There is no other turning point and no point of inflexion.

(2) $y = \sinh x$; $\frac{dy}{dx} = \cosh x$, $\frac{d^2y}{dx^2} = \sinh x$.

$\frac{dy}{dx}$, *i.e.*, $\cosh x$ is always positive and does not vanish. Consequently $\sinh x$ is **always increasing** and has no turning point. When $x = 0$, $\frac{d^2y}{dx^2} = 0$, and is negative before and positive after. Therefore there is a **point of inflexion** when $x = 0$; since $\frac{dy}{dx}$, *i.e.*, $\cosh x = 1$ when $x = 0$, the gradient at O is unity and the slope $\frac{\pi}{4}$.

(3) $y = \tanh x$; $\frac{dy}{dx} = \text{sech}^2\, x$.

Since $\text{sech}^2\, x$ is always positive, $\tanh x$ is always increasing between $-\infty$ and $+\infty$. Also since $\sinh x$ and $\cosh x$ are always continuous and $\cosh x$ never vanishes, $\tanh x$ must be a **continuous function.**

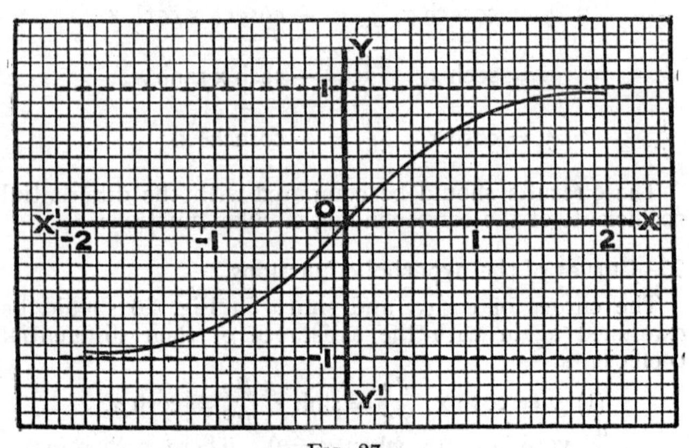

FIG. 27.

As was shown in § 91, $\tanh x$ can be written in the form:

$$\tanh x = \frac{e^{2x} - 1}{e^{2x} + 1}$$

$$= 1 - \frac{2}{e^{2x} + 1}.$$

From this form it is evident that while x increases from $-\infty$ to 0, e^{2x} increases from 0 to 1.

$$\therefore \quad 1 - \frac{2}{e^{2x}+1} \text{ or } \tanh x \text{ increases from } -1 \text{ to } 0.$$

Similarly, while x increases from 0 to $+\infty$, $\tanh x$ increases from 0 to $+1$.

The curve therefore has the lines $y = \pm 1$ as its asymptotes and is as shown in Fig. 27.

96. Differentiation of the inverse hyperbolic functions.

Inverse hyperbolic functions correspond to inverse circular functions, and their differential coefficients are found by similar methods.

(1) **Differential coefficient of $\sinh^{-1} x$.**

Let $\qquad y = \sinh^{-1} x$

Then $\qquad x = \sinh y.$

$$\therefore \quad \frac{dx}{dy} = \cosh y$$

or
$$\frac{dy}{dx} = \frac{1}{\cosh y} = \frac{1}{\sqrt{1+\sinh^2 y}} \qquad (\S\ 93)$$

$$= \frac{1}{\sqrt{1+x^2}} \quad \text{or} \quad \frac{1}{\sqrt{x^2+1}}.$$

(2) **Differential coefficient of $\cosh^{-1} x$.**

Using the same method as above we get:

$$\frac{dy}{dx} = \frac{1}{\sqrt{x^2-1}}.$$

(3) **Differential coefficient of $\tanh^{-1} x$.**

If $\qquad y = \tanh^{-1} x$

$$x = \tanh y$$

$$\therefore \quad \frac{dx}{dy} = \text{sech}^2 y$$

and
$$\frac{dy}{dx} = \frac{1}{\text{sech}^2 y}$$

$$= \frac{1}{1-\tanh^2 y} \qquad (\S\ 93)$$

$$= \frac{1}{1-x^2}.$$

(4) The differential coefficients of the reciprocals of the above can be found by the same methods. They are:

$$y = \operatorname{sech}^{-1} x, \quad \frac{dy}{dx} = -\frac{1}{x\sqrt{1 - x^2}}.$$

$$y = \operatorname{cosech}^{-1} x, \quad \frac{dy}{dx} = -\frac{1}{x\sqrt{1 + x^2}}.$$

$$y = \coth^{-1} x, \quad \frac{dy}{dx} = -\frac{1}{x^2 - 1}.$$

The following forms will be found of importance later :

(1) If $y = \sinh^{-1} \frac{x}{a}$,

$$\frac{dy}{dx} = \frac{1}{\sqrt{\left(1 + \frac{x^2}{a^2}\right)}} \times \frac{1}{a}$$

$$= \frac{1}{\sqrt{a^2 + x^2}} \quad \text{or} \quad \frac{1}{\sqrt{x^2 + a^2}}.$$

(2) Similarly, if $y = \cosh^{-1} \frac{x}{a}$,

$$\frac{dy}{dx} = \frac{1}{\sqrt{x^2 - a^2}}$$

97. Logarithm equivalents of the inverse hyperbolic functions.

(1) $\sinh^{-1} x = \log \{x + \sqrt{1 + x^2}\}$.

Let $\qquad y = \sinh^{-1} x.$

Then $\qquad x = \sinh y.$

But
$$\cosh^2 y = 1 + \sinh^2 y \qquad (\S\ 93)$$
$$= 1 + x^2.$$
$$\cosh y = \sqrt{1 + x^2} \qquad \text{(A)}$$
$$\therefore \quad \sinh y + \cosh y = x + \sqrt{1 + x^2}$$

but
$$\sinh y + \cosh y = e^y \qquad (\S\ 91)$$
$$\therefore \quad e^y = x + \sqrt{1 + x^2}.$$

Taking logs $\qquad y = \log \{x + \sqrt{1 + x^2}\}$

i.e., $\qquad \sinh^{-1} x = \log \{x + \sqrt{x^2 + 1}\}.$

Note.—Since $\cosh y$ is always positive, the plus sign only is taken in A.

(2) $\cosh^{-1} x = \log\{x + \sqrt{x^2 - 1}\}$.

Let $y = \cosh^{-1} x$.

$\therefore \quad x = \cosh y,$

but $\sinh^2 y = \cosh^2 y - 1$ (§ 93)

$= x^2 - 1.$

$\therefore \quad \sinh y = \sqrt{x^2 - 1}$ (both signs applicable).

As above $e^y = \cosh y + \sinh y$

$= x \pm \sqrt{x^2 - 1}.$

$\therefore \quad y = \log\{x \pm \sqrt{x^2 - 1}\}$

or $\cosh^{-1} x = \log\{x \pm \sqrt{x^2 - 1}\}.$

The two values thus obtained are:

$$\log\{x + \sqrt{x^2 - 1}\} \text{ and } \log\{x - \sqrt{x^2 - 1}\}.$$

Their sum :

$$\begin{aligned}&\log\{x + \sqrt{x^2 - 1}\} + \log\{x - \sqrt{x^2 - 1}\}\\ &= \log\{(x + \sqrt{x^2 - 1}) \times (x - \sqrt{x^2 - 1})\}\\ &= \log\{x^2 - (x^2 - 1)\}\\ &= \log 1\\ &= 0.\end{aligned}$$

$\therefore$ these two values of $\cosh^{-1} x$ are equal, differing only in their sign. Hence we may write:

$$\cosh^{-1} x = \pm \log\{x + \sqrt{x^2 - 1}\}$$

Note.—x must lie between 1 and $+\infty$.

(3) $\tanh^{-1} x = \frac{1}{2} \log \frac{1 + x}{1 - x}$.

Let $y = \tanh^{-1} x$

then $x = \tanh y$

(and x lies between $+1$ and -1 (§ 95))

$$= \frac{e^{2y} - 1}{e^{2y} + 1} \qquad (\S\ 91)$$

$\therefore \quad x(e^{2y} + 1) = e^{2y} - 1.$

Whence $$e^{2y} = \frac{1+x}{1-x}.$$

$$\therefore \quad 2y = \log \frac{1+x}{1-x}$$

and $$y = \tfrac{1}{2} \log \frac{1+x}{1-x},$$

i.e., $$\tanh^{-1} x = \tfrac{1}{2} \log \frac{1+x}{1-x}.$$

98. Summary of formulae of inverse functions.

Function.	Diff. Coeff.
$\sinh^{-1} x$	$\dfrac{1}{\sqrt{x^2+1}}$
$\cosh^{-1} x$	$\dfrac{1}{\sqrt{x^2-1}}$
$\tanh^{-1} x$	$\dfrac{1}{1-x^2}$
$\operatorname{cosech}^{-1} x$	$\dfrac{-1}{x\sqrt{1+x^2}}$
$\operatorname{sech}^{-1} x$	$-\dfrac{1}{x\sqrt{1-x^2}}$
$\coth^{-1} x$	$-\dfrac{1}{x^2-1}$

The following additional forms are important. When

$$y = \sinh^{-1} \frac{x}{a}, \quad \frac{dy}{dx} = \frac{1}{\sqrt{x^2+a^2}}$$

$$y = \cosh^{-1} \frac{x}{a}, \quad \frac{dy}{dx} = \frac{1}{\sqrt{x^2-a^2}}$$

$$y = \tanh^{-1} \frac{x}{a}, \quad \frac{dy}{dx} = \frac{a}{a^2-x^2}.$$

Logarithm equivalents.

$$\sinh^{-1} x = \log\{x + \sqrt{x^2 + 1}\}$$
$$\cosh^{-1} x = \pm \log\{x + \sqrt{x^2 - 1}\}$$
$$\tanh^{-1} x = \tfrac{1}{2}\log\frac{1 + x}{1 - x}.$$

Also

$$\sinh^{-1}\frac{x}{a} = \log\left\{\frac{x + \sqrt{x^2 + a^2}}{a}\right\}$$
$$\cosh^{-1}\frac{x}{a} = \pm\log\left\{\frac{x + \sqrt{x^2 - a^2}}{a}\right\}$$
$$\tanh^{-1}\frac{x}{a} = \tfrac{1}{2}\log\frac{a + x}{a - x}.$$

Exercise 15.

Differentiate the following functions:

1. (*a*) $\sinh\frac{x}{2}$; (*b*) $\sinh 2x$; (*c*) $\cosh\frac{x}{3}$.
2. (*a*) $\tanh ax$; (*b*) $\tanh\frac{x}{4}$; (*c*) $\sinh ax + \cosh ax$.
3. (*a*) $\sinh\frac{1}{x}$; (*b*) $\sinh^2 x$; (*c*) $\cosh^3 x$.
4. (*a*) $\sinh(ax + b)$; (*b*) $\cosh 2x^2$; (*c*) $\sinh^n ax$.
5. (*a*) $\sinh x\cosh x$; (*b*) $\sinh^2 x + \cosh^2 x$; (*c*) $\tanh^2 x$.
6. (*a*) $\log\tanh x$; (*b*) $x\sinh x - \cosh x$; (*c*) $\log\cosh x$.
7. (*a*) $x^3\sinh 3x$; (*b*) $\log(\sinh x + \cosh x)$; (*c*) $e^{\sinh x}$.
8. (*a*) $\sqrt{\sinh x}$; (*b*) $\log\frac{1 + \tanh x}{1 - \tanh x}$; (*c*) $e^{\tanh x}$.
9. (*a*) $\sinh^{-1}\frac{x}{2}$; (*b*) $\cosh^{-1}\frac{x}{5}$; (*c*) $\sinh^{-1}\frac{1 - x}{1 + x}$.
10. (*a*) $\sinh^{-1}\tan x$; (*b*) $\tan^{-1}\sinh x$; (*c*) $\tanh^{-1}\sin x$.
11. (*a*) $\sin^{-1}\tanh x$; (*b*) $\cosh^{-1}\sec x$; (*c*) $\tanh^{-1}\frac{2x}{1 + x^2}$.
12. (*a*) $\cosh^{-1}(4x + 1)$; (*b*) $\sinh^{-1} 2x\sqrt{1 + x^2}$; (*c*) $\tanh^{-1}\frac{1}{1 + x}$.
13. (*a*) $\tan^{-1} x + \tanh^{-1} x$; (*b*) $\tanh^{-1}(\tan\frac{1}{2}x)$; (*c*) $\tan^{-1}(\tanh\frac{1}{2}x)$.

14. Write the logarithmic equivalents of :

(*a*) $\sinh^{-1}\frac{x}{2}$; (*b*) $\cosh^{-1}\frac{x}{3}$; (*c*) $\sinh^{-1}\frac{2x}{3}$

(*d*) $\cosh^{-1}\frac{3x}{2}$; (*e*) $\tanh^{-1}\frac{x}{4}$.

15. Differentiate:

(*a*) $\log\left\{\frac{x+\sqrt{x^2+a^2}}{a}\right\}$;

(*b*) $\log\left\{\frac{x+\sqrt{x^2-a^2}}{a}\right\}$; (*c*) $\frac{1}{2}\log\frac{a+x}{a-x}$.

CHAPTER X

INTEGRATION. STANDARD INTEGRALS

99. Meaning of integration.

The integral calculus is concerned with **the operation of integration,** which, in one of its aspects, is the **converse of differentiation.**

From this point of view the problem to be solved in integration is: *What is the function which on being differentiated produces a given function?* For example, what is the function which, being differentiated, produces $\cos x$? In this case we **know** from the work on the previous chapters on differentiation, that $\sin x$ is the function required. We therefore conclude that **$\sin x$ is the integral of $\cos x$.**

Generally **if $f'(x)$ represents the differential coefficient of $f(x)$, then the problem of integration is, given $f'(x)$, find $f(x)$, or given $\frac{dy}{dx}$, find y.**

But the process of finding the integral is seldom as simple as in the example above. A converse operation is usually more difficult than the direct one, and integration is no exception. A sound knowledge of differentiation will help in many cases, such as that above, but, even when the **type** of function is known, there may arise minor complications of signs and constants.

For example, if the integral of $\sin x$ is required, we know that $\cos x$, when differentiated, produces $-\sin x$. We therefore conclude that the function which produces $+\sin x$ on differentiation must be $-\cos x$. Thus **the integral of $\sin x$ is $-\cos x$.**

Again, **suppose the integral of x is required.** We know that the function which produces this on differentiation must be of the form x^2. But $\frac{d}{dx}(x^2) = 2x$. If therefore x is to be the result of the differentiation, the integral must contain a constant factor of x such that it cancels with the 2 in $2x$. Clearly this constant must be $\frac{1}{2}$. Hence the **integral must be $\frac{1}{2}x^2$.**

These two simple examples may help the student to realise some of the difficulties which face him in the integral calculus. In the differential calculus, with a knowledge of the rules which have been formulated in previous chapters, it is possible to differentiate not only all the ordinary types of functions, but also complicated expressions formed by products, powers, quotients, logs, etc., of these functions. But simplifications, cancellings and other operations occur before the final form of the differential coefficient is reached. When reversing the process, as in integration, we want to know the original function, it is usually impossible to reverse through these changes, and in very many cases the integration cannot be effected.

It is not possible, therefore, to formulate a set of rules by which **any** function may be integrated. Methods have been devised, however, for integrating certain types of functions, and these will be stated in succeeding chapters. With a knowledge of these and much practice, the student, if he possesses a good grasp of differentiation, will be able to integrate most of the commonly occurring functions.

These methods, in general, consist of transposing and manipulating the functions so that they assume the known form of standard functions of which the integrals are known. The final solution becomes a matter of recognition and inspection.

Integration has one advantage—the result can always be checked. If the function obtained by integration be differentiated we should get the original function. *The student should not omit this check.*

100. The constant of integration.

When a function containing a constant term is differentiated, the constant term disappears, since its differential coefficient is zero.

When the process is reversed and we integrate, the constant cannot be determined without further information.

For example, let $y = x^2 + 3.$

Then $\frac{dy}{dx} = 2x$

If the process is now reversed and $2x$ is integrated as it

stands, the result is x^2. Consequently **to get a complete integral an unknown constant must be added.**

In the above example let C denote the constant. Then we may state that the integral of $2x$ is $x^2 + C$, where **C is an undetermined constant.** Consequently the integral is called an **Indefinite Integral.**

This may be illustrated graphically as follows.

In Fig. 28 there are represented the graphs of $y = x^2$, $y = x^2 + 2$, and $y = x^2 - 3$, all of which are included in

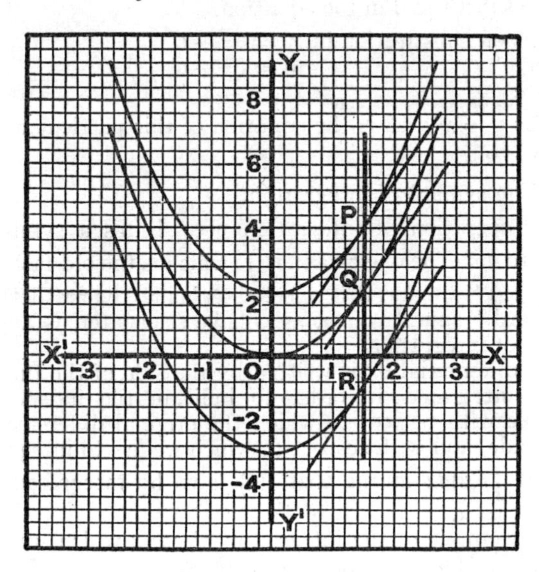

Fig. 28.

the general form $y = x^2 + C$. They are termed **integral curves,** since they represent the curves of the integral $x^2 + C$, when the values 0, + 2, and − 3 are assigned to C. Evidently there is an infinite number of such curves.

Let P, Q, R be points on these curves where they are cut by the ordinate 2 = 1·5.

At all three points **the gradient is the same.** They have

the same coefficient, $2x$, which for these points has the value 3.

The integral $y = x^2 + C$ therefore represents a series of corresponding curves having the **same gradient at points with the same abscissa.**

The equation of any particular curve in the series can be found when a pair of corresponding values of x and y is known. These enable us to find C. If, for example, a curve passes through the point (3, 6) these values of x and y can be substituted in the equation.

Thus on substitution in $y = x^2 + C$
we have $6 = 3^2 + C$
whence $C = -3.$

Thus $y = x^2 - 3$ is the equation of this particular curve in the set.

101. The symbol of integration.

The operation of integration necessitates a symbol to indicate it. The one chosen is $\int$, which is the old-fashioned elongated " s," and it is selected as being the first letter of the word " sum," which, as will be seen later, is another aspect of integration.

The differential dx is written by the side of the function to be integrated in order to indicate the independent variable with respect to which the original differentiation was made, and with respect to which we are to integrate.

Thus $\int f(x)dx$ means that $f(x)$ is to be integrated with respect to x.

The example of the integration of $\cos x$ which was considered in § 99 would be written thus:

$$\int \cos x dx = \sin x + C.$$

It is important to remember that the variables in the function to be integrated and in the differential must be the same. Thus $\int \cos y dx$ could not be obtained as it stands. It would first be necessary if possible to express $\cos y$ as a function of x.

Note.—Any other letter may be used to represent the

independent variable besides x. Thus $\int t dt$ indicates that t is the independent variable and we need to integrate with respect to it.

102. Integration of a constant factor.

It was shown in § 39 that when a function contains a constant number as a factor, this number will be a factor of the differential coefficient of the function. Thus if

$$y = ax^n$$
$$\frac{dy}{dx} = a(nx^{n-1}).$$

It will be obvious from § 39 that when the operation is reversed, and we integrate a function containing a constant factor, **this factor must also be a factor of the final integral.**

When finding an integral it is better to transfer such a factor to the left side of the integration sign before proceeding with the integration of the function. Thus:

$$\int 5x dx = 5\int x dx$$
$$= 5(\tfrac{1}{2}x^2) + C$$
$$= \tfrac{5}{2}x^2 + C.$$

Generally $$\int af'(x)dx = a\int f'(x)dx.$$

It should be noted that **no factor which involves the variable can thus be transferred** to the other side of the integration sign.

103. The integration of x^n.

Simple examples of this can be obtained by inspection, viz.:

$$\int x dx = \tfrac{1}{2}x^2 + C$$
$$\int x^2 dx = \tfrac{1}{3}x^3 + C$$
$$\int x^3 dx = \tfrac{1}{4}x^4 + C$$
$$\int x^4 dx = \tfrac{1}{5}x^5 + C.$$

From these examples we may readily deduce that:

$$\int x^n dx = \frac{1}{n+1} x^{n+1} + C.$$

Also, in accordance with the rule of § 102:

$$\int ax^n dx = a\int x^n dx$$
$$= \frac{a}{n+1} x^{n+1} + C.$$

Remembering the rule for the differentiation of a function of a function, we can also deduce that

$$\int (ax + b)^n dx = \frac{1}{a(n+1)} (ax + b)^{n+1} + C.$$

If a student has any difficulty in realising such a result as this last, he will see the reason for it by differentiating the integral obtained.

It was seen in § 38 that the rule for the differentiation of x^n holds for all values of n. The formula above for the integration of the function similarly holds for all values of the index.

Note.—It should be noted that $\int dx = x + C$.

104. Worked examples.

(1) $\int 3x^7 dx = 3\int x^7 dx = 3 \times \frac{x^8}{8} + C = \frac{3}{8}x^8 + C$

(2) $\int 4\sqrt{x}dx = 4\int x^{\frac{1}{2}}dx = 4 \times \frac{x^{\frac{1}{2}+1}}{\frac{1}{2}+1} + C = 4 \times \frac{2}{3}x^{\frac{3}{2}} + C$

$$= \frac{8}{3}x^{\frac{3}{2}} + C.$$

(3) $\int \frac{dx}{\sqrt{x}} = \int \frac{dx}{x^{\frac{1}{2}}} = \int x^{-\frac{1}{2}}dx$

$$= \frac{x^{-\frac{1}{2}+1}}{-\frac{1}{2}+1} + C$$
$$= 2x^{\frac{1}{2}} + C$$
$$= 2\sqrt{x} + C.$$

Note.—This last integral and those of the following allied functions should be carefully noted:

$$\int \frac{dx}{\sqrt{x+b}} = 2\sqrt{x+b} + C.$$

$$\int \frac{dx}{\sqrt{ax+b}} = \frac{2}{a}\sqrt{ax+b} + C$$

105. Integration of a sum.

It is evident from a consideration of the differentiation of a sum of a number of functions (§ 41), that on reversing the process the same rule must hold for integration—*i.e.*, **the integral of a sum of a number of functions is equal to the sum of the integrals of these functions.**

Examples.

(1) $\int (x^3 - 5x^2 + 7x - 11)dx$

$$= \int x^3 dx - 5\int x^2 dx + 7\int x dx - 11\int dx$$

$$= \frac{x^4}{4} - \tfrac{5}{3}x^3 + \tfrac{7}{2}x^2 - 11x + C.$$

Note.—The constants which would arise from the integration of the separate terms can all be included in one constant, since this constant is arbitrary and undetermined.

(2) $\int \left(\sqrt[3]{x} - \frac{1}{\sqrt[3]{x}}\right) dx$

$$= \int x^{\frac{1}{3}} dx - \int x^{-\frac{1}{3}} dx$$

$$= \frac{1}{\frac{1}{3}+1} x^{\frac{1}{3}+1} - \frac{1}{-\frac{1}{3}+1} x^{-\frac{1}{3}+1} + C$$

$$= \tfrac{3}{4}x^{\frac{4}{3}} - \tfrac{3}{2}x^{\frac{2}{3}} + C.$$

106. The integration of $\frac{1}{x}$.

If the rule for the integration of x^n be applied to the case of $\frac{1}{x}$ or x^{-1}, we get:

$$\int \frac{dx}{x} = \int x^{-1} dx = \frac{x^{-1+1}}{-1+1} + C$$

$$= \frac{x^0}{0} + C = \frac{1}{0} + C.$$

This result is apparently infinite, and the rule does not seem to apply. The apparent contradiction must be left for future consideration, but it should be remembered that in these processes we are dealing with limits.

We know, however, that by the rule for the differentiation of a logarithmic function (§ 87) that the differential coefficient of $\log_e x$ is $\frac{1}{x}$.

Hence we conclude that

$$\int \frac{dx}{x} = \log_e x.$$

107. A useful rule for integration.

By combining with this last result the rule in differentiation for the function of a function we know that:

if
$$y = \log \{f(x)\}$$
$$\frac{dy}{dx} = \frac{1}{f(x)} \times f'(x) = \frac{f'(x)}{f(x)}$$

consequently
$$\int \frac{f'(x)}{f(x)} dx = \log f(x) + C.$$

Hence—*when integrating a fractional function in which, after a suitable adjustment of constants, if necessary, it is seen that the numerator is the differential coefficient of the denominator, then the integral is the logarithm of the denominator.*

Clearly **all** fractional functions of x in which the denominator is a function of the first degree can be integrated by this rule by a suitable adjustment of constants.

108. Worked examples.

(1) $$\int \frac{dx}{ax} = \frac{1}{a} \int \frac{a}{ax} dx$$
$$= \frac{1}{a} \log ax + C.$$

(2) $$\int \frac{dx}{ax + b} = \frac{1}{a} \log (ax + b) + C.$$

(3) $$\int \frac{xdx}{2x^2 + 3} = \frac{1}{4} \int \frac{4xdx}{2x^2 + 3}$$
$$= \frac{1}{4} \log (2x^2 + 3) + C.$$

(4) $\int \frac{2(x+1)dx}{x^2+2x+7} = \int \frac{(2x+2)dx}{x^2+2x+7}$
$= \log(x^2+2x+7).$

(5) $\int \tan x dx = \int \frac{\sin x}{\cos x} dx$
$= -\int \frac{-\sin x}{\cos x} dx$
$= -\log \cos x + C$

i.e., $\int \tan x dx = \log \sec x + C.$

(6) $\int \cot x dx = \int \frac{\cos x}{\sin x} dx.$

$\therefore \quad \int \cot x dx = \log \sin x + C.$

(7) $\frac{6x+5}{3x^2+5x+1} dx = \log(3x^2+5x+1) + C.$

(8) $\int (x+2)(2x-1)dx.$

Although there is a definite rule for the differentiation of the product of two functions, there is none for the integration of a product as in the above example. In such a case the factors must be multiplied.

Then $\int (x+2)(2x-1)dx = \int (2x^2+3x-2)dx$
$= 2\int x^2 dx + 3\int x dx - 2\int dx$
$= \frac{2}{3}x^3 + \frac{3}{2}x^2 - 2x + C.$

(9) $\int \frac{x^4+3x^2+1}{x^3} dx.$

In this example we employ a device which will be used later in more complicated cases; the fraction is split up into its component fractions. This we do by dividing each term of the numerator by the denominator.

Then $\int \frac{x^4+3x^2+1}{x^3} dx = \int \left(x + \frac{3}{x} + \frac{1}{x^3}\right) dx$
$= \int x dx + 3\int \frac{1}{x} dx + \int \frac{dx}{x^3}$
$= \frac{1}{2}x^2 + 3\log x - \frac{1}{2x^2} + C.$

109. If $\frac{d^2y}{dx^2} = x^3$ express y in terms of x.

Since $\frac{d^2y}{dx^2}$ is the differential coefficient of $\frac{dy}{dx}$, it follows that by integrating $\frac{d^2y}{dx^2}$ we obtain $\frac{dy}{dx}$. Having thus found $\frac{dy}{dx}$, a second integration will give the equation connecting y and x.

Since $\frac{d^2y}{dx^2} = x^3$.

Integrating $\frac{dy}{dx} = \int x^3 dx$

$= \frac{1}{4}x^4 + C_1$.

Integrating again $y = \int (\frac{1}{4}x^4 + C_1)dx$

$= \int \frac{1}{4}x^4 dx + \int C_1 dx$

$= \frac{1}{4} \times \frac{1}{5}x^5 + C_1x + C_2$.

$\therefore \quad y = \frac{1}{20}x^5 + C_1x + C_2$.

As a result of integrating twice, two constants are introduced, and these are distinguished as C_1 and C_2.

To find these it is necessary to have two pairs of corresponding values of x and y. On substituting these, we get two simultaneous equations involving C_1 and C_2 as the two unknowns. Solving these, the values found are substituted in the equation

$$y = \frac{1}{20}x^5 + C_1x + C_2,$$

and so the equation connecting x and y is found completely.

Exercise 16.

Find the following integrals.

Constants of integration may always be omitted unless they are going to have a bearing on any further answers; e.g., *the constant C_2 above may be omitted, but not C_1.*

1. $\int 3x dx$.
2. $\int 5x^2 dx$.
3. $\int \frac{1}{2}x^3 dx$.
4. $\int 0{\cdot}4x^4 dx$.

5. $\int 12x^8 dx.$

6. $\int 15t^3 dt.$

7. $\int \frac{dx}{2}.$

8. $\int d\theta.$

9. $\int (4x^2 - 5x + 1)dx.$

10. $\int (3x^4 - 5x^3)dx.$

11. $\int x(8x - \frac{1}{2})dx.$

12. $\int 6x^2(x^2 + x)dx.$

13. $\int \{(x - 3)(x + 3)\}dx.$

14. $\int \{(2x - 3)(x + 4)\}dx.$

15. $\int \frac{dx}{x^2}.$

16. $\int 3x^{-4} dx.$

17. $\int \frac{dx}{x^{1\cdot 4}}.$

18. $\int \sqrt[3]{x} \,.\, dx.$

19. $\int \frac{1}{2}x^{-\frac{1}{2}} dx.$

20. $\int \frac{dx}{\sqrt[3]{x}}.$

21. $\int (x^{\frac{1}{2}} + x^{-\frac{1}{2}})dx.$

22. $\int (x^{\frac{2}{3}} + 1 + x^{-\frac{2}{3}})dx.$

23. $\int \frac{1}{2\sqrt{2x^3}}\, dx.$

24. $\int \left(\frac{\pi}{2} - 5x^{-0\cdot 5}\right) dx.$

25. $\int g\,dt.$

26. $\int \left(\frac{1}{x^3} - \frac{1}{x^2} + \frac{1}{x} - 1\right) dx.$

27. $\int \sqrt{t} \,.\, dt.$

28. $\int \left(1 - \frac{1}{3}x^2 - \frac{1}{2\sqrt{x}}\right) dx.$

29. $\int \frac{1\cdot 4}{x}\, dx.$

30. $\int \frac{dx}{x + 3}.$

31. $\int \frac{dx}{ax + b}.$

32. $\int \left(\frac{3}{x - 1} - \frac{4}{x - 2}\right) dx.$

33. $\int \frac{2x\,dx}{x^2 + 4}.$

34. $\int \frac{dx}{3 - 2x}.$

35. $\int \frac{x + 3}{x}\, dx.$

36. $\int \frac{x^3 - 7}{x}\, dx.$

37. $\int \frac{x^2 - x + 1}{x^3}\, dx.$

38. $\int \sqrt{ax + b}\,dx.$

39. $\int \sqrt{2x + 3}\,dx.$

40. $\int \sqrt{1 + \frac{x}{2}}\, dx.$

41. $\int \frac{dx}{\sqrt{ax+b}}$.

42. $\int \frac{dx}{\sqrt{1-x}}$.

43. $\int (ax+b)^2 dx$.

44. $\int x(1+x)(1+x^2)dx$.

45. $\int \frac{xdx}{x^2-1}$.

46. $\int \frac{\sin axdx}{1+\cos ax}$.

47. $\int \frac{e^{3x}dx}{e^{3x}+6}$.

48. $\int \frac{1+\cos 2xdx}{2x+\sin 2x}$.

---

49. If $\frac{d^2y}{dx^2} = 3x^2$, find y in terms of x.

50. If $\frac{dy}{dx} = 6x^2$, find y in terms of x, when $y = 5$ if $x = 1$.

51. If $\frac{d^2y}{dx^2} = 5x$, find y in terms of x when it is known that if $x = 2$, $\frac{dy}{dx} = 12$, and when $x = 1$, $y = 1$.

52. The gradient of a curve is given by $\frac{dy}{dx} = 4x - 5$. When $x = 1$ it is known that $y = 3$. Find the equation of the curve.

53. The gradient of a curve is given by $\frac{dy}{dx} = 9x^2 - 10x + 4$. If the curve passes through the point (1, 6), find its equation.

54. If $\frac{d^2s}{dt^2} = 8t$, find s in terms of t, when it is known that if $t = 0$, $s = 10$, and $\frac{ds}{dt} = 8$.

110. Integrals of Standard Forms.

We collect below a number of integrals known as **Standard Forms**, which are obtained mainly by inspection as being the known differential coefficients of functions; a few of them were employed above.

(*a*) **Algebraic functions.**

(1) $\int x^n dx = \frac{1}{n+1} x^{n+1}$.

(2) $\int \frac{dx}{x} = \log_e x$.

(3) $\int a^x dx = a^x \times \log_a e.$

(4) $\int e^x dx = e^x.$

(*b*) Trigonometrical functions.

(5) $\int \sin x dx = -\cos x.$

(6) $\int \cos x dx = \sin x.$

(7) $\int \tan x dx = -\log \cos x = \log \sec x.$ (§ 107)

(8) $\int \cot x dx = \log \sin x.$ (§ 107)

Note.—The differential coefficients of sec x and cosec x—viz., sec x tan x and cosec x cot x—do not give rise to standard forms, but to products of these. They are not therefore included in the list above, but follow below. The integrals of sec x and cosec x do not arise by direct differentiation. They will be given later (§120).

(*c*) Hyperbolic functions.

(9) $\int \sinh x dx = \cosh x.$

(10) $\int \cosh x dx = \sinh x.$

(11) $\int \tanh x dx = \log \cosh x.$ (using method of § 107)

(12) $\int \coth x dx = \log \sinh x.$ (using method of § 107)

Note.—The following variations of the above should be carefully noted:

$$\int \sin ax dx = -\frac{1}{a} \cos ax$$

$$\int \sin (ax + b) dx = -\frac{1}{a} \cos (ax + b)$$

$$\int \cos ax dx = \frac{1}{a} \sin ax$$

$$\int \cos (ax + b)dx = \frac{1}{a} \sin (ax + b)$$

$$\int \tan axdx = \frac{1}{a} \log \sec ax$$

$$\int \sinh axdx = \frac{1}{a} \cosh ax$$

$$\int \cosh axdx = \frac{1}{a} \sinh ax.$$

Exercise 17.

Find the following integrals:

1. $\int 3e^{2x}dx.$
2. $\int e^{3x-1}dx.$
3. $\int (e^x + e^{-x})^2dx.$
4. $\int e^{\frac{x}{a}}dx.$
5. $\int (e^{\frac{x}{2}} + e^{-\frac{x}{2}})dx.$
6. $\int (e^{ax} - e^{-ax})dx.$
7. $\int (e^{3x} + a^{3x})dx.$
8. $\int 2^x dx.$
9. $\int 10^{3x}dx.$
10. $\int (a^x + a^{-x})dx.$
11. $\int xe^{x^2}dx.$
12. $\int e^{\cos x} \sin xdx.$
13. $\int \sin 3xdx.$
14. $\int \cos 5xdx.$
15. $\int \sin \tfrac{1}{2}\left(x + \frac{\pi}{3}\right) dx.$
16. $\int \cos(2x + \alpha)dx.$
17. $\int \sin \tfrac{1}{3}xdx.$
18. $\int \sin (\alpha - 3x)dx.$
19. $\int (\cos ax + \sin bx)dx.$
20. $\int \sin 2axdx.$
21. $\int \left(\cos 3x - \sin \frac{x}{3}\right) dx.$
22. $\int \frac{1 + \cos x}{x + \sin x} dx.$
23. $\int \sin^3 x \cos xdx.$
24. $\int \sec^2 xe^{\tan x} dx.$
25. $\int (\tan ax + \cot bx)dx.$
26. $\int \frac{\sin 2x}{1 + \sin^2 x} dx.$

27. $\int \cosh 2x dx.$

28. $\int \sinh \frac{ax}{2} dx.$

29. $\int \tanh 3x dx.$

30. $\int \{\sin (a + bx) - \cos (a - bx)\} dx.$

31. $\int \frac{(e^x + 1)^2}{\sqrt{e^x}} dx.$

32. $\int \tan \frac{3x}{2} dx.$

33. $\int \sec^2 \frac{x}{3} dx.$

34. $\int \frac{e^x}{1 + e^x} dx.$

35. $\int \frac{\sec^2 x}{1 + \tan x} dx.$

36. $\int \cos x \sqrt{\sin x} \, . \, dx.$

111. Additional standard integrals.

In addition to the above integrals of standard forms, the following additional integrals, which are obtained by the differentiation of standard forms, are of importance, especially Nos. 17–25:

(*a*) Trigonometrical.

(13) $\int \sec x \tan x dx = \sec x.$

(14) $\int \operatorname{cosec} x \cot x dx = -\operatorname{cosec} x.$

(15) $\int \operatorname{cosec}^2 x dx = -\cot x.$

(16) $\int \sec^2 x = \tan x.$

(*b*) Inverse trigonometrical.

(17) $\int \frac{dx}{\sqrt{a^2 - x^2}} = \sin^{-1} \frac{x}{a} \text{ or } -\cos^{-1} \frac{x}{a}.$

(18) $\int \frac{dx}{a^2 + x^2} = \frac{1}{a} \tan^{-1} \frac{x}{a} \text{ or } -\frac{1}{a} \cot^{-1} \frac{x}{a}.$

(19) $\int \frac{dx}{x\sqrt{x^2 - a^2}} = \frac{1}{a} \sec^{-1} \frac{x}{a} \text{ or } -\frac{1}{a} \operatorname{cosec}^{-1} \frac{x}{a}.$

(*c*) Inverse hyperbolic functions.

(20) $\int \frac{dx}{\sqrt{x^2 + a^2}} = \sinh^{-1} \frac{x}{a}$ or

$$\log \left\{ \frac{x + \sqrt{x^2 + a^2}}{a} \right\}.$$

(21) $\int \frac{dx}{\sqrt{x^2 - a^2}} = \cosh^{-1} \frac{x}{a}$ or $\log \left\{ \frac{x + \sqrt{x^2 - a^2}}{a} \right\}$.

(22) $\int \frac{dx}{a^2 - x^2} = \frac{1}{a} \tanh^{-1} \frac{x}{a}$ or $\frac{1}{2a} \log \frac{a + x}{a - x}$.

(23) $\int \frac{dx}{x^2 - a^2} = -\frac{1}{a} \coth^{-1} \frac{x}{a}$ or $\frac{1}{2a} \log \frac{x - a}{x + a}$.

(24) $\int \frac{dx}{x\sqrt{a^2 - x^2}} = -\frac{1}{a} \operatorname{sech}^{-1} \frac{x}{a}$ or $-\frac{1}{a} \log \frac{a + \sqrt{a^2 - x^2}}{x}$.

(25) $\int \frac{dx}{x\sqrt{a^2 + x^2}} = -\frac{1}{a} \operatorname{cosech}^{-1} \frac{x}{a}$ or $-\frac{1}{a} \log \frac{a + \sqrt{a^2 + x^2}}{x}$.

The following variations of Nos. 20–25 will be found useful, especially in some of the applications in the next chapter:

20 (a) $\int \frac{dx}{\sqrt{b^2x^2 + a^2}} = \frac{1}{b} \sinh^{-1} \frac{bx}{a}$
$$= \frac{1}{b} \log \left\{ \frac{bx + \sqrt{b^2x^2 + a^2}}{a} \right\}.$$

21 (a) $\int \frac{dx}{\sqrt{b^2x^2 - a^2}} = \frac{1}{b} \cosh^{-1} \frac{bx}{a}$
$$= \frac{1}{b} \log \left\{ \frac{bx + \sqrt{b^2x^2 - a^2}}{a} \right\}.$$

22 (a) $\int \frac{dx}{a^2 - b^2x^2} = \frac{1}{ba} \tanh^{-1} \frac{bx}{a} = \frac{1}{2ba} \log \frac{a + bx}{a - bx}$.

23 (a) $\int \frac{dx}{b^2x^2 - a^2} = \frac{1}{ba} \coth^{-1} \frac{bx}{a} = \frac{1}{2ba} \log \frac{bx - a}{bx + a}$.

24 (a) $$\int \frac{dx}{x\sqrt{a^2 - b^2x^2}} = -\frac{1}{a}\,\text{sech}^{-1}\frac{bx}{a}$$
$$= -\frac{1}{a}\log\frac{a + \sqrt{a^2 - b^2x^2}}{bx}.$$

25 (a) $$\int \frac{dx}{x\sqrt{a^2 + b^2x^2}} = -\frac{1}{a}\,\text{cosech}^{-1}\frac{bx}{a}$$
$$= -\frac{1}{a}\log\frac{a + \sqrt{a^2 + b^2x^2}}{bx}.$$

Notes.

(1) In Formulae 20, 21, 20(a), 21(a) the "a" which appears in the denominator of the logarithm is omitted. This means that the $-\log a$ is merged in the constant of integration.

(2) In Formulae 17–25, if $a = 1$, we get the simpler form stated in §§ 78 and 95.

(3) The Formulae 17–25 will be proved directly in a later chapter.

(4) In the trigonometrical integrals it will assist the memory if it be noted that whenever the name of the function in the resulting integral begins with "co" the function is negative.

112. Worked examples.

Example 1. *Evaluate the integral* $\int \frac{dx}{\sqrt{16 - 9x^2}}$.

The form of this integral can be transformed to that of No. 17:

$$\int \frac{dx}{\sqrt{16 - 9x^2}} = \int \frac{dx}{3\sqrt{\frac{16}{9} - x^2}} = \frac{1}{3}\int \frac{dx}{\sqrt{\frac{16}{9} - x^2}}.$$

This is now in the form of No. 17, where $a = \frac{4}{3}$.

$$\therefore \text{ Integral } = \frac{1}{3}\sin^{-1}\left(x \div \frac{4}{3}\right).$$
$$= \frac{1}{3}\sin^{-1}\frac{3x}{4}.$$

Example 2. *Evaluate the integral* $\int \frac{dx}{\sqrt{9x^2 - 1}}$

The form is that of No. 21 (a), where $b = 3$, $a = 1$.

Hence

$$\int \frac{dx}{\sqrt{9x^2 - 1}} = \frac{1}{3}\cosh^{-1} 3x = \frac{1}{3}\{\log 3x + \sqrt{9x^2 - 1}\}$$

Example 3. *Find the integral* $\int \frac{dx}{9x^2 + 4}$.

This can be transformed into No. 18.

Thus
$$\int \frac{dx}{9x^2 + 4} = \int \frac{dx}{9(x^2 + \frac{4}{9})}$$
$$= \frac{1}{9}\int \frac{dx}{x^2 + (\frac{2}{3})^2}.$$

$\therefore$ by No. 18 integral
$$= \left(\frac{1}{9} \times \frac{1}{\frac{2}{3}}\right) \tan^{-1} \frac{x}{\frac{2}{3}}$$
$$= \left(\frac{1}{9} \times \frac{3}{2}\right) \tan^{-1} \frac{3x}{2}$$
$$= \frac{1}{6} \tan^{-1} \frac{3x}{2}.$$

Exercise 18.

Find the following integrals:

1. (*a*) $\int \frac{dx}{\sqrt{9 - x^2}}$; (*b*) $\int \frac{dx}{\sqrt{x^2 - 9}}$; (*c*) $\int \frac{dx}{\sqrt{x^2 + 9}}$.
2. (*a*) $\int \frac{dx}{9 + x^2}$; (*b*) $\int \frac{dx}{9 - x^2}$; (*c*) $\int \frac{dx}{x^2 - 9}$.
3. (*a*) $\int \frac{dx}{\sqrt{16 - x^2}}$; (*b*) $\int \frac{dx}{16 - x^2}$.
4. (*a*) $\int \frac{dx}{\sqrt{x^2 - 16}}$; (*b*) $\int \frac{dx}{x^2 - 16}$.
5. (*a*) $\int \frac{dx}{\sqrt{x^2 + 16}}$; (*b*) $\int \frac{dx}{x^2 + 16}$.
6. (*a*) $\int \frac{dx}{\sqrt{25 - 9x^2}}$; (*b*) $\int \frac{dx}{\sqrt{9x^2 - 25}}$; (*c*) $\int \frac{dx}{\sqrt{9x^2 + 25}}$.
7. (*a*) $\int \frac{dx}{4x^2 + 9}$; (*b*) $\int \frac{dx}{9 - 4x^2}$; (*c*) $\int \frac{dx}{4x^2 - 9}$.

8. (a) $\int \frac{dx}{9x^2+4}$; (b) $\int \frac{dx}{\sqrt{9x^2+4}}$; (c) $\int \frac{dx}{\sqrt{9x^2-4}}$.

9. (a) $\int \frac{dx}{\sqrt{49x^2+25}}$; (b) $\int \frac{dx}{\sqrt{2x^2+5}}$.

10. (a) $\int \frac{dx}{\sqrt{5-x^2}}$; (b) $\int \frac{dx}{\sqrt{5-4x^2}}$.

11. (a) $\int \frac{dx}{\sqrt{5+4x^2}}$; (b) $\int \frac{dx}{25-4x^2}$.

12. (a) $\int \frac{dx}{\sqrt{7x^2+36}}$; (b) $\int \frac{dx}{x\sqrt{1+x^2}}$.

13. (a) $\int \frac{dx}{x\sqrt{x^2-4}}$; (b) $\int \frac{dx}{x\sqrt{x^2+4}}$.

14. (a) $\int \frac{dx}{x\sqrt{4-x^2}}$; (b) $\int \frac{(x+1)dx}{x\sqrt{x^2-9}}$.

CHAPTER XI

SOME ELEMENTARY METHODS OF INTEGRATION

113. This chapter will contain some of the rules and devices for integration which were referred to in § 99. The general aim of these will be, not direct integration, but transformations of the function to be integrated so that it takes the form of one of the known standard integrals which were given in the last chapter.

Transformations of Trigonometric Functions.

114. Certain trigonometrical formulae may frequently be used with advantage to change products or powers of trigonometric functions into sums of other functions when the rules of § 105 or § 107 may be employed to effect a solution. Examples of this were given in § 108, Nos. 5 and 6, where, by changing $\tan x$ to $\frac{\sin x}{\cos x}$ and $\cot x$ to $\frac{\cos x}{\sin x}$, the integrals $\int \tan x dx$ and $\int \cot x dx$ were found.

Among the formulae which are commonly employed are the following:

(1) $\sin^2 x = \frac{1}{2}(1 - \cos 2x)$.

(2) $\cos^2 x = \frac{1}{2}(1 + \cos 2x)$. (*Trigonometry*, § 83.)

Hence, $\int \sin^2 x dx = \int \frac{1}{2}(1 - \cos 2x)dx$

$= \frac{1}{2}(x - \frac{1}{2}\sin 2x)$.

Similarly, $\int \cos^2 x dx = \frac{1}{2}(x + \frac{1}{2}\sin 2x)$.

It will be noticed that the formula employed in each case enabled us to change a power of this function into a sum, when integration was immediately possible. The following are two further examples:

(3) $\tan^2 x = \sec^2 x - 1$.

$\therefore \int \tan^2 x dx = \int (\sec^2 x - 1)dx$

$= \tan x - x$.

(4) $\cot^2 x$. By the same device

$$\int \cot^2 x dx = \int (\operatorname{cosec}^2 x - 1) dx.$$
$$= -(\cot x + x).$$

(5) $\int \sin^3 x dx$. This can be found by employing the rule:

$$\sin 3A = 3 \sin A - 4 \sin^3 A$$

whence $\sin^3 A = \frac{1}{4}(3 \sin A - \sin 3A)$.

The integral can now be written down:

(6) $\int \cos^3 x dx$. The method is the same as in No. 5, using

$$\cos 3A = 4 \cos^3 A - 3 \cos A.$$

The following formulae are useful for changing products of sines and cosines into sums of these functions:

(*a*) $\sin A \cos B = \frac{1}{2}\{\sin (A + B) + \sin (A - B)\}$
(*b*) $\cos A \sin B = \frac{1}{2}\{\sin (A + B) - \sin (A - B)\}$
(*c*) $\cos A \cos B = \frac{1}{2}\{\cos (A + B) + \cos (A - B)\}$
(*d*) $\sin A \sin B = \frac{1}{2}\{\cos (A - B) - \cos (A + B)\}$

(*Trigonometry*, § 86.)

115. Worked examples.

Example 1. *Evaluate the integral* $\int \frac{\sin^3 x}{\cos^2 x} dx$.

Rearranging

$$\int \frac{\sin^3 x}{\cos^2 x} dx = \int \frac{\sin^2 x \sin x}{\cos^2 x} dx$$
$$= \int \frac{(1 - \cos^2 x) \sin x}{\cos^2 x} dx$$
$$= \int \frac{\sin x}{\cos^2 x} dx - \int \frac{\sin x \cos^2 x}{\cos^2 x} dx$$
$$= \int \sec x \tan x dx - \int \sin x dx$$
$$= \sec x + \cos x.$$

Example 2. *Integrate* $\int \sin 3x \cos 4x dx$.

Using formula (*b*) above

$$\int \cos 4x \sin 3x dx = \int \tfrac{1}{2}\{(\sin (4x + 3x) - \sin (4x - 3x)\}$$
$$= \tfrac{1}{2}\int \sin 7x dx - \tfrac{1}{2}\int \sin x dx$$
$$= \tfrac{1}{2}\{- \tfrac{1}{7} \cos 7x + \cos x\}$$
$$= \tfrac{1}{2}(\cos x - \tfrac{1}{7} \cos 7x).$$

Exercise 19.

Evaluate the following integrals:

1. $\int \sin^2 \frac{x}{2} dx.$
2. $\int \cos^2 \frac{x}{2} dx.$
3. $\int \tan^2 \frac{x}{2} dx.$
4. $\int \cos^4 x dx.$
5. $\int \sin^4 x dx.$
6. $\int \cot^2 2x dx.$
7. $\int \sin^2 2x dx.$
8. $\int \cos^2 3x dx.$
9. $\int \cos^2 (ax + b) dx.$
10. $\int \sin^3 x dx.$
11. $\int \cos^3 x dx.$
12. $\int \sin 2x \sin 3x dx.$
13. $\int \cos 3x \cos x dx.$
14. $\int \sin 4x \cos 2x dx.$
15. $\int \sin 4x \cos \frac{3x}{2} dx.$
16. $\int \sin ax \cos bx dx.$
17. $\int \sin \theta \cos \theta \, d\theta.$
18. $\int \sin^2 x \cos^2 x dx.$
19. $\int \frac{dx}{\sin^2 x \cos^2 x}.$
20. $\int \frac{1 + \sin^2 x}{\cos^2 x} dx.$
21. $\int \tan^3 x dx.$
22. $\int \sin^4 x \cos^2 x dx.$
23. $\int \sqrt{1 + \cos x} dx.$
24. $\int \sec^4 x dx.$

Integration by Substitution.

116. It is sometimes possible, by changing the independent variable, to transform a function into another which can be readily integrated. Experience will suggest the particular form of substitution which is likely to be effective, but there are some easily recognised forms in which certain known substitutions can be employed.

Irrational functions can frequently be treated in this way, as will be seen in the following examples, and those employed serve to prove some of the standard integrals given in § 111.

A. Some trigonometrical and hyperbolic substitutions.

117. $\int\sqrt{a^2 - x^2}dx$.

The form of this suggests that if x be replaced by $a \sin \theta$, we get $a^2 - a^2 \sin^2 \theta$, *i.e.*, $a^2(1 - \sin^2 \theta)$. This is equal to $a^2 \cos^2 \theta$, and on taking the square root the irrational quantity disappears.

It will be seen that we are then left with **two** independent variables—viz., x and θ, since dx remains as part of the integral. But we must have the same variable throughout the integral. Consequently

dx must be expressed in terms of θ.

Since $x = a \sin \theta$.

Differentiating with respect to θ

$$\frac{dx}{d\theta} = a \cos \theta$$

which, for this purpose, we can write as

$$dx = a \cos \theta d\theta.$$

The solution will therefore be as follows:

To integrate $\int\sqrt{a^2 - x^2} . dx$.

Let $x = a \sin \theta$.

Then $dx = a \cos \theta d\theta$.

Substituting in the integral

$$\begin{aligned}\int\sqrt{a^2 - x^2}dx &= \int\sqrt{a^2 - a^2 \sin^2 \theta} \times a \cos \theta . d\theta \\ &= \int a\sqrt{1 - \sin^2 \theta} \times a \cos \theta d\theta \\ &= a^2\int \cos^2 \theta d\theta \\ &= a^2\{\tfrac{1}{2}(\theta + \tfrac{1}{2} \sin 2\theta)\} \qquad \text{(See § 114.)} \\ &= \tfrac{1}{2}a^2\theta + \tfrac{1}{2}a \sin \theta \times a \cos \theta \\ &\qquad \text{(since } \sin 2\theta = 2 \sin \theta \cos \theta).\end{aligned}$$

It is now necessary to change the variable from θ to x.

Since $x = a \sin \theta$ and $\sin \theta = \frac{x}{a}$

$$\theta = \sin^{-1} \frac{x}{a}$$

also

$$a \cos \theta = a\sqrt{1 - \sin^2 \theta}$$
$$= \sqrt{a^2 - a^2 \sin^2 \theta}$$
$$= \sqrt{a^2 - x^2}.$$

$\therefore$ Substituting in

$$\int \sqrt{a^2 - x^2}dx = \tfrac{1}{2}a^2\theta + \tfrac{1}{2}a \sin \theta \times a \cos \theta$$

we get

$$\int \sqrt{a^2 - x^2}dx = \frac{a^2}{2} \sin^{-1}\frac{x}{a} + \tfrac{1}{2}x\sqrt{a^2 - x^2}.$$

Note.—Instead of substituting $x = a \sin \theta$ we could equally well put $x = a \cos \theta$. The student should work this through for practice.

$$\int \frac{dx}{\sqrt{a^2 - x^2}}.$$

Using the same substitution as in the previous case—

viz., $x = a \sin \theta$

we have: $\sqrt{a^2 - x^2} = a \cos \theta$

and $dx = a \cos \theta d\theta$

$$\therefore \int \frac{dx}{\sqrt{a^2 - x^2}} = \int \frac{a \cos \theta d\theta}{a \cos \theta}$$
$$= \int d\theta$$
$$= \theta$$
$$= \sin^{-1} \frac{x}{a}$$

$$\therefore \int \frac{dx}{\sqrt{a^2 - x^2}} = \sin^{-1} \frac{x}{a}.$$ (See § 111.)

$\int \sqrt{x^2 - a^2}dx.$

For this integral we employ hyperbolic functions.

Let $$x = a \cosh z.$$

Then $$z = \cosh^{-1} \frac{x}{a}.$$

From $\cosh^2 z - \sinh^2 z = 1$ (See § 92.)

and
$$\sinh z = \sqrt{\cosh^2 z - 1}$$
$$= \sqrt{\frac{x^2}{a^2} - 1} = \frac{1}{a}\sqrt{x^2 - a^2}.$$

Also since
$$x = a \cosh z$$
$$dx = a \sinh z \,.\, dz$$

$$\therefore \int \sqrt{x^2 - a^2}dx = \int \sqrt{a^2 \cosh^2 z - a^2} \times a \sinh z \,.\, dz$$
$$= \int a\sqrt{\sinh^2 z} \times a \sinh z \,.\, dz$$
$$= a^2 \int \sinh^2 z dz$$
$$= a^2 \int \tfrac{1}{2}(\cosh 2z - 1)dz \qquad \text{(See § 93.)}$$
$$= \frac{a^2}{2}\left(\tfrac{1}{2} \sinh 2z - z\right)$$
$$= \frac{a^2}{4} \sinh 2z - \frac{a^2}{2} z$$
$$= \frac{a^2}{4} 2 \sinh z \cosh z - \frac{a^2}{2} z$$
$$= \tfrac{1}{2}(a \sinh z \times a \cosh z) - \frac{a^2}{2} z$$
$$= \tfrac{1}{2}(\sqrt{x^2 - a^2} \times x) - \frac{a^2}{2} \cosh^{-1} \frac{x}{a}$$

(from above)

$$\therefore \int \sqrt{x^2 - a^2}dx = \tfrac{1}{2}x\sqrt{x^2 - a^2} - \frac{a^2}{2} \cosh^{-1} \frac{x}{a} \text{ or}$$

$$\tfrac{1}{2}x\sqrt{x^2 - a^2} - \frac{a^2}{2}\left(\log \left\{\frac{x + \sqrt{x^2 - a^2}}{a}\right\}\right).$$

(See § 97.)

$$\int \frac{dx}{\sqrt{x^2 - a^2}}.$$

As in the case of the preceding integral

Let $\quad x = a \cosh z.$

Using the equivalents found above:

$$\int \frac{dx}{\sqrt{x^2 - a^2}} = \int \frac{1}{a \sinh z} \times a \sinh z \,.\, dz$$
$$= \int dz$$
$$= z$$

$$\therefore \quad \int \frac{dx}{\sqrt{x^2 - a^2}} = \cosh^{-1} \frac{x}{a} \quad \text{or}$$
$$\log \frac{x + \sqrt{x^2 - a^2}}{a}. \quad \text{(See § 111, No. 21.)}$$

$$\int \sqrt{x^2 + a^2} dx.$$

Let $\quad x = a \sinh z$

$\therefore \quad dx = a \cosh z dz$

and $\quad z = \sinh^{-1} \frac{x}{a}$ and $\cosh z = \frac{1}{a}\sqrt{x^2 + a^2}.$

Substituting

$$\int \sqrt{(x^2 + a^2)}dx = \int a^2\sqrt{\sinh^2 z + 1} \times a \cosh z dz$$
$$= \int a \cosh z \times a \cosh z dz$$
$$= a^2 \int \cosh^2 z dz$$
$$= \frac{a^2}{2}\int (\cosh 2z + 1) dz$$
$$= \frac{a^2}{2} (\tfrac{1}{2} \sinh 2z + z)$$
$$= \frac{a^2}{4} \times 2 \sinh z \cosh z + \frac{a^2}{2} z$$
$$= \tfrac{1}{2} a \sinh z \times a \cosh z + \frac{a^2}{2} z$$

$$\therefore \int \sqrt{x^2 + a^2}dx = \tfrac{1}{2}x\sqrt{x^2 + a^2} + \frac{a^2}{2}\sinh^{-1}\frac{x}{a} \text{ or}$$

$$\tfrac{1}{2}x\sqrt{x^2 + a^2} + \frac{a^2}{2}\log\frac{x + \sqrt{x^2 + a^2}}{a}.$$

$$\int\frac{dx}{\sqrt{x^2 + a^2}}.$$

As above, let $x = a \sinh z.$

Then
$$\int\frac{dx}{\sqrt{x^2 + a^2}} = \int\frac{a \cosh z dz}{a \cosh z}$$
$$= \int dz$$
$$= z$$

$$\therefore \int\frac{dx}{\sqrt{x^2 + a^2}} = \sinh^{-1}\frac{x}{a} \text{ or}$$

$$\log\frac{x + \sqrt{x^2 + a^2}}{a}.$$ (See § 111, No. 20.)

118. $\displaystyle\int\frac{dx}{x^2 + a^2}.$

The form of this suggests the substitution

$$\tan^2\theta + 1 = \sec^2\theta.$$

Accordingly, let $x = a \tan\theta.$

$$\therefore \quad \theta = \tan^{-1}\frac{x}{a}$$

then $dx = a \sec^2\theta d\theta.$

Substituting

$$\int\frac{dx}{x^2 + a^2} = \int\frac{a \sec^2\theta d\theta}{a^2(\tan^2\theta + 1)}$$
$$= \int\frac{a \sec^2\theta}{a^2 \sec^2\theta}d\theta$$
$$= \frac{1}{a}\int d\theta$$
$$= \frac{1}{a}\theta$$

$$\therefore \int \frac{dx}{x^2 + a^2} = \frac{1}{a} \tan^{-1} \frac{x}{a}.$$

(See § 111, No. 18.)

119. Summary of the above formulae.

	Integral.	Substitution.	Result.
1.	$\int \sqrt{a^2 - x^2}dx$ $x < a$	$x = a \sin \theta$	$\frac{a^2}{2} \sin^{-1}\frac{x}{a} + \frac{1}{2}x\sqrt{a^2 - x^2}$
2.	$\int \frac{dx}{\sqrt{a^2 - x^2}}$	$x = a \sin \theta$	$\sin^{-1} \frac{x}{a}$
3.	$\int \sqrt{x^2 - a^2}dx$ $x > a$	$x = a \cosh z$	$\frac{1}{2}x\sqrt{x^2 - a^2} - \frac{a^2}{2} \cosh^{-1} \frac{x}{a}$ *or* $\frac{1}{2}x\sqrt{x^2 - a^2} -$ $\frac{a^2}{2} \log \frac{x + \sqrt{x^2 - a^2}}{a}$
4.	$\int \frac{dx}{\sqrt{x^2 - a^2}}$	$x = a \cosh z$	$\cosh^{-1} \frac{x}{a}$ *or* $\log \frac{x + \sqrt{x^2 - a^2}}{a}$
5.	$\int \sqrt{x^2 + a^2}dx$	$x = a \sinh z$	$\frac{1}{2}x\sqrt{x^2 + a^2} + \frac{a^2}{2}\sinh^{-1} \frac{x}{a}$ *or* $\frac{1}{2}x\sqrt{x^2 + a^2} +$ $\frac{a^2}{2} \log \frac{x + \sqrt{x^2 + a^2}}{a}$
6.	$\int \frac{dx}{\sqrt{x^2 + a^2}}$	$x = a \sinh z$	$\sinh^{-1} \frac{x}{a}$ *or* $\log \frac{x + \sqrt{x^2 + a^2}}{a}$
7.	$\int \frac{dx}{x^2 + a^2}$	$x = a \tan \theta$	$\frac{1}{a} \tan^{-1} \frac{x}{a}$

Note.—$\int \frac{dx}{x^2 - a^2}$ and $\int \frac{dx}{a^2 - x^2}$ are solved by a method which will be given later (§ 129).

120. A useful trigonometrical substitution is given by means of the following formulae, in which $\sin x$ and $\cos x$ are expressed in terms of $\tan \frac{x}{2}$. The formulae are:

$$\sin x = \frac{2 \tan \frac{1}{2}x}{1 + \tan^2 \frac{1}{2}x}$$

$$\cos x = \frac{1 - \tan^2 \frac{1}{2}x}{1 + \tan^2 \frac{1}{2}x}.$$

In using these formulae it is convenient to proceed as follows:

Let $$t = \tan \tfrac{1}{2}x$$

then $$\sin x = \frac{2t}{1 + t^2}$$

$$\cos x = \frac{1 - t^2}{1 + t^2}.$$

Since $$t = \tan\tfrac{1}{2}x$$

$$dt = \tfrac{1}{2} \sec^2 \tfrac{1}{2}x dx$$

$$\therefore \quad dx = \frac{2dt}{\sec^2 \frac{1}{2}x} = \frac{2dt}{1 + \tan^2 \frac{1}{2}x}$$

$$\therefore \quad dx = \frac{2dt}{1 + t^2}.$$

This substitution can be used to find the following integral:

$$\int \operatorname{cosec} x dx = \int \frac{dx}{\sin x}$$

$$= \int \frac{2dt}{1 + t^2} \div \frac{2t}{1 + t^2}$$

$$= \int \frac{dt}{t}$$

$$= \log t$$

$$\therefore \quad \int \operatorname{cosec} x dx = \log \tan \frac{x}{2}.$$

$\int \sec x dx$ can be found similarly or may be derived from the above thus:

From Trigonometry $\sec x = \operatorname{cosec}\left(\frac{\pi}{2} + x\right)$

$$\therefore \quad \int \sec x dx = \int \operatorname{cosec}\left(\frac{\pi}{2} + x\right)$$

$$\therefore \quad \int \sec x dx = \log \tan\left(\frac{\pi}{4} + \frac{x}{2}\right).$$

It may also be shown that this is equal to

$$\log(\sec x + \tan x).$$

The integrals $\int \frac{dx}{a + b\cos x}$ and $\int \frac{dx}{a + b\sin x}$ can be solved by the above substitution. The following example will illustrate the method.

Find the integral $\int \frac{dx}{5 + 4\cos x}$.

Let $\qquad dx = \frac{2dt}{1 + t^2}$, where $t = \tan \frac{1}{2}x$

then $\qquad \cos x = \frac{1 - t^2}{1 + t^2}$.

On simplication the integral becomes:

$$\int \frac{2dt}{5(1 + t^2) + 4(1 - t^2)} = 2\int \frac{dt}{9 + t^2}.$$

This is of the form of integral (18) of § 111.

$$\therefore \quad \text{integral} = 2\left\{\tfrac{1}{3}\tan^{-1}\frac{t}{3}\right\}$$

$$= \tfrac{2}{3}\tan^{-1}\left(\tfrac{1}{3}\tan\frac{1}{2}x\right).$$

The resulting integral may take one of the forms 18, 22, or 23 of the standard integrals of § 111, according to the relative values of a and b. Or, it may require methods given in Chapter 12.

Worked examples.

The following worked examples are numerical variations of the above.

Example 1. *Integrate* $\int\sqrt{16 - 9x^2}dx$.

Let $3x = 4\sin\theta$.

then $x = \frac{4}{3}\sin\theta$ and $\theta = \sin^{-1}\frac{3}{4}x$

$dx = \frac{4}{3}\cos\theta d\theta$

$$\cos\theta = \sqrt{1 - \sin^2\theta} = \sqrt{1 - \frac{9x^2}{16}} = \frac{1}{4}\sqrt{16 - 9x^2}.$$

Substituting

$$\begin{aligned}\int\sqrt{16 - 9x^2}dx &= \int\sqrt{16 - 16\sin^2\theta} \times \tfrac{4}{3}\cos\theta d\theta\\ &= 4\int\cos\theta \times \tfrac{4}{3}\cos\theta d\theta\\ &= \tfrac{16}{3}\int\cos^2\theta d\theta\\ &= \tfrac{16}{3}\int\frac{1 + \cos 2\theta}{2}d\theta\\ &= \tfrac{8}{3}(\theta + \tfrac{1}{2}\sin 2\theta)\\ &= \tfrac{8}{3}\{\sin^{-1}\tfrac{3}{4}x + \sin\theta\cos\theta\}\\ &= \tfrac{8}{3}\{\sin^{-1}\tfrac{3}{4}x + \frac{3x}{4} \times \tfrac{1}{4}\sqrt{16 - 9x^2}\}\\ &= \tfrac{8}{3}\sin^{-1}\tfrac{3}{4}x + \tfrac{1}{2}x\sqrt{16 - 9x^2}.\end{aligned}$$

Example 2. *Integrate* $\int\frac{dx}{\sqrt{9x^2 + 1}}$.

Put $x = \frac{1}{3}\sinh z$; then $z = \sinh^{-1}3x$

$\therefore\ dx = \frac{1}{3}\cosh z dz$

and $\cosh z = \sqrt{1 + \sinh^2 z} = \sqrt{1 + 9x^2}$.

Then

$$\begin{aligned}\int\frac{dx}{\sqrt{9x^2 + 1}} &= \int\frac{\frac{1}{3}\cosh z \,.\, dz}{\sqrt{\sinh^2 z + 1}}\\ &= \tfrac{1}{3}\int\frac{\cosh z dz}{\cosh z}\\ &= \tfrac{1}{3}\int dz\\ &= \tfrac{1}{3}z\\ &= \tfrac{1}{3}\sinh^{-1}3x.\end{aligned}$$

Example 3. *Integrate* $\int \frac{dx}{\sqrt{2 - 3x^2}}$.

Put $\quad x = \sqrt{\tfrac{2}{3}} \sin \theta, \quad$ then $\quad \theta = \sin^{-1} \sqrt{\tfrac{3}{2}}x.$

$$\therefore \quad dx = \sqrt{\tfrac{2}{3}} \cos \theta d\theta$$

$$\cos \theta = \sqrt{1 - \sin^2 \theta} = \sqrt{1 - \tfrac{3}{2}x^2} = \frac{1}{\sqrt{2}} \sqrt{2 - 3x^2}$$

$$\begin{aligned}
\int \frac{dx}{\sqrt{2 - 3x^2}} &= \int \frac{\sqrt{\tfrac{2}{3}} \cos \theta d\theta}{\sqrt{2 - 2 \sin^2 \theta}} \\
&= \int \frac{\sqrt{\tfrac{2}{3}} \cos \theta d\theta}{\sqrt{2} \cos \theta} \\
&= \frac{1}{\sqrt{3}} \int d\theta \\
&= \frac{1}{\sqrt{3}} \theta \\
&= \frac{1}{\sqrt{3}} \sin^{-1} \sqrt{\tfrac{3}{2}}x.
\end{aligned}$$

Example 4. *Integrate* $\int \frac{dx}{x^2\sqrt{1 + x^2}}$.

Let $\quad x = \tan \theta.$

Then $\quad dx = \sec^2 \theta d\theta$

and $\quad \sec \theta = \sqrt{1 + x^2}.$

Then

$$\begin{aligned}
\int \frac{dx}{x^2\sqrt{1 + x^2}} &= \int \frac{\sec^2 \theta d\theta}{\tan^2 \theta \sqrt{1 + \tan^2 \theta}} \\
&= \int \frac{\sec^2 \theta d\theta}{\tan^2 \theta \sec \theta} \\
&= \int \frac{1}{\cos \theta} \times \frac{\cos^2 \theta}{\sin^2 \theta} d\theta \\
&= \int \frac{\cos \theta d\theta}{\sin^2 \theta} \\
&= -\frac{1}{\sin \theta} \qquad \text{(by inspection or by putting } \sin \theta = z) \\
&= -\frac{\sec \theta}{\tan \theta} \\
&= -\frac{\sqrt{1 + x^2}}{x}.
\end{aligned}$$

Exercise 20.

Use the methods given above to find the following Integrals by using suitable substitutions.

Note.—For other examples analogous to 1–10 but involving the irrational quantities as the denominators of fractions, the student is recommended to solve some of the examples of Exercise 18 by the method of substitution.

1. $\int\sqrt{9 - x^2}dx$ (put $x = 3\sin\theta$).
2. $\int\sqrt{25 - x^2}dx$.
3. $\int\sqrt{1 - 4x^2}dx$.
4. $\int\sqrt{9 - 4x^2}dx$ (put $x = \frac{3}{2}\sin\theta$).
5. $\int\sqrt{x^2 - 4}dx$.
6. $\int\sqrt{x^2 - 25}dx$.
7. $\int\sqrt{x^2 + 49}dx$.
8. $\int\sqrt{x^2 + 5}dx$.
9. $\int\sqrt{25x^2 + 16}dx$.
10. $\int\sqrt{x^2 - 3}dx$.
11. $\int\frac{dx}{x^2\sqrt{1 - x^2}}$.
12. $\int\frac{x^2dx}{\sqrt{x^2 + 1}}$.
13. $\int\frac{\sqrt{1 + x^2}}{x^2}dx$.
14. $\int\frac{dx}{x^2\sqrt{a^2 + x^2}}$.
15. $\int\frac{x^2dx}{(1 - x^2)^{\frac{3}{2}}}$.
16. $\int\frac{dx}{(1 - x)\sqrt{1 - x^2}}$ (put $x = \cos\theta$).
17. $\int\operatorname{cosec}\frac{1}{2}xdx$.
18. $\int\sec\frac{1}{2}xdx$.
19. $\int\operatorname{cosec}3xdx$.
20. $\int\sec x\operatorname{cosec}xdx$.
21. $\int\frac{dx}{1 + \cos x}$.
22. $\int\frac{dx}{1 + \sin x}$.
23. $\int\frac{dx}{1 - \sin x}$.
24. $\int(\sec x + \tan x)dx$.
25. $\int\frac{dx}{5 + 3\cos x}$.
26. $\int\frac{dx}{5 - 3\cos x}$.
27. $\int\frac{dx}{4 + 5\cos x}$.
28. $\int\frac{dx}{4 - 5\sin x}$.

B. Algebraic Substitutions.

121. Transformation of a function into a form in which it can readily be integrated can be effected by suitable algebraical substitutions in which the independent variable is changed. The forms these take will depend upon the kind of function to be integrated and, in general, experience and experiment must guide the student. The general aim will be to simplify the function so that it may become easier to integrate.

A frequent example of this method is in the cases of irrational functions in which the expression under the radical sign is of the first degree, that is of the form $ax + b$. These can be integrated by substitution.

Let $\quad ax + b = u^2$

or $\quad u = \sqrt{ax + b}.$

The following examples are typical of the use of algebraical substitution.

122. Worked examples.

Example 1. *Integrate* $\int x\sqrt{2x + 1}dx.$

Let $\quad 2x + 1 = u^2$

or $\quad u = \sqrt{2x + 1}.$

Then $\quad x = \frac{1}{2}(u^2 - 1)$

and $\quad dx = udu.$

Substituting

$$\begin{aligned}\int x\sqrt{2x + 1}dx &= \int \tfrac{1}{2}(u^2 - 1) \times u \times udu \\ &= \tfrac{1}{2}\int u^2(u^2 - 1)du \\ &= \tfrac{1}{2}\int (u^4 - u^2)du \\ &= \tfrac{1}{2}\left(\frac{u^5}{5} - \frac{u^3}{3}\right) \\ &= \tfrac{1}{30}(3u^5 - 5u^3) \\ &= \tfrac{1}{30}\{3(2x + 1)^{\frac{5}{2}} - 5(2x + 1)^{\frac{3}{2}}\}.\end{aligned}$$

Example 2. *Integrate* $\int \frac{xdx}{\sqrt{x + 3}}.$

We rationalise the denominator by the substitution,

$$z = \sqrt{x+3} \quad \text{or} \quad x + 3 = z^2.$$

Then $\quad x = z^2 - 3$

and $\quad dx = 2zdz.$

Substituting

$$\int \frac{xdx}{\sqrt{x+3}} = \int \frac{(z^2-3)2zdz}{\sqrt{z^2}}$$
$$= 2\int \frac{(z^2-3)zdz}{z}$$
$$= 2\int (z^2-3)dz$$
$$= 2\left(\frac{z^3}{3} - 3z\right)$$
$$= \frac{2z}{3}(z^2-9)$$
$$= \tfrac{2}{3}(x-6)\sqrt{x+3}.$$

Example 3. *Integrate* $\int x^3\sqrt{1-x^2}dx.$

Let $\quad u^2 = 1 - x^2 \quad \text{and} \quad x^2 = 1 - u^2.$

Then $\quad x = \sqrt{1-u^2}$

and $\quad dx = -\dfrac{u}{\sqrt{1-u^2}}du.$

Substituting

$$\int x^3\sqrt{(1-x^2)}dx = \int (1-u^2)\sqrt{1-u^2} \times u \times \frac{-u}{\sqrt{1-u^2}}du$$
$$= -\int u^2(1-u^2)du$$
$$= -\left(\frac{5u^3 - 3u^5}{15}\right)$$
$$= -\tfrac{1}{15}u^3(5-3u^2)$$
$$= -\tfrac{1}{15}(1-x^2)\sqrt{1-x^2}\{5-3(1-x^2)\}$$
$$= -\tfrac{1}{15}\{(1-x^2)^{\frac{3}{2}}(2+3x^2)\}.$$

Example 4. *Evaluate* $\int \dfrac{dx}{e^x + e^{-x}}.$

In this case no rationalisation is needed, but we try a substitution which will simplify the exponential form, thus:

Let $$u = e^x$$

then $$e^{-x} = \frac{1}{u}$$

$$du = e^x dx$$

or $$dx = \frac{du}{e^x} \text{ or } \frac{du}{u}.$$

Substituting

$$\int \frac{dx}{e^x + e^{-x}} = \int \frac{du}{u} \div \left(u + \frac{1}{u}\right)$$

$$= \int \frac{du}{u^2 + 1}.$$

Thus we have reached a standard form, viz., No. 18 (§ 111).

$\therefore$ Integral $= \tan^{-1} u$
$= \tan^{-1} e^x.$

Example 5. *Integrate* $\int \frac{\cos^3 x}{\sqrt[3]{\sin x}} dx.$

This example illustrates the advantage in certain cases of changing trigonometrical forms into algebraical, the reverse of the method employed in §§ 117–120. It will then be easier to operate with the indices.

Let $$u = \sin x.$$

$$\therefore \cos x = \sqrt{1 - u^2}.$$

Then $$du = \cos x dx.$$

$$\therefore \int \frac{\cos^3 x dx}{\sqrt[3]{\sin x}} = \int \frac{\cos^2 x \times \cos x dx}{(\sin x)^{\frac{1}{3}}}$$

$$= \int \frac{(1 - u^2) \times du}{u^{\frac{1}{3}}}$$

$$= \int (u^{-\frac{1}{3}} - u^{\frac{5}{3}}) du$$

$$= \tfrac{3}{2} u^{\frac{2}{3}} - \tfrac{3}{8} u^{\frac{8}{3}}$$

$$= 3u^{\frac{2}{3}} \left(\tfrac{1}{2} - \frac{u^2}{8}\right)$$

$$= 3 \sin^{\frac{2}{3}} x \left(\tfrac{1}{2} - \frac{\sin^2 x}{8}\right)$$

Example 6. *Find the value of the integral* $\int \sin^3 x \cos^4 x dx.$

The form suggests trying the same substitution as that of the preceding example.

Let $\cos x = u$

then $\sin x = \sqrt{1 - \cos^2 x}$.

Also $-\sin x dx = du$.

Splitting the factor $\sin^3 x$ into $\sin^2 x . \sin x$ and substituting

$$\begin{aligned}\int \sin^3 x \cos^4 x dx &= \int \sin^2 x . \cos^4 x . \sin x dx \\ &= \int (1 - u^2) \times u^4 \times (-du) \\ &= -\int (u^4 - u^6) du \\ &= -\left(\frac{u^5}{5} - \frac{u^7}{7}\right) \\ &= \tfrac{1}{7}\cos^7 x - \tfrac{1}{5}\cos^5 x.\end{aligned}$$

Example 7. *Integrate* $\int \frac{dx}{\sqrt{x} + 2}$.

Let $x = u^2$

then $dx = 2udu$

$$\begin{aligned}\therefore \int \frac{dx}{\sqrt{x} + 2} &= \int \frac{2udu}{u + 2} \\ &= 2\int \frac{(u + 2) - 2}{u + 2} du \\ &= 2\int \left(1 - \frac{2}{u + 2}\right) du \\ &= 2\{u - 2 \log (u + 2)\} \\ &= 2\{\sqrt{x} - 2 \log (\sqrt{x} + 2)\}.\end{aligned}$$

Exercise 21.

Note.—Some of the following examples may be solved by inspection, remembering the rule for the differentiation of a function of a function. The student is advised, however, if only for the sake of practice, to solve by the method of substitution.

Integrate the following functions:—

1. $\int x^2 \cos x^3 dx$ (put $x^3 = u$).
2. $\int \frac{x^2 dx}{1 - 2x^3}$ (put $2x^3 = u$).
3. $\int \frac{xdx}{\sqrt{1 + x^2}}$.
4. $\int \frac{dx}{\sqrt{2 - 5x}}$.

5. $\int \frac{1}{\sqrt{x}} \sin \sqrt{x} \,.\, dx.$

6. $\int \frac{x^2 dx}{\sqrt{1 + x^3}}.$

7. $\int \frac{\sin x dx}{1 + 2 \cos x}.$

8. $\int \frac{\log x dx}{x}.$

9. $\int x\sqrt{5 + x^2} dx.$

10. $\int \frac{2x dx}{1 + x^4}.$

11. $\int x(x - 2)^4 dx.$

12. $\int \frac{x^2 dx}{(x + 1)^3}.$

13. $\int \frac{x dx}{\sqrt{x - 1}}.$

14. $\int x\sqrt{x - 1} dx.$

15. $\int \frac{x dx}{\sqrt{5 - x^2}}.$

16. $\int \frac{x^3 dx}{\sqrt{x^2 - 1}}.$

17. $\int x\sqrt[3]{x - 2} dx.$

18. $\int \frac{x^3 dx}{x - 1}$ (put $x - 1 = u$).

19. $\int x^3\sqrt{x^2 - 2} dx.$

20. $\int \frac{dx}{\sqrt{x} - 3}.$

21. $\int \frac{\sqrt{x}}{\sqrt{x} + 1} dx.$

22. $\int \frac{x dx}{\sqrt{x} + 1}.$

23. $\int \sin^3 x \cos^2 x dx.$

24. $\int \sin^2 x \cos^5 x dx.$

25. $\int \frac{x^3 dx}{(x^2 + 1)^{\frac{3}{2}}}.$

26. $\int \frac{dx}{e^{2x} - 2e^x}.$

27. $\int x^5(1 + 2x^3)^{\frac{1}{2}} dx.$

28. $\int \frac{dx}{x^2\sqrt{1 + x^2}}.$

29. $\int \frac{\sqrt{1 - x^2}}{x^4} dx$ $\left(\text{put } x = \frac{1}{u}\right).$

30. $\int \frac{\sqrt{1 + \log x}}{x} dx$ (put $1 + \log x = z$).

Integration by parts.

123. This method of integration is derived from the rule for the differentiation of a product of two functions (§ 43), viz.:—

$$\frac{d(uv)}{dx} = u \,.\, \frac{dv}{dx} + v \,.\, \frac{du}{dx}$$

in which u and v are functions of x.

Integrating throughout with respect to x, we get:

$$uv = \int u \cdot \frac{dv}{dx} dx + \int v \cdot \frac{du}{dx} dx.$$

Since u and v are functions of x, this may be written more conveniently in the form:

$$uv = \int u \cdot dv + \int v \cdot du.$$

Thus if either of the integrals on the right side is known, the other can be found. We thus have a choice of solving either of two integrals, whichever is possible or the easier. If, for example, it is decided that $\int vdu$ can readily be determined, then the other integral—viz., $\int udv$—can be found, thus:

$$\int udv = uv - \int v \cdot du \quad . \quad . \quad . \quad (A)$$

The method to be employed will be better understood by studying an example. Suppose it is required to find the integral

$$\int x \cos xdx.$$

Let $u = x$ and $dv = \cos xdx$.

Then $du = dx$

since $dv = \cos xdx$

$$v = \int \cos xdx = \sin x.$$

Substituting in the formulae

$$\int udv = uv - \int vdu$$

we get

$$\int x \cdot \cos xdx = x \sin x + \int \sin xdx.$$

Thus instead of finding the original integral, we have now to find the simpler one of $\int \sin xdx$, which we know to be $-\cos x$.

$$\therefore \quad \int x \cos xdx = x \sin x + \cos x.$$

If u and v had been selected as follows:—

$u = \cos x$ then $du = -\sin xdx$

$dv = xdx$ and $v = \int xdx = \frac{1}{2}x^2$.

Substituting in the formula we get:

$$\int x \cos x dx = \tfrac{1}{2}x^2 \cos x - \int \tfrac{1}{2}x^2(-\sin x).$$

Thus the integral to be found is more difficult than the original.

Formula (A) above could of course be written in the form:

$$\int v du = uv - \int u dv \quad . \quad . \quad . \quad . \quad \text{(B)}$$

The choice is arbitrary, but the student will probably find it better **always** to use one of the two forms. If the form selected is (A), then u will always stand for the function which is to be differentiated and dv as the one to be integrated to complete the formula. In determining which of the functions is thus to be represented by u and which by v, trial must be made as to which will produce the easier final integral.

The following worked examples will perhaps serve to make these points clear.

124. Worked examples.

Example 1. *Evaluate the integral* $\int \log x dx$.

Evidently since $\log x$ produces a simple expression on being differentiated, we put:

$$u = \log x. \qquad \therefore \quad du = \frac{dx}{x}$$

$$dv = dx. \qquad \therefore \quad v = \int dx = x.$$

$\therefore$ substituting in

$$\int u dv = uv - \int v du$$

$$\int \log x dx = x \log x - \int x \times \frac{1}{x} dx$$

$$= x \log x - \int dx$$

$$= x \log x - x$$

or $$\int \log x dx = x(\log x - 1).$$

This important integral should be carefully noted.

Example 2. *Evaluate* $\int xe^{ax}dx$.

We know that e^{ax} produces the same result, except for constants, whether it be differentiated or integrated. But x has a simple form for its differential coefficient.

Hence let $u = x$. $\therefore \quad du = dx$

$dv = e^{ax}dx \quad \therefore \quad v = \int e^{ax}dx = \frac{1}{a}e^{ax}$.

Substituting in

$$\int udv = uv - \int vdu$$

$$\int xe^{ax}dx = x \times \frac{1}{a}e^{ax} - \frac{1}{a} \times \frac{1}{a}e^{ax}$$

$$= \frac{1}{a}e^{ax}\left(x - \frac{1}{a}\right).$$

Example 3. *Integrate* $\int x^2 \sin xdx$.

For the reasons given in § 123, we choose

$u = x^2$. $\therefore \quad du = 2xdx$

and $dv = \sin xdx$ and $v = \int \sin xdx = -\cos x$.

Substituting in Formula A, we get:

$$\int x^2 \sin xdx = -x^2 \cos x + 2\int x \cos xdx.$$

In this example we arrive at an integral which cannot be evaluated by inspection, but is the one evaluated in § 123 and requires itself to be "integrated by parts."

As was shown above

$$\int x \cos xdx = x \sin x + \cos x.$$

Substituting this in the result obtained above, we get

$$\int x^2 \sin xdx = -x^2 \cos x + 2\{x \sin x + \cos x\}$$
$$= -x^2 \cos x + 2x \sin x + 2 \cos x.$$

This repetition of the process will occur in many other cases. For example, if $\int x^3 \sin xdx$ were required, the integration process would have to be applied three times.

Example 4. *Integrate* $\int \sin^{-1} x dx$.

As in Example 2, we must represent dx by dv and u by $\sin^{-1} x$.

Let $u = \sin^{-1} x$. $\therefore \quad du = \frac{dx}{\sqrt{1 - x^2}}$.

$dv = dx$. $\therefore \quad v = \int dx = x$.

Substituting in

$$\int u dv = uv - \int v du$$

we get $\int \sin^{-1} x dx = x \sin^{-1} x - \int \frac{x dx}{\sqrt{1 - x^2}}$.

Noting that the numerator with adjustment of sign is the differential coefficient of $\sqrt{(1 - x^2)}$ in the denominator

$$\int \frac{x dx}{\sqrt{1 - x^2}} = - \sqrt{1 - x^2}.$$

Hence $\int \sin^{-1} x = x \sin^{-1} x + \sqrt{1 - x^2}$.

Example 5. *Evaluate* $\int e^x \cos x dx$.

Take $u = e^x$ $\therefore \quad du = e^x dx$.

Take $dv = \cos x dx$ $\therefore \quad v = \int \cos x dx = \sin x$.

Substituting in

$$\int u dv = uv - \int v du$$

we get

$$\int e^x \cos x dx = e^x \sin x - \int e^x \sin x dx \quad . \quad \text{(A)}$$

Thus we are left with an integral of the same type as the original.

Now try $u = \cos x$. $\therefore \quad du = - \sin x dx$.
and $dv = e^x dx$. $\therefore \quad v = e^x$.

Substituting the formula above

$$\int e^x \cos x dx = e^x \cos x - \int e^x (- \sin x dx).$$

$$\therefore \int e^x \cos x dx = e^x \cos x + \int e^x \sin x dx \quad . \quad . \quad \text{(B)}$$

By addition of (A) and (B)

$$2\int e^x \cos x dx = e^x \sin x + e^x \cos x.$$

$$\therefore \quad \int e^x \cos x dx = \tfrac{1}{2}e^x (\sin x + \cos x).$$

In the same way we may find the general form of these integrals:

$$\int e^{ax} \cos bx dx = \frac{e^{ax}}{a^2 + b^2}\{a \cos bx + b \sin bx\}$$

and

$$\int e^{ax} \sin bx dx = \frac{e^{ax}}{a^2 + b^2}\{a \sin bx - b \cos bx\}$$

or more generally

$$\int e^{ax} \cos (bx + c) dx = \frac{e^{ax}}{a^2 + b^2}\{a \cos (bx + c) + b \sin (bx + c)\}$$

$$\int e^{ax} \sin (bx + c) dx = \frac{e^{ax}}{a^2 + b^2}\{a \sin (bx + c) - b \sin (bx + c)\}.$$

Exercise 22.

Evaluate the following integrals:—

1. $\int x \sin x dx.$ 2. $\int x \sin 3x dx.$

2. $\int x^2 \cos x dx.$ 4. $\int x^3 \cos x dx.$

5. $\int x \log x dx.$ 6. $\int x^2 \log x dx.$

7. $\int x^3 \log x dx.$ 8. $\int \sqrt{x} \log x dx.$

9. $\int x e^x dx.$ 10. $\int x^2 e^x dx.$

11. $\int x e^{-ax} dx.$ 12. $\int e^x \cos 2x dx.$

13. $\int \cos^{-1} x dx.$ 14. $\int \tan^{-1} x dx$

15. $\int x \tan^{-1} x dx.$ 16. $\int e^x \sin x dx.$

17. $\int x \sin^2 x dx.$ 18. $\int x \sin x \cos x dx.$

19. $\int x \sec^2 x dx.$ 20. $\int x \sinh x dx.$

21. $\int x^2 \sin^{-1} x dx.$ 22. $\int x^3 (\log x)^2 dx.$

CHAPTER XII

INTEGRATION OF ALGEBRAIC FRACTIONS

I. Rational fractions.

125. Fractions of certain types have occurred frequently among the functions which have been integrated in previous work. One of the commonest is that in which the numerator can be expressed as the differential coefficient of the denominator. As stated in § 107

$$\int \frac{f'(x)}{f(x)}\,dx = \log f(x) + C.$$

A special form of this which will constantly appear in the work which follows is that in which the denominator is of the first degree, the general form of which is:

$$\int \frac{dx}{ax+b} = \frac{1}{a}\int \frac{a\,dx}{ax+b}$$
$$= \frac{1}{a}\log(ax+b) + C.$$

126. Variants of the above include fractions in which the numerator is of the same as or of higher dimensions than the denominator, simple examples of which have already occurred. Such fractions can often be transposed so that the rule quoted above may be applied. Worked examples illustrating this follow.

127. Worked examples.

Example I. *Evaluate* $\int \frac{x^2}{x+1}\,dx.$

The process employed in transforming such a fraction is similar to that employed in arithmetic. Thus the fraction $\frac{11}{8} = \frac{8+3}{8} = 1 + \frac{3}{8}.$

Similarly, in the example above

$$\begin{aligned}\int \frac{x^2dx}{x+1} &= \int \frac{(x^2-1)+1}{x+1}dx \\ &= \int \frac{x^2-1}{x+1}dx + \int \frac{1}{x+1}dx \\ &= \int (x-1)dx + \int \frac{1}{x+1}dx \\ &= \tfrac{1}{2}x^2 - x + \log(x+1).\end{aligned}$$

Example 2. *Evaluate* $\int \frac{3x+1}{2x-3}dx.$

Then

$$\begin{aligned}\int \frac{3x+1}{2x-3}dx &= \int \frac{\{\frac{3}{2}(2x-3)+\frac{9}{2}\}+1}{2x-3}dx \\ &= \int \frac{\frac{3}{2}(2x-3)+\frac{11}{2}}{2x-3}dx \\ &= \int \tfrac{3}{2}dx + \int \frac{\frac{11}{2}}{2x-3}dx \\ &= \tfrac{3}{2}x + \tfrac{11}{2} \times \tfrac{1}{2}\log(2x-3) \quad (\S\,125.) \\ &= \tfrac{3}{2}x + \tfrac{11}{4}\log(2x-3).\end{aligned}$$

Exercise 23.

Integrate the following:

1. $\int \frac{xdx}{x+2}$.
2. $\int \frac{xdx}{1-x}$.
3. $\int \frac{xdx}{a+bx}$.
4. $\int \frac{x+1}{x-1}dx$.
5. $\int \frac{1-x}{1+x}dx$.
6. $\int \frac{2x-1}{2x+3}dx$.
7. $\int \frac{x^2}{x+2}dx$.
8. $\int \frac{x^2dx}{1-x}$.
9. $\int \frac{x^2dx}{3x-1}$.
10. $\int \frac{x^2dx}{a+bx}$.
11. $\int \frac{3x^3dx}{x+2}$.
12. $\int \frac{x^2dx}{x-1}$.

128. Method of partial fractions.

In the fractions above the denominators are of the first degree. We next proceed to consider fractions of which the donominators are of the second or higher degree.

When adding two such fractions as

$$\frac{2}{x+3} - \frac{1}{x+5}$$

we get $\dfrac{2(x+5)-(x+3)}{(x+3)(x+5)} = \dfrac{x+7}{x^2+8x+15}$.

By reversing this process, $\dfrac{x+7}{x^2+8x+15}$ can be resolved into the two fractions $\dfrac{2}{x+3}$ and $\dfrac{-1}{x+5}$, which are called its "**Partial Fractions**", and these can be integrated directly. By this device we obtain the integral of $\dfrac{x+7}{x^2+8x+15}$. In proceeding to develop this method we will, for the present, consider those cases in which **the denominator of the fraction to be integrated can be resolved into linear factors which are different.**

If in the fraction to be integrated the numerator is of the same or higher dimensions than the denominator, the fraction can first be simplified by the process given in § 127.

The following examples will indicate how the partial fractions are obtained.

129. Worked examples.

Example 1. *Integrate* $\displaystyle\int \frac{x+35}{x^2-25}\,dx$.

Factorising the denominator

$$\frac{x+35}{x^2-25} = \frac{x+35}{(x+5)(x-5)}.$$

From what has been stated above this is resolvable into two partial fractions with denominators $(x+5)$ and $(x-5)$. Since the numerator of the given fraction is of lower dimensions than the denominator, it is evident that the numerators of the partial fractions will be numbers, *i.e.*, not containing x.

Let the numerators be A and B, so that

$$\frac{x+35}{(x+5)(x-5)} = \frac{A}{x-5} + \frac{B}{x+5} \quad . \quad . \quad (1)$$

Clearing the fractions

$$x + 35 = A(x+5) + B(x-5) \quad . \quad . \quad (2)$$

This is an identity and therefore true for any values of x. Let $x = 5$, by which means the coefficient of B vanishes.

Then
$$5 + 35 = 10A + 0.$$
$$\therefore \quad 10A = 40$$
$$A = 4.$$

Substitution of this value of A in (2) would give a equation which could be solved for B. But in this, and in most such cases, it is more simple to substitute a value of x in (2) so that the coefficient of A vanishes.

$\therefore$ let
$$x = -5.$$

Substituting in (2)

$$-5 + 35 = 0 + B(-5-5).$$
$$\therefore \quad 10B = -30$$

and
$$B = -3.$$

Substituting for A and B in (1)

$$\frac{x+35}{x^2-25} = \frac{4}{x-5} - \frac{3}{x+5}.$$

Hence

$$\int \frac{x+35}{x^2-25}\,dx = \int \frac{4dx}{x-5} - \frac{3dx}{x+5}.$$
$$= 4 \log (x-5) - 3 \log (x+5).$$

Example 2. *Integrate* $\dfrac{dx}{x^2-a^2}$.

This is a generalised form of Example 1, and is No. 23 of the Standard Integrals (§ 111).

Factorising

$$\frac{1}{x^2-a^2} = \frac{1}{(x+a)(x-a)}.$$

Let

$$\frac{1}{(x+a)(x-a)} = \frac{A}{x-a} + \frac{B}{x+a}.$$

$$\therefore \quad 1 = A(x+a) + B(x-a).$$

(1) Let $x = a$

then $1 = A(2a) + B(0). \quad \therefore \quad A = \frac{1}{2a}.$

(2) Let $x = -a$

then $1 = A(0) + B(-2a). \quad \therefore \quad B = -\frac{1}{2a}.$

$$\therefore \quad \frac{1}{x^2 - a^2} = \frac{1}{2a}\cdot\frac{1}{x-a} - \frac{1}{2a}\cdot\frac{1}{x+a}.$$

$$\therefore \quad \int\frac{dx}{x^2-a^2} = \frac{1}{2a}\int\left\{\frac{dx}{x-a} - \frac{dx}{x+a}\right\}$$

$$= \frac{1}{2a}\{\log(x-a) - \log(x+a)\}$$

or

$$\int\frac{dx}{x^2-a^2} = \frac{1}{2a}\log\frac{x-a}{x+a} = -\frac{1}{a}\coth^{-1}\frac{x}{a}.$$

Similarly

$$\int\frac{dx}{a^2-x^2} = \frac{1}{2a}\log\frac{a+x}{a-x} = \frac{1}{a}\tanh^{-1}\frac{x}{a}.$$

Example 3. *Integrate* $\int\frac{23-2x}{2x^2+9x-5}dx.$

Factorising the denominator

$$\frac{23-2x}{2x^2+9x-5} = \frac{23-2x}{(2x-1)(x+5)}.$$

Let

$$\frac{23-2x}{(2x-1)(x+5)} = \frac{A}{2x-1} + \frac{B}{x+5}.$$

$$\therefore \quad 23 - 2x = A(x+5) + B(2x-1).$$

(1) Let $x = -5;$

then $23 + 10 = A(0) + B(-11). \quad \therefore \quad B = -3.$

(2) Let $x = \frac{1}{2};$

then $23 - 1 = A(\frac{11}{2}) + B(0). \quad \therefore \quad A = 4.$

Hence

$$\frac{23-2x}{2x^2+9x-5}=\frac{4}{2x-1}-\frac{3}{x+5}.$$

$$\therefore \int\frac{23-2x}{2x^2+9x-5}dx=\int\frac{4dx}{2x-1}-\int\frac{3dx}{x+5}.$$

$$=2\log(2x-1)-3\log(x+5).$$

Example 4. *Integrate* $\int\frac{x^2+10x+6}{x^2+2x-8}dx.$

The numerator being of the same dimensions as the denominator we proceed as shown in § 127, Example 1.

Then $\frac{x^2+10x+6}{x^2+2x-8}=\frac{(x^2+2x-8)+(8x+14)}{x^2+2x-8}$

$$=1+\frac{8x+14}{x^2+2x-8}.$$

The fraction thus obtained is now resolved into partial fractions. Factorising the denominator

$$\frac{8x+14}{x^2+2x-8}=\frac{8x+14}{(x-2)(x+4)}.$$

Let $\frac{8x+14}{x^2+2x-8}=\frac{A}{x-2}+\frac{B}{x+4}.$

$$\therefore \quad 8x+14=A(x+4)+B(x-2).$$

(1) Let $x=-4$;

then $-18=A(0)+B(-6). \quad \therefore \quad B=3.$

(2) Let $x=2$;

then $30=A(6)+B(0). \quad \therefore \quad A=5.$

$$\therefore \quad \frac{8x+14}{(x-2)(x+4)}=\frac{5}{x-2}+\frac{3}{x+4}.$$

$$\therefore \int\frac{x^2+10x+6}{x^2+2x-8}dx=\int\left(1+\frac{5}{x-2}+\frac{3}{x+4}\right)dx$$

$$=x+5\log(x-2)+3\log(x+4).$$

130. When the denominator is the square of a binomial, as, for example, $(x+a)^2$.

In this case the fraction may be the sum of two fractions of which the denominators are $(x+a)$ and $(x+a)^2$ with constants as numerators.

Example. *Integrate* $\int \frac{3x + 1}{(x + 1)^2} dx$.

Let
$$\frac{3x + 1}{(x + 1)^2} = \frac{A}{x + 1} + \frac{B}{(x + 1)^2}.$$

$$\therefore \quad 3x + 1 = A(x + 1) + B \quad . \quad . \quad . \quad . \quad (1)$$

Let
$$x = -1;$$
then
$$-2 = A(0) + B. \quad \therefore \quad B = -2.$$

A may be found by using the property of an identity, viz., the coefficients of like terms on the two sides of the identity are equal. Comparing the coefficients of x in (1), above, we get:

$$3 = A.$$

$$\therefore \quad \frac{3x + 1}{(x + 1)^2} = \frac{3}{x + 1} - \frac{2}{(x + 1)^2}.$$

$$\therefore \quad \int \frac{3x + 1}{(x + 1)^2} dx = \int \frac{3dx}{x + 1} - \int \frac{2dx}{(x + 1)^2}$$

$$= 3 \log (x + 1) + \frac{2}{x + 1}.$$

The second integral, *i.e.*, $\int \frac{2dx}{(x + 1)^2}$, is found by inspection, remembering that

$$\int \frac{dx}{x^2} = -\frac{1}{x}.$$

Exercise 24.

Find the following integrals:

1. $\int \frac{dx}{x^2 - 1}$.
2. $\int \frac{dx}{1 - x^2}$.
3. $\int \frac{x^2 dx}{x^2 - 4}$.
4. $\int \frac{dx}{4x^2 - 9}$.
5. $\int \frac{x + 8}{x^2 + 6x + 8} dx$.
6. $\int \frac{3x - 1}{x^2 + x - 6} dx$.
7. $\int \frac{8x + 1}{2x^2 - 9x - 35} dx$.
8. $\int \frac{x + 1}{3x^2 - x - 2} dx$.
9. $\int \frac{7x - 8}{4x^2 + 3x - 1} dx$.
10. $\int \frac{1 + x}{(1 - x)^2} dx$.

11. $\int \frac{2x - 1}{(x + 2)^2} dx.$

12. $\int \frac{2x + 1}{(2x + 3)^2} dx.$

13. $\int \frac{x^2 - 2}{x^2 - x - 12} dx.$

14. $\int \frac{x^2 + 1}{x^2 - x - 2} dx.$

15. $\int \frac{2x^3 - 2x^2 - 11x - 8}{x^2 - x - 6} dx.$

16. $\int \frac{x^3 - 2x^2 - 1}{x^2 - 1} dx.$

131. Denominator of higher degree than the second and resolvable into factors.

(a) When the denominator is entirely resolvable into different linear factors.

The method is the same as when there are only two factors, but the number of partial fractions will correspond to the number of factors.

Example. *Integrate* $\frac{3 - 4x - x^2}{x(x^2 - 4x + 3)}$.

Factorising the denominator we get:

$$\frac{3 - 4x - x^2}{x(x - 1)(x - 3)}.$$

Let

$$\frac{-x^2 - 4x + 3}{x(x - 1)(x - 3)} = \frac{A}{x} + \frac{B}{x - 1} + \frac{C}{x - 3}.$$

Then

$$-x^2 - 4x + 3 = A(x - 1)(x - 3) + Bx(x - 3) + Cx(x - 1).$$

(1) Let $x = 0$;

then $3 = 3A + B(0) + C(0)$. $\therefore A = 1$.

(2) Let $x = 1$;

then $-2 = A(0) - 2B + C(0)$. $\therefore B = 1$.

(3) Let $x = 3$;

then $-18 = A(0) + B(0) + 6C$. $\therefore C = -3$.

$$\therefore \int \frac{-x^2 - 4x + 3}{x(x - 1)(x - 3)} dx = \int \left(\frac{1}{x} + \frac{1}{x - 1} - \frac{3}{x - 3}\right) dx$$

$$= \log x + \log (x - 1) - 3 \log (x - 3).$$

(b) When the denominator can be resolved into linear factors, one or more of which may be repeated.

Example. *Integrate* $\int \frac{-2dx}{(x - 1)^2(x - 2)}$.

The procedure is the same as that of § 130.

Let

$$\frac{-1}{(x-1)^2(x-2)} = \frac{A}{x-1} + \frac{B}{(x-1)^2} + \frac{C}{x-2}.$$

then

$$-1 = A(x-1)(x-2) + B(x-2) + C(x-1)^2.$$

(1) Let $x = 1$;

then $\quad -1 = A(0) - B + C(0). \qquad \therefore B = 1.$

(2) Let $x = 2$;

then $\quad -1 = A(0) + B(0) + C. \qquad \therefore C = -1.$

(3) Let $x = 0$;

then $\quad -1 = 2A - 2 - 2. \qquad \therefore A = 1.$

(on substituting the values already found for B and C).

$$\therefore \int \frac{-dx}{(x-1)^2(x-2)} = \int\left(\frac{1}{x-1} + \frac{1}{(x-1)^2} - \frac{1}{x-2}\right) dx$$

$$= \log(x-1) - \frac{1}{x-1} - \log(x-2).$$

132. When the denominator contains a quadratic factor which cannot itself be factorised.

The method adopted in cases already considered can be employed.

Example. *Integrate* $\int \frac{(x-1)dx}{(x+1)(x^2+1)}$.

The factor $(x^2 + 1)$ cannot itself be resolved into real factors. However, two partial fractions with the denominators $(x + 1)$ and $(x^2 + 1)$ can be obtained. But the numerator of the fraction in which the denominator is of the second degree, viz. $(x^2 + 1)$ may be of the first degree in x. The general form of this can be expressed by $(Bx + C)$.

$\therefore$ Let

$$\frac{x-1}{(x+1)(x^2+1)} = \frac{A}{x+1} + \frac{Bx+C}{x^2+1}.$$

Then $\quad x - 1 = A(x^2 + 1) + (Bx + C)(x + 1) \qquad (1)$

Let $\quad x = -1$

then $\quad -2 = A(2) + 0. \qquad \therefore A = -1.$

Substituting this value of A in (1), we get:

$$x - 1 = -(x^2 + 1) + (Bx + C)(x + 1)$$

or

$$x^2 + x = (Bx + C)(x + 1).$$

Equating coefficients of x^2

$$1 = B \qquad \therefore \quad B = 1. \qquad (\S\ 130.)$$

Equating coefficients of x

$$1 = B + C. \qquad \therefore \quad C = 0.$$

Hence

$$\frac{x - 1}{(x - 1)(x^2 + 1)} = \frac{-1}{x + 1} + \frac{x}{x^2 + 1}.$$

$$\therefore \quad \int \frac{x - 1}{(x + 1)(x^2 + 1)} dx = -\int \frac{dx}{x + 1} + \int \frac{xdx}{x^2 + 1}$$

$$= -\log(x + 1) + \tfrac{1}{2}\log(x^2 + 1).$$

Note.—The integral $\int \frac{xdx}{x^2 + 1}$ is one which can be found by the application of the rule in § 107, but more difficult cases will require the methods given in the next section.

Exercise 25.

Integrate the following:

1. $\int \frac{dx}{x(x^2 - 1)}$.
2. $\int \frac{dx}{x^2(x + 2)}$.
3. $\int \frac{(2x + 3)dx}{x(x - 1)(x + 2)}$.
4. $\int \frac{(x^2 - 3)dx}{(x - 1)(x - 2)(x + 3)}$.
5. $\int \frac{(2x + 1)dx}{(x + 2)(x - 3)^2}$.
6. $\int \frac{xdx}{(x + 1)^2(x - 1)}$.
7. $\int \frac{(x^3 + 1)dx}{x(x - 1)^3}$.
8. $\int \frac{dx}{x(x^2 + 1)}$.
9. $\int \frac{dx}{(x^2 + 1)(x - 2)}$.
10. $\int \frac{xdx}{(x + 1)(x^2 + 4)}$.
11. $\int \frac{(x + 2)dx}{(x^2 + 4)(1 - x)}$.
12. $\int \frac{xdx}{x^4 - 1}$.
13. $\int \frac{x^2dx}{x^4 - 1}$.
14. $\int \frac{(x + 1)^2}{x^3 + x} dx$.

133. Denominator of the form $ax^2 + bx + c$ and not resolvable into factors.

The student will have learnt from *Algebra* that the expression $ax^2 + bx + c$ can always be expressed as the sum or difference of two squares. The following examples illustrate this.

$$\begin{aligned}
x^2 + 4x + 2 &= \{x^2 + 4x + (2)^2\} - 2^2 + 2 \\
&= (x + 2)^2 - 2 \quad or \quad (x + 2)^2 - (\sqrt{2})^2. \\
2x^2 - 3x + 1 &= 2(x^2 - \tfrac{3}{2}x) + 1 \\
&= 2\{x^2 - \tfrac{3}{2}x + (\tfrac{3}{4})^2\} - 2 \times (\tfrac{3}{4})^2 + 1 \\
&= 2(x - \tfrac{3}{4})^2 - \tfrac{1}{8} = \{\sqrt{2}(x - \tfrac{3}{4})\}^2 - (\sqrt{\tfrac{1}{8}})^2 \\
x^2 + 6x + 14 &= \{x^2 + 6x + (3)^2\} - (3)^2 + 14 \\
&= (x + 3)^2 + 5 = (x + 3)^2 + (\sqrt{5})^2 \\
12 + 5x - x^2 &= 12 - (x^2 - 5x) \\
&= 12 - \{x^2 - 5x + (\tfrac{5}{2})^2\} + \tfrac{25}{4} \\
&= \tfrac{73}{4} - (x - \tfrac{5}{2})^2 \\
&= \left(\frac{\sqrt{73}}{2}\right)^2 - (x - \tfrac{5}{2})^2.
\end{aligned}$$

All of these are expressions for which there are no rational factors. They are all included in the three types:

(1) $x^2 - a^2$
(2) $x^2 + a^2$
(3) $a^2 - x^2$

We have seen that fractions of which these are denominators are of standard form (*see* § 111, Nos. 18, 22, 23). Consequently the denominator of a fraction which is of the form $ax^2 + bx + c$ can be transformed into one of these three types. For convenience these three integrals are repeated, as they will be in constant use in work which follows.

$$\text{A} \quad \int \frac{dx}{x^2 - a^2} = -\frac{1}{a} \coth^{-1} \frac{x}{a} \quad or \quad \frac{1}{2a} \log \frac{x - a}{x + a}.$$

$$\text{B} \quad \int \frac{dx}{a^2 + x^2} = \frac{1}{a} \tan^{-1} \frac{x}{a}.$$

$$\text{C} \quad \int \frac{dx}{a^2 - x^2} = \frac{1}{a} \tanh^{-1} \frac{x}{a} \quad or \quad \frac{1}{2a} \log \frac{a + x}{a - x}.$$

If $(x + b)$ is substituted for x in each of the above, since the differential coefficients of $x + b$ and x are the same, we have:

$$A \quad \int \frac{dx}{(x+b)^2 - a^2} = -\frac{1}{a}\coth^{-1}\frac{x+b}{a} \quad or \quad \frac{1}{2a}\log\frac{(x+b)-a}{(x+b)+a}.$$

$$B \quad \int \frac{dx}{(x+b)^2 + a^2} = \frac{1}{a}\tan^{-1}\frac{x+b}{a}.$$

$$C \quad \int \frac{dx}{a^2 - (x+b)^2} = \frac{1}{a}\tanh^{-1}\frac{x+b}{a} \quad or \quad \frac{1}{2a}\log\frac{a+(x+b)}{a-(x+b)}.$$

Two cases may occur in the integration of such fractional functions. They will be illustrated by the following:

(1) When the numerator is constant.

Type $$\int \frac{dx}{ax^2 + bx + c}.$$

134. Worked examples.

Example 1. *Integrate* $\int \frac{dx}{x^2 + 6x + 2}$.

We first express the denominator in the form $x^2 \pm a^2$.

$$x^2 + 6x + 2 = \{x^2 + 6x + (\tfrac{6}{2})^2\} - 9 + 2.$$
$$= (x + 3)^2 - 7.$$

$$\therefore \quad \int \frac{dx}{x^2 + 6x + 2} = \int \frac{dx}{(x+3)^2 - (\sqrt{7})^2}$$

which is of form A above.

$$\therefore \quad \int \frac{dx}{x^2 + 6x + 2} = -\frac{1}{\sqrt{7}}\coth^{-1}\frac{x+3}{\sqrt{7}} \quad or$$
$$\frac{1}{2\sqrt{7}}\log\frac{x+3-\sqrt{7}}{x+3+\sqrt{7}}.$$

Example 2. *Integrate* $\int \frac{dx}{2 + 3x - x^2}$.

Since $2 + 3x - x^2 = 2 - \{x^2 - 3x + \frac{9}{4}\} + \frac{9}{4}$
$$= \tfrac{17}{4} - (x - \tfrac{3}{2})^2.$$

Using formula C

$$\int\frac{dx}{2+3x-x^2}=\int\frac{dx}{\frac{17}{4}-(x-\frac{3}{2})^2}$$

$$=\frac{2}{\sqrt{17}}\tanh^{-1}\frac{x-\frac{3}{2}}{\sqrt{17}}\ \text{or}\ \frac{1}{2\sqrt{17}}\log\frac{\sqrt{17}+(2x-3)}{\sqrt{17}-(2x-3)}.$$

Example 3. *Integrate* $\int\frac{dx}{2x^2+4x+3}$.

Rearranging the denominator

$$2x^2+4x+3=2\{(x^2+2x+1)-1+\tfrac{3}{2}\}$$

$$=2\left\{(x+1)^2+\left(\frac{1}{\sqrt{2}}\right)^2\right\}.$$

Using formula B and substituting

$$\int\frac{dx}{2x^2+4x+3}=\tfrac{1}{2}\int\frac{dx}{(x+1)^2+\frac{1}{2}}.$$

Using (B) as integral

$$\int\frac{dx}{2x^2+4x+3}=\tfrac{1}{2}\int\frac{dx}{(x+1)^2+\frac{1}{2}}$$

$$=\tfrac{1}{2}\left\{\frac{1}{\sqrt{\frac{1}{2}}}\tan^{-1}\frac{x+1}{\sqrt{\frac{1}{2}}}\right\}$$

$$=\frac{\sqrt{2}}{2}\tan^{-1}\sqrt{2}(x+1).$$

(2) When the numerator is of the first degree in the variable.

Type $$\int\frac{Ax+B}{ax^2+bx+c}dx.$$

135. To solve this integral a combination of the devices previously used is required as shown in the following examples.

Example 1. *Integrate* $\int\frac{6x+7}{x^2-x-1}dx$.

The numerator must be first rearranged so that part of it is the differential coefficient of the denominator.

Now $$\frac{d}{dx}(x^2-x-1)=2x-1.$$

Re-arranging numerator

$$6x + 7 = 3(2x - 1) + 10.$$

denominator $\qquad x^2 - x - 1 = (x - \tfrac{1}{2})^2 - \tfrac{5}{4}.$

$$\therefore \quad \int \frac{(6x+7)dx}{x^2 - x - 1} = \int \frac{3(2x-1)+10}{x^2-x-1}dx.$$

$$= \int \frac{3(2x-1)}{x^2-x-1}dx + \int \frac{10dx}{(x-\frac{1}{2})^2 - \frac{5}{4}}.$$

The first integral is found by the rule of § 107 and the second by using the standard form (*A*) above.

$$\therefore \int \frac{(6x+7)dx}{x^2-x-1} = 3\log(x^2 - x + 1) + \frac{10}{\sqrt{5}}\log\frac{(x-\frac{1}{2}) - \frac{\sqrt{5}}{2}}{(x-\frac{1}{2}) + \frac{\sqrt{5}}{2}}.$$

Verification.—The student will find it a very useful exercise to verify some of these results, by differentiating the integral obtained. The verification of the exercise above, follows as an example.

Let

$$y = 3\log(x^2 - x + 1) + \frac{10}{\sqrt{5}}\log\frac{(x-\frac{1}{2}) - \frac{\sqrt{5}}{2}}{(x-\frac{1}{2}) + \frac{\sqrt{5}}{2}}.$$

Then

$$\frac{dy}{dx} = \frac{3(2x-1)}{x^2-x+1} + \frac{10}{\sqrt{5}}\left\{\frac{1}{(x-\frac{1}{2}) - \frac{\sqrt{5}}{2}} - \frac{1}{(x-\frac{1}{2}) + \frac{\sqrt{5}}{2}}\right\}$$

$$= \frac{6x-3}{x^2-x+1} + \frac{10}{\sqrt{5}}\left\{\frac{\left(x - \frac{1}{2} + \frac{\sqrt{5}}{2}\right) - \left(x - \frac{1}{2} - \frac{\sqrt{5}}{2}\right)}{\left(x - \frac{1}{2} - \frac{\sqrt{5}}{2}\right)\left(x - \frac{1}{2} + \frac{\sqrt{5}}{2}\right)}\right\}$$

$$= \frac{6x-3}{x^2-x+1} + \frac{10}{\sqrt{5}}\left\{\frac{\sqrt{5}}{(x-\frac{1}{2})^2 - \left(\frac{\sqrt{5}}{2}\right)^2}\right\}$$

$$= \frac{6x-3}{x^2-x+1} + \frac{10}{x^2-x+1}$$

$$= \frac{6x+7}{x^2-x+1}.$$

Example 2. *Integrate* $\int \frac{5x+1}{2x^2+4x+3} dx.$

$$\frac{d}{dx}(2x^2+4x+3) = 4x+4.$$

Re-arranging numerator

$$5x+1 = \tfrac{5}{4}(4x+4) - 4.$$

Re-arranging denominator

$$\begin{aligned} 2x^2+4x+3 &= 2\{x^2+2x+\tfrac{3}{2}\} \\ &= 2\{(x+1)^2+\tfrac{1}{2}\}. \end{aligned}$$

$$\begin{aligned} \therefore \quad \int \frac{(5x+1)dx}{2x^2+4x+3} \\ &= \int \frac{\frac{5}{4}(4x+4)-4}{2x^2+4x+3} dx \\ &= \tfrac{5}{4}\int \frac{(4x+4)dx}{2x^2+4x+3} - 4\int \frac{dx}{2\{(x+1)^2+\frac{1}{2}\}} \\ &= \tfrac{5}{4}\log(2x^2+4x+3) - (2 \div \sqrt{\tfrac{1}{2}})\tan^{-1}\frac{x+1}{\sqrt{\frac{1}{2}}} \\ &= \tfrac{5}{4}\log(2x^2+4x+3) - 2\sqrt{2}\tan^{-1}\sqrt{2}(x+1). \end{aligned}$$

Example 3. *Integrate* $\int \frac{2x+1}{x^3-1} dx.$

First we must resolve the fraction into partial fractions.

Since $\quad x^3 - 1 = (x-1)(x^2+x+1).$

Let $\quad \frac{2x+1}{x^3-1} = \frac{A}{x-1} + \frac{Bx+C}{x^2+x+1}.$

$$\therefore \quad 2x+1 = A(x^2+x+1) + (Bx+C)(x-1).$$

Let $\quad x = 1$; then $3 = A(3) + 0$; $\therefore$ A = 1.

Comparing coefficients

(1) x^2 $\quad 0 = A + B = 1 + B. \quad \therefore B = -1.$
(2) Constants $\quad 1 = -C + 1. \quad \therefore C = 0.$

$$\therefore \quad \int \frac{(2x+1)dx}{x^3-1} = \int \frac{dx}{x-1} - \int \frac{xdx}{x^2+x+1}.$$

(1) $\int \frac{dx}{x-1} = \log(x-1).$

(2) $\int \frac{xdx}{x^2 + x + 1}$

$$= \int \frac{\frac{1}{2}(2x + 1) - \frac{1}{2}}{x^2 + x + 1} dx$$

$$= \frac{1}{2}\int \frac{2x + 1}{x^2 + x + 1} - \frac{1}{2}\int \frac{dx}{(x + \frac{1}{2})^2 + \frac{3}{4}}$$

$$= \frac{1}{2} \log (x^2 + x + 1) - \left(\frac{1}{2} \times \frac{2}{\sqrt{3}}\right) \tan^{-1} \frac{x + \frac{1}{2}}{\frac{\sqrt{3}}{2}}$$

$$= \frac{1}{2} \log (x^2 + x + 1) - \frac{1}{\sqrt{3}} \tan^{-1} \frac{2x + 1}{\sqrt{3}}.$$

Adding (1) and (2)

$$\int \frac{2x + 1}{x^3 + 1}$$
$$= \log (x - 1) + \frac{1}{2} \log (x^2 + x + 1) - \frac{1}{\sqrt{3}} \tan^{-1} \frac{2x + 1}{\sqrt{3}}.$$

Exercise 26.

Integrate the following:

1. $\int \frac{dx}{x^2 + 6x + 17}$.
2. $\int \frac{dx}{x^2 + 6x - 4}$.
3. $\int \frac{dx}{x^2 + 4x + 6}$.
4. $\int \frac{dx}{2x^2 + 2x + 7}$.
5. $\int \frac{(1 - 3x)dx}{3x^2 + 4x + 2}$.
6. $\int \frac{(4x - 5)dx}{x^2 - 2x - 1}$.
7. $\int \frac{(2x + 5)dx}{x^2 + 4x + 5}$.
8. $\int \frac{dx}{x^3 + 1}$.
9. $\int \frac{(x - 1)^2 dx}{x^2 + 2x + 2}$.
10. $\int \frac{(3x + 5)dx}{1 - 2x - x^2}$.
11. $\int \frac{(4x + 5)dx}{3x^2 + x + 3}$.
12. $\int \frac{x^2 + 1}{x^3 + 1} dx$.

II. Fractions with irrational denominators.

136. Type $\int \frac{dx}{\sqrt{ax^2 + bx + c}}$.

By the use of methods similar to those employed in

previous sections, integrals of this type can be transformed to one of the following standard forms (see § 111).

$$\text{(A)} \int \frac{dx}{\sqrt{x^2 - a^2}} = \cosh^{-1}\frac{x}{a} = \log\{x + \sqrt{x^2 - a^2}\}$$

$$\text{(B)} \int \frac{dx}{\sqrt{a^2 - x^2}} = \sin^{-1}\frac{x}{a}.$$

$$\text{(C)} \int \frac{dx}{\sqrt{x^2 + a^2}} = \sinh^{-1}\frac{x}{a} = \log\{x + \sqrt{x^2 + a^2}\}.$$

In this type the numerator is a constant and does not involve the variable. Consequently it is necessary only to transform the denominator into one of the three forms (A), (B), or (C).

The method of doing this is illustrated in the following examples:

137. Worked examples.

Example 1. *Integrate* $\int \frac{dx}{\sqrt{x^2 + 6x + 10}}$.

Now

$$x^2 + 6x + 10 = x^2 + 6x + (3)^2 - 9 + 10 = (x + 3)^2 + 1.$$

$$\therefore \quad \int \frac{dx}{\sqrt{x^2 + 6x + 10}} = \int \frac{dx}{\sqrt{(x + 3)^2 + 1}}.$$

This is of type (C) above, in which x is replaced by $x + 3$, which has the same differential coefficient.

$$\therefore \quad \int \frac{dx}{\sqrt{x^2 + 6x + 10}}$$
$$= \sinh^{-1}(x + 3) \text{ or } \log\{(x + 3) + \sqrt{x^2 + 6x + 10}\}.$$

Example 2. *Integrate* $\int \frac{dx}{\sqrt{2x^2 + 3x - 2}}$.

Now, $\quad (2x^2 + 3x - 2) = 2(x^2 + \frac{3}{2}x - 1)$
$\qquad\qquad = 2\{(x + \frac{3}{4})^2 - \frac{25}{16}\}.$

$$\therefore \quad \int \frac{dx}{\sqrt{2x^2 + 3x - 2}} = \frac{1}{\sqrt{2}} \int \frac{dx}{\sqrt{(x + \frac{3}{4})^2 - \frac{25}{16}}}.$$

$\therefore$ Using Type A,

$$\int \frac{dx}{\sqrt{2x^2 + 3x - 2}} = \frac{1}{\sqrt{2}} \cosh^{-1} \frac{x + \frac{3}{4}}{\frac{5}{4}}$$

$$= \frac{1}{\sqrt{2}} \cosh^{-1} \frac{4x + 3}{5}.$$

Example 3. *Integrate* $\int \frac{dx}{\sqrt{4 + 8x - 5x^2}}$.

$$4 + 8x - 5x^2 = 5\{\tfrac{4}{5} - (x^2 - \tfrac{8}{5}x)\}$$
$$= 5\{\tfrac{36}{25} - (x - \tfrac{4}{5})^2\}.$$

$$\therefore \int \frac{dx}{\sqrt{4 + 8x - 5x^2}} = \int \frac{dx}{\sqrt{5\{\frac{36}{25} - (x - \frac{4}{5})^2}}$$

$$= \frac{1}{\sqrt{5}} \int \frac{dx}{\sqrt{\frac{36}{25} - (x - \frac{4}{5})^2}}$$

$$= \frac{1}{\sqrt{5}} \sin^{-1} \frac{x - \frac{4}{5}}{\frac{6}{5}}$$

$$= \frac{1}{\sqrt{5}} \sin^{-1} \frac{5x - 4}{6}.$$

138. Type $\int \frac{(Ax + B)dx}{\sqrt{ax^2 + bx + c}}$.

Let us consider a special case in which the denominator is, say, $\sqrt{2x^2 + 7x + 8}$, *i.e.*, $(2x^2 + 7x + 8)^{\frac{1}{2}}$.

Then $\frac{d}{dx}(\sqrt{2x^2 + 7x + 8})$

$$= \tfrac{1}{2}(2x^2 + 7x - 8)^{-\frac{1}{2}} \times \frac{d}{dx}(2x^2 + 7x - 8)$$

$$= \tfrac{1}{2}(2x^2 + 7x - 8)^{-\frac{1}{2}} \times (4x + 7)$$

$$= \frac{\frac{1}{2}(4x + 7)}{\sqrt{2x^2 + 7x - 8}}.$$

From this it is evident that, *if the numerator of a fraction, of this type, is one half of the differential coefficient of the expression under the root sign in the denominator, then the integral of the fraction is equal to the denominator, i.e.,*

$$\int \frac{\frac{1}{2} \cdot \frac{d}{dx}(ax^2 + bx + c)}{\sqrt{ax^2 + bx + c}} dx = \sqrt{ax^2 + bx + c}.$$

Consequently, when evaluating an integral of this type, arrange the numerator so that part of it is the differential coefficient of the expression under the root sign in the denominator.

In general, this results in a constant being left over in the numerator, as in the corresponding type in § 135. The expression can then be divided, as in § 135, into two fractions, in which the first will be as above, and the second as in § 137.

A worked example will make this clear.

139. Worked example.

Integrate $\int \frac{(x+1)dx}{\sqrt{2x^2+x-3}}$.

Now $$\frac{d}{dx}(2x^2+x-3) = 4x+1.$$

Re-arranging the numerator

$$\begin{aligned} x+1 &= \tfrac{1}{4}(4x+1) + \tfrac{3}{4} \\ &= \tfrac{1}{2}\{\tfrac{1}{2}(4x+1)\} + \tfrac{3}{4}. \end{aligned}$$

$$\therefore \quad \int \frac{(x+1)dx}{\sqrt{2x^2+x-3}}$$

$$= \int \frac{\tfrac{1}{2}\{\tfrac{1}{2}(4x+1)\} + \tfrac{3}{4}}{\sqrt{2x^2+x-3}}dx$$

$$= \tfrac{1}{2}\int \frac{\tfrac{1}{2}(4x+1)dx}{\sqrt{2x^2+x-3}} + \int \frac{\tfrac{3}{4}dx}{\sqrt{2x^2+x-3}}.$$

As shown above $\tfrac{1}{2}\int \frac{\tfrac{1}{2}(4x+1)dx}{\sqrt{2x^2+x-3}} = \tfrac{1}{2}\sqrt{2x^2+x-3}.$

Also using the methods of §§ 135, 136

$$\int \frac{\tfrac{3}{4}dx}{\sqrt{2x^2+x-3}} = \tfrac{3}{4}\int \frac{dx}{\sqrt{2x^2+x-3}}$$

$$= \tfrac{3}{4}\left\{\frac{1}{\sqrt{2}}\cosh^{-1}\frac{4x+1}{5}\right\}$$

$$\therefore \quad \int \frac{(x+1)dx}{\sqrt{2x^2+x-3}}$$

$$= \tfrac{1}{2}\sqrt{2x^2+x-3} + \frac{3}{4\sqrt{2}}\cosh^{-1}\frac{4x+1}{5}.$$

Exercise 27.

Integrate the following:

1. $\displaystyle\int \frac{dx}{\sqrt{x^2 + 6x + 10}}$.

2. $\displaystyle\int \frac{dx}{\sqrt{x^2 + 2x + 4}}$.

3. $\displaystyle\int \frac{dx}{\sqrt{x^2 - 4x + 2}}$.

4. $\displaystyle\int \frac{dx}{\sqrt{1 - x - x^2}}$.

5. $\displaystyle\int \frac{dx}{\sqrt{5x^2 - 12x + 4}}$.

6. $\displaystyle\int \frac{dx}{\sqrt{x(4 - x)}}$.

7. $\displaystyle\int \frac{xdx}{\sqrt{x^2 + 1}}$.

8. $\displaystyle\int \frac{(x + 1)dx}{\sqrt{x^2 + 1}}$.

9. $\displaystyle\int \frac{(x + 1)dx}{\sqrt{x^2 - 1}}$.

10. $\displaystyle\int \frac{xdx}{\sqrt{x^2 - x + 1}}$.

11. $\displaystyle\int \frac{(2x - 3)dx}{\sqrt{x^2 - 2x + 5}}$.

12. $\displaystyle\int \frac{(2x + 1)dx}{\sqrt{3 - 4x - x^2}}$.

13. $\displaystyle\int \frac{(2x + 3)dx}{\sqrt{x^2 + x + 1}}$.

14. $\displaystyle\int \frac{(x + 2)dx}{\sqrt{x^2 + 2x - 1}}$.

140. Some useful devices.

Other irrational functions can sometimes be transformed so as to admit of the use of the above methods.

(*a*) **Rationalisation.** In certain cases the rationalisation of the numerator enables the integration to be performed.

Example. *Integrate* $\displaystyle\int \sqrt{\frac{x - 1}{x + 1}}\, dx.$

Rationalising the numerator

$$\frac{\sqrt{x - 1}}{\sqrt{x + 1}} = \frac{\sqrt{x - 1} \times \sqrt{x - 1}}{\sqrt{x + 1} \times \sqrt{x - 1}}$$

$$= \frac{x - 1}{\sqrt{x^2 - 1}}.$$

$$\therefore \quad \int \sqrt{\frac{x - 1}{x + 1}} \cdot dx = \int \frac{(x - 1)dx}{\sqrt{x^2 - 1}}$$

$$= \int \frac{xdx}{\sqrt{x^2 - 1}} - \int \frac{dx}{\sqrt{x^2 - 1}}$$

$$= \sqrt{x^2 - 1} - \cosh^{-1} x.$$

(*b*) **Substitution.** By substitution for the irrational expression a new variable, such as u, the Integral can be simplified as shown in the following examples:

Example 1. *Integrate* $\int \frac{dx}{x\sqrt{x^2+4}}$.

Let $\quad x = \frac{1}{u}$ or $u = \frac{1}{x}$.

Then $\quad dx = -\frac{1}{u^2}\,du$.

$$\therefore \int \frac{dx}{x\sqrt{x^2+4}} = \int \frac{-\frac{1}{u^2}\,du}{\frac{1}{u}\sqrt{\frac{1}{u^2}+4}}$$

$$= -\int \frac{\frac{1}{u^2}\,du}{\frac{1}{u^2}\sqrt{1+4u^2}}$$

$$= -\int \frac{du}{\sqrt{1+4u^2}}$$

$$= -\tfrac{1}{2}\sinh^{-1} 2u$$

$$= -\tfrac{1}{2}\sinh^{-1}\frac{2}{x}.$$

Example 2. *Integrate* $\int \frac{dx}{x\sqrt{x^2-x+1}}$.

Let $\quad x = \frac{1}{u}$ and $u = \frac{1}{x}$.

Then $\quad dx = -\frac{du}{u^2}$.

$$\therefore \int \frac{dx}{x\sqrt{x^2-x+1}} = \int \frac{-\frac{1}{u^2}\,du}{\frac{1}{u}\sqrt{\frac{1}{u^2}-\frac{1}{u}+1}}$$

$$= \int \frac{-\frac{1}{u^2}\,du}{\frac{1}{u^2}\sqrt{1-u+u^2}}$$

$$= -\int \frac{du}{\sqrt{1 - u + u^2}}$$

$$= -\sinh^{-1}\frac{2u - 1}{\sqrt{3}} \quad \text{(using method of § 137)}$$

$$= -\sinh^{-1}\frac{\frac{2}{x} - 1}{\sqrt{3}}$$

$$= -\sinh^{-1}\frac{2 - x}{x\sqrt{3}}.$$

Exercise 28.

Integrate the following functions:

1. $\int\sqrt{\frac{x-2}{x+2}}\,dx.$

2. $\int\sqrt{\frac{2x+3}{2x-3}}\,dx.$

3. $\int\sqrt{\frac{x}{x+3}}\,dx.$

4. $\int\sqrt{\frac{x+1}{2x-3}}\,dx.$

5. $\int\frac{dx}{x+\sqrt{x^2-1}}$ (rationalise the denominator).

6. $\int\frac{dx}{x\sqrt{x^2+6x+10}}.$

7. $\int\frac{dx}{x\sqrt{1+x+x^2}}.$

8. $\int\frac{dx}{x\sqrt{3x^2+4x+1}}.$

9. $\int\frac{dx}{(x+1)\sqrt{x^2+4x+2}}$ $\left(\text{put } x+1=\frac{1}{u}\right).$

10. $\int\frac{dx}{x\sqrt{1+x^2}}.$

11. $\int\frac{\sqrt{1+x^2}}{x}\,dx$ (rationalise numerator).

12. $\int\frac{dx}{x^2\sqrt{1+x^2}}$ (put $x = \tan u$).

13. $\int\frac{\sqrt{x^2+1}}{x^2}\,dx.$

14. $\int\frac{x^2dx}{\sqrt{x^2+1}}.$

15. $\int\frac{x\,.\,dx}{\sqrt{1-x}}.$

16. $\int\frac{dx}{(x+1)\sqrt{x+2}}$ (put $\sqrt{x+2} = u$).

17. $\int\frac{dx}{x\sqrt{x+1}}.$

CHAPTER XIII

AREAS BY INTEGRAL CALCULUS. DEFINITE INTEGRALS

141. Areas by integration.

THE integral calculus had its origin in the endeavour to find a general method for the determination of the areas of regular figures. When these figures are bounded by straight lines, elementary geometry supplies the means of obtaining formulae for their areas; but when the boundaries are wholly, or in part, regular curves, such as the circle, ellipse, semi-circle, etc., then, unless we have the help of the integral calculus, we must depend upon experimental or approximate methods. We proceed therefore to investigate how integration can be employed to determine any such area.

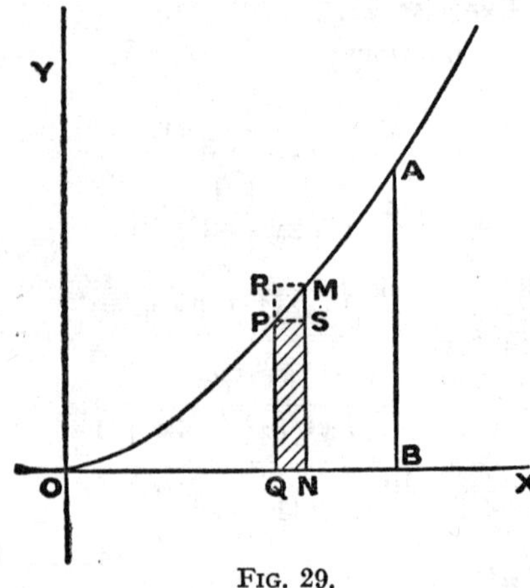

FIG. 29.

Let us consider, as an example, the parabola

$$y = x^2.$$

In Fig. 29, OA represents a part of this curve.

Let A be any point on the curve and AB the corresponding ordinate.

Let $OB = a$ units.

Suppose it is required to find the area under OA, that is, the area of OAB, which is bounded by the curve, OX, and AB.

Let the area be A square units.

Let P be any point on the curve OA.

Let its co-ordinates be (x, y).

Drawing the ordinate PQ we have $OQ = x$, $PQ = y$.

Suppose the area to be increased by a small amount δA, due to the point P moving along the curve to M, and Q moving along the axis to N.

Draw PS and MR parallel to OX and produce QP to meet MR at R.

Then we can represent QN by δx

and MS by δy.

$$\therefore \quad ON = x + \delta x$$
$$MN = y + \delta y.$$

Also δA is represented by the figure $QPMN$.

The area $QPMN$ lies between the areas of $QRMN$ and $QPSN$,

and area of $QRMN$ is $(y + \delta y)\delta x$

,, $QPSN$ is $y\delta x$.

$\therefore$ δA lies between $y\delta x$ and $(y + \delta y)\delta x$

and $\dfrac{\delta A}{\delta x}$,, y and $y + \delta y$.

Now suppose δx to be decreased indefinitely.

Then as $\delta x \longrightarrow 0$, $\delta y \longrightarrow 0$, and $\dfrac{\delta A}{\delta x}$ becomes $\dfrac{dA}{dx}$ in the limit, *i.e.*, in the limit

$$\frac{dA}{dx} = y$$
$$= x^2.$$
$$\therefore \quad dA = x^2dx.$$

Integrating $\qquad A = \frac{1}{3}x^3 + C.$

This result provides a formula for the area A in terms of any abscissa x and the undetermined constant C.

But when $\qquad x = 0,\ A = 0,$

then $\qquad C = 0.$

$\therefore$ for any value of x, when measured from O.

$$A = \tfrac{1}{3}x^3.$$

When $x = a$ as in the figure for the area of OAB

$$A = \tfrac{1}{3}a^3 \text{ square units.}$$

If now another value of x, say b, be taken, so that OD in Fig. 30 $= b$. Then by the above result

$$\text{Area of } OCD = \tfrac{1}{3}b^3.$$

$$\therefore \text{Area of } CDBA = \tfrac{1}{3}(a^3 - b^3).$$

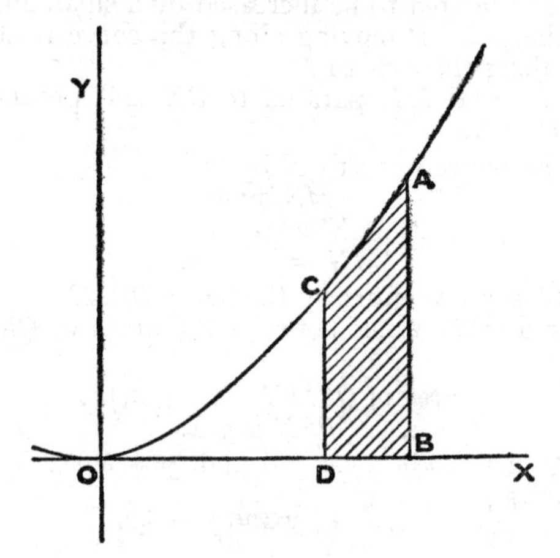

FIG. 30.

We will now proceed to establish a general rule which will apply to any function.

142. Definite Integrals.

Let the curve drawn in Fig. 31 represent part of the function

$$y = \phi(x).$$

Let AB and CD be fixed ordinates such that

$$OB = a, \ OD = b.$$

Let $ABDC$ be the area which we require to find and let it be A sq. inch.

Let PQ be a variable ordinate corresponding to any point, (x, y), so that $OQ = x$, $PQ = y = \phi(x)$.

Then, if Q moves along OX so that x be increased by δx (*i.e.*, QN), P will in consequence move along the curve to M (say).

Draw PS and MR parallel to OX.

Then $\qquad MS = \delta y$
and $\qquad MN = y + \delta y$
also $\qquad ON = x + \delta x.$

Let the area be increased by δA, where δA is represented by the figure $PQNM$.

Then the area of $PQNM$ lies between the areas of $PQNS$ and $QRMN$.

$\therefore$ δA lies between $y\delta x$ and $(y + \delta y)\delta x$,

i.e., $\frac{\delta A}{\delta x}$ lies between y and $y + \delta y$.

Let δx be decreased indefinitely.

Then, as $\delta x \longrightarrow 0$, $\delta y \longrightarrow 0$, and $\frac{\delta A}{\delta x}$ approaches $\frac{dA}{dx}$ as its limit.

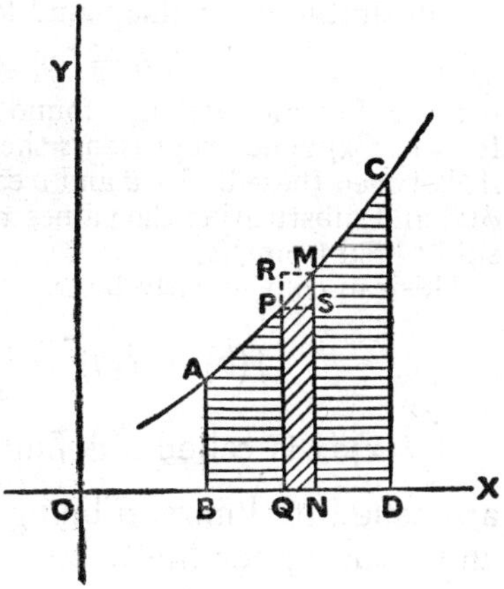

Fig. 31.

$\therefore$ in the limit
$$\frac{dA}{dx} = y$$
$$= \phi(x).$$
$$\therefore \quad dA = \phi(x)dx.$$

Integrating, and representing the integral of $\phi(x)$ by $f(x)$.
$$\int dA = \int \phi(x)dx$$
and
$$A = f(x) + C \quad . \quad . \quad . \quad . \quad . \quad \text{(I)}$$
where C is an undetermined constant. Its value can be determined when the value of A is known for some value of x.

Now A has been taken to represent the area $ABDC$, *i.e.*, between the ordinates where $x = a$, and $x = b$ respectively, and the variable ordinate moves from $x = a$ to $x = b$.

But when $x = a, A = 0$.
Substituting in I, $0 = f(a) + C$,

$\therefore \quad C = -f(a)$.

When $x = b$, *i.e.*, at D

$A = f(b) + C$.

Substituting the value found for C.

$$A = f(b) - f(a) \quad . \quad . \quad . \quad . \quad \text{(II)}$$

Since $f(a)$ and $f(b)$ are found by substituting a and b for x in $f(x)$ which represents the integral of $\phi(x)$, the area, A, between these limits a and b can be found by integrating $\phi(x)$ and substituting the values $x = a$ and $x = b$, $f(a)$ being subtracted from $f(b)$.

This can conveniently be expressed by the notation

$$f(b) - f(a) = \int_a^b \varphi(x)dx.$$

$\int_a^b \phi(x)dx$ is called a definite integral and a and b are called its limits, a being called the lower limit and b the upper limit.

$\therefore$ To evaluate a definite integral such as $\int_a^b \phi(x)dx$

(1) *Find the indefinite integral* $\int \phi(x)dx$, *viz.* $f(x)$.
(2) *Substitute for x in this the* upper limit b, *i.e.*, $f(b)$.
(3) ,, ,, ,, lower limit a, *i.e.*, $f(a)$.
(4) *Subtract* $f(a)$ *from* $f(b)$.

In practice the following notation and arrangement is found convenient:

$$\int_a^b \phi(x)dx = \Big[f(x)\Big]_a^b$$
$$= f(b) - f(a).$$

143. Characteristics of a definite integral.

The following points about a definite integral should be carefully noted :

(*a*) The results of substituting the limits in the integral are respectively $f(a) + C$ and $f(b) + C$. Consequently on subtraction the constant C disappears, hence the term "definite." If a and b are numbers the integral will also be a number.

(*b*) The variable is assumed to be increasing from the lower limit to the upper limit, *i.e.*, in the above from a to b. This must be carefully remembered when dealing with negative limits. If, for example, the limits are -2 and 0, then the variable x is increasing from -2 to 0. Consequently the upper limit is 0 and the lower limit -2.

This definite integral would therefore be written

$$\int_{-2}^{0} \phi(x)dx.$$

(*c*) The term "limit" in this connection has not the meaning attached to it previously in § 15. It denotes the values of the variable x at the ends of the range of values a to b over which we are finding the value of the definite integral.

144. Worked examples.

Example 1. *Evaluate the definite integral* $\int_2^5 3x \, . \, dx$.

Now $$\int 3xdx = \tfrac{3}{2}x^2 + C.$$

$$\therefore \quad \int_2^5 3xdx = \tfrac{3}{2}\Big[x^2\Big]_2^5$$
$$= \tfrac{3}{2}\{(5)^2 - (2)^2\}$$
$$= \tfrac{3}{2} \times 21$$
$$= \frac{63}{2}.$$

The student will find it a useful exercise to check this by drawing the graph of $y = 3x$, the ordinates at $x = 2$ and $x = 5$ and finding the area of the trapezium by the ordinary geometrical rule.

Example 2. *Evaluate the definite integral* $\int_0^{\frac{\pi}{2}} \sin x dx$.

Now $\int \sin x dx = -\cos x + C.$

$$\therefore \int_0^{\frac{\pi}{2}} \sin x dx = \Big[-\cos x\Big]_0^{\frac{\pi}{2}}$$
$$= \left\{\left(-\cos\frac{\pi}{2}\right) - (-\cos 0)\right\}$$
$$= 0 + 1$$
$$= 1.$$

Note.—This gives, in square units, the area beneath the curve of $y = \sin x$ between 0 and $\frac{\pi}{2}$. A graph of this function between 0 and π is shown in Fig. 32. Clearly, from symmetry the area under this curve between 0 and π must be twice that between 0 and $\frac{\pi}{2}$, *i.e.*, 2 square units. This can be checked by evaluating $\int_0^{\pi} \sin x dx$.

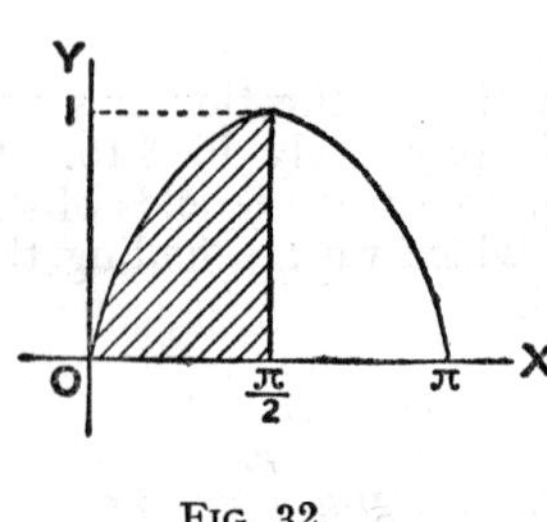

FIG. 32.

Example 3. *Evaluate* $\int_0^1 xe^{x^2}dx$.

Now $\int xe^{x^2}dx = \frac{1}{2}e^{x^2} + C$ (by inspection)

$$\therefore \int_0^1 xe^{x^2}dx = \Big[\tfrac{1}{2}e^{x^2}\Big]_0^1$$
$$= \tfrac{1}{2}(e^1 - e^0)$$
$$= \tfrac{1}{2}(e - 1).$$

Example 4. *Evaluate the definite integral*

$$\int_{-1}^{0} (1 + 3x - 2x^2)dx.$$

Since $\int (1 + 3x - 2x^2)dx = x + \frac{3x^2}{2} - \frac{2}{3}x^3$

$$\therefore \quad \int_{-1}^{0} (1 + 3x - 2x^2)dx = \left[x + \frac{3x^2}{2} - \frac{2}{3}x^3\right]_{-1}^{0}$$
$$= 0 - (-1 + \tfrac{3}{2} + \tfrac{2}{3})$$
$$= -\frac{7}{6}.$$

Example 5. *Evaluate the definite integral* $\int_2^3 \frac{dx}{\sqrt{x^2 - 1}}$.

Since $\int \frac{dx}{\sqrt{x^2 - 1}} = \cosh^{-1} x.$

$$\therefore \quad \int_2^3 \frac{dx}{\sqrt{x^2 - 1}} = \left[\cosh^{-1} x\right]_2^3$$
$$= \cosh^{-1}(3) - \cosh^{-1}(2)$$
$$= 1{\cdot}763 - 1{\cdot}316 \quad \text{(both approx.)}$$
$$= 0{\cdot}447.$$

Rough values for $\cosh^{-1}(3)$ and $\cosh^{-1}(2)$ can be found from the tables on p. 379.

More exact values can be found by using the algebraical equivalent of $\cosh^{-1} x$, viz. $\log_e\{x + \sqrt{x^2 + 1}\}$ using the hyperbolic logs on p. 377.

Example 6. *Evaluate the definite integral* $\int_1^e x \log x dx$.

Using the result of Exercise 22, No. 5, we get:

$$\int x \log x dx = \frac{x^2}{2}(\log x - \tfrac{1}{2}).$$

$$\therefore \quad \int_1^e x \log x dx = \left[\frac{x^2}{2}(\log x - \tfrac{1}{2})\right]_1^e$$
$$= \frac{e^2}{2}(\log_e e - \tfrac{1}{2}) - \tfrac{1}{2}(\log_e 1 - \tfrac{1}{2})$$
$$= \left(\frac{e^2}{2} \times \tfrac{1}{2}\right) - \tfrac{1}{2}(0 - \tfrac{1}{2})$$
$$= \frac{e^2}{4} + \frac{1}{4}.$$

Exercise 29.

Evaluate the following definite integrals:

1. $\int_1^3 x^n dx.$
2. $\int_0^1 (x^2 + 4)dx.$
3. $\int_1^2 (x^2 + 3x - 5)dx.$
4. $\int_{-2}^1 (2x + 1)^2 dx.$
5. $\int_1^{10} x^{-0\cdot 8} dx.$
6. $\int_1^4 \sqrt{x}\, dx.$
7. $\int_0^4 (x^{\frac{1}{2}} + x^{-\frac{1}{2}})dx.$
8. $\int_0^{\frac{\pi}{6}} \cos 3x dx.$
9. $\int_0^{\frac{\pi}{2}} (\cos \theta - \sin 2\theta)d\theta.$
10. $\int_0^{\frac{\pi}{4}} \cos\left(2\theta + \frac{\pi}{4}\right)d\theta.$
11. $\int_{-1}^1 2^x dx.$
12. $\int_0^2 e^{\frac{1}{2}x} dx.$
13. $\int_0^{\frac{\pi}{2}} \frac{1}{2}r^2 d\theta.$
14. $\frac{1}{2}\pi p \int_{-a}^a (a^2 - x^2)^2 dx.$
15. $\int_a^b e^{kx} dx.$
16. $\int_0^{\frac{\pi}{2}} \sin^2 x dx.$
17. $\int_2^3 \frac{x dx}{1 + x^2}.$
18. $\int_0^{\frac{\pi}{2}} x \sin x dx.$
19. $\int_0^1 x \log x dx.$
20. $\int_0^1 x^2 \log x dx.$
21. $\int_0^1 \sin^{-1} x dx.$
22. $\int_0^1 \tan^{-1} x dx.$
23. $\int_0^1 \frac{dx}{\sqrt{1 - x^2}}.$
24. $\int_1^2 \sqrt{1 + 3x} dx.$
25. $\int_0^3 \frac{x dx}{\sqrt{4 - x}}.$
26. $\int_0^4 \frac{dx}{\sqrt{x} + 1}.$
27. $\int_{-1}^1 \frac{dx}{1 + x^2}.$
28. $\int_0^a \frac{dx}{a^2 + x^2}.$
29. $\int_0^a \frac{dx}{\sqrt{a^2 - x^2}}.$
30. $\int_0^1 \frac{x dx}{\sqrt{1 - x^2}}.$

31. $\int_1^2 \frac{dx}{\sqrt{x^2-1}}$. 32. $\int_0^2 \frac{dx}{\sqrt{x^2+2x+2}}$.

33. $\int_0^1 \sqrt{\frac{1-x}{1+x}}\,dx$. 34. $\int_0^{\frac{\pi}{4}} \tan^2 x dx$.

35. $\int_0^1 \frac{dx}{\sqrt{12-4x-x^2}}$. 36. $\int_0^1 \frac{dx}{\sqrt{x(1-x)}}$.

37. $\int_{-2}^{-1} \frac{dx}{(x-2)^3}$.

145. Some properties of definite integrals.

(1) Interchange of limits.

Let $\phi(x)$ be the indefinite integral of $f(x)$.

Then, if the limits of the definite integral are a and b

$$\int_a^b f(x)dx = \phi(b) - \phi(a).$$

If the limits be interchanged

$$\int_b^a f(x)dx = \phi(a) - \phi(b)$$

i.e.

$$\int_a^b f(x)dx = - \int_b^a f(x)dx.$$

Thus, the interchange of the limits of integration changes only the sign of the definite integral.

(2)

$$\int_a^b f(x)dx = \int_c^b f(x)dx + \int_a^c f(x)dx.$$

Let $\phi(x)$ be the indefinite integral of $f(x)$.

$$\therefore \quad \int_a^b f(x)dx = \phi(b) - \phi(a)$$

also

$$\int_c^b f(x)dx = \phi(b) - \phi(c)$$

and

$$\int_a^c f(x)dx = \phi(c) - \phi(a).$$

$$\therefore \quad \int_c^b f(x)dx + \int_a^c f(x)dx = \{\phi(b) - \phi(c)\} + \{\phi(c) - \phi(a)\}$$
$$= \phi(b) - \phi(a)$$
$$= \int_a^b f(x)dx.$$

In Fig. 30 there is a graphical illustration of this theorem. Clearly,

Area of OAB = Area of $ABDC$ + Area of OCD,

i.e.,
$$\int_0^a f(x) = \int_b^a f(x) + \int_0^b f(x).$$

(3) Since $\int_a^b f(x)dx = \phi(b) - \phi(a)$, where $\phi(x)$ is the indefinite integral of $\int f(x)dx$, then as the definite integral $\phi(b) - \phi(a)$ does not contain x, any other letter could be used in the integral, provided the function of each of the two letters in the sum is the same.

For,
$$\int_a^b f(y)dy = \phi(b) - \phi(a)$$

but
$$\int_a^b f(x)dx = \phi(b) - \phi(a).$$

$$\therefore \quad \int_a^b f(x)dx = \int_a^b f(y)dy.$$

(4) $\int_0^a f(x)dx = \int_0^a f(a - x)dx.$

Let $\qquad x = a - u \quad$ or $\quad a - x = u.$
Then $\qquad dx = - du.$

Now, if in definite integration the variable is changed, the limits will also be changed and the new limits must be determined.

$\therefore$ in the above, when $x = a$,
$$u = a - x = a - a = 0.$$
when $x = 0$
$$u = a - x = a - 0 = a.$$

Thus when $x = a$, $u = 0$, and when $x = 0$, $u = a$.

$\therefore$ When x is replaced by $a - u$ in $\int_0^a f(x)dy$, the limits must be changed to those found above.

$$\therefore \quad \int_0^a f(x)dx = -\int_a^0 (a-u)du$$

$$= \int_0^a (a-u)dx \qquad \text{(by (1) above)}$$

$$= \int_0^a (a-x)dx \qquad \text{(by (3) above)}$$

Examples.

$$\int_0^{\frac{\pi}{2}} \sin x dx = \int_0^{\frac{\pi}{2}} \sin\left(\frac{\pi}{2} - x\right)dx$$

$$= \int_0^{\frac{\pi}{2}} \cos x dx.$$

In general

$$\int_0^{\frac{\pi}{2}} f(\sin x)dx = \int_0^{\frac{\pi}{2}} f(\cos x)dx.$$

146. Infinite limits and infinite integrals.

In the calculation of definite integrals between two limits a and b it has been assumed

(1) That these limits are finite.

(2) That all the values of the function between them are also finite, *i.e.*, the function is continuous.

We must, however, consider cases when one or both of these conditions is not satisfied.

147. Infinite limits.

The problems which arise when one of the limits is infinite can be illustrated by considering the case of $y = \frac{1}{x^2}$.

In studying this function it will be helpful to refer to its graph, part of which is shown in Fig. 33. The values of $\frac{1}{x^2}$ being always positive, the curve of the function lies entirely above OX. It consists of two parts, corresponding to positive and negative values of x. These two parts are clearly symmetrical about OY.

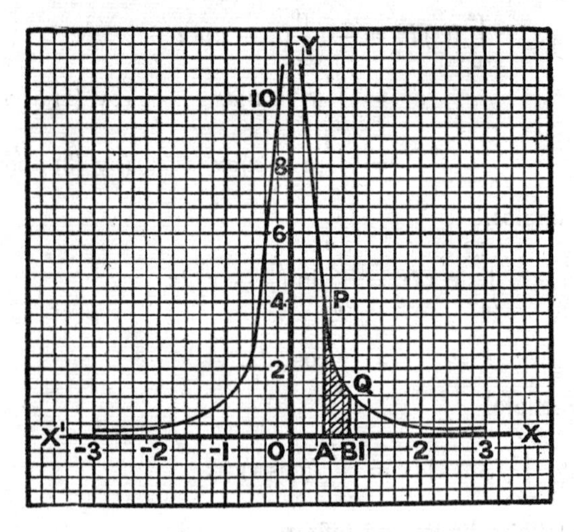

FIG. 33.

Let P, Q be two points on the curve.

Let PA, QB be the corresponding ordinate.

Let $OA = a$, $OB = b$.

Then, as shown in § 142, the area beneath the part of the curve PQ and bounded by PA, QB and OX, is as shown by the shaded part of the figure and is represented by

$$\int_a^b \frac{dx}{x^2} = \left[-\frac{1}{x}\right]_a^b = -\left(\frac{1}{b} - \frac{1}{a}\right).$$

(1) Suppose the ordinate QB to move indefinitely away from OY, so that OB—*i.e.*, b—is increased indefinitely.

Then the ordinate QB decreases indefinitely and in the limit OX is an asymptote to the curve (§ 14),

$$i.e., \text{ as } b \longrightarrow \infty,\ QB \longrightarrow 0.$$

The definite integral can now be written:

$$\int_a^{b \to \infty} \frac{dx}{x^2}, \text{ or more conveniently,}$$

$$\int_a^{\infty} \frac{dx}{x^2} = \left[-\frac{1}{x}\right]_a^{\infty}.$$

Its value in the limit becomes $-\left(\frac{1}{\infty} - \frac{1}{a}\right)$.

But the limit of $\frac{1}{\infty}$ is zero.

$\therefore$ **The value of the definite integral is $\frac{1}{a}$ and is finite.**

(2) Next, suppose the ordinate PA to move towards OY; then PA increases rapidly, and when OA—*i.e.*, a—is decreased without limit, the ordinate, *i.e.*, the value of y, increases without limit,

$\therefore$ as $\qquad a \longrightarrow 0, \quad y \longrightarrow \infty.$

The definite integral can now be written:

$$\int_{a \to 0}^{b} \frac{dx}{x^2} \quad \text{or} \quad \int_0^b \frac{dx}{x^2} = \left[-\frac{1}{x}\right]_0^b$$

$$= -\left[\frac{1}{b} - \frac{1}{0}\right].$$

In the limit $\frac{1}{0}$ becomes infinite.

Thus the definite integral becomes infinite and cannot be found numerically.

At the same time OY becomes an asymptote to the curve.

We therefore conclude that in the definite integral $\int_a^b \frac{dx}{x^2}$,

(*a*) If x becomes infinitely great, while y becomes indefinitely small, the integral will have **a finite value.**

(*b*) If x becomes indefinitely small, while y becomes infinitely large, the integral has **no finite value.**

It is clear therefore that in all such cases we must investigate and determine whether the definite integral can have a finite value or not.

Next we will consider an example in which both limits become infinite.

$$\int_a^b \frac{dx}{1 + x^2} = \left[\tan^{-1} x\right]_a^b$$

$$= \tan^{-1} b - \tan^{-1} a.$$

If $\quad b \longrightarrow \infty, \quad$ then $\quad \tan^{-1} b \longrightarrow \frac{\pi}{2}.$

If $\quad a \longrightarrow -\infty, \quad$ then $\quad \tan^{-1} a \longrightarrow -\frac{\pi}{2}.$

$\therefore$ In the limit

$$\int_{-\infty}^{+\infty} \frac{dx}{1+x^2} \quad \text{becomes} \quad \left\{\frac{\pi}{2} - \left(-\frac{\pi}{2}\right)\right\} = \pi.$$

Therefore there is a finite value of the integral $\int_{-\infty}^{+\infty} \frac{dx}{1+x^2}$.

148. Functions with infinite values.

We next consider functions which become infinite for some value, or values, of the variable between the limits of the definite integral, *i.e.*, the function is not continuous.

The function $\frac{1}{x^2}$, which was considered above, is an example; it becomes infinite when $x = 0$, as shown above. If therefore it is required to find the value of the integral $\int_{-2}^{+2} \frac{dx}{x^2}$ it is evident that the function becomes infinite for a value of x between the limits, viz. $x = 0$.

If $\int_{-2}^{+2} \frac{dx}{x^2}$ be evaluated as usual, disregarding this infinity value, the result is as follows:

$$\int_{-2}^{+2} \frac{dx}{x^2} = \left[-\frac{1}{x}\right]_{-2}^{+2} = -\{\tfrac{1}{2} - (-\tfrac{1}{2})\} = -1.$$

But this result is at variance with that obtained above when it was shown that $\int_{0}^{+2} \frac{dx}{x^2}$ becomes infinite as x approaches zero. Similarly, it can also be shown that $\int_{-2}^{0} \frac{dx}{x^2}$ is infinite.

As $\int_{-2}^{+2} \frac{dx}{x^2}$ must be the sum of these (§ 145), it must be infinite.

It is therefore necessary, before evaluating certain

integrals, to ascertain if the function is continuous between the assigned limits, or whether it becomes infinite for some value of x.

This is specially necessary in the case of fractional functions in which, while the numerator remains finite, the denominator vanishes for one or more values of x.

Thus $\frac{x}{(x-1)(x-2)}$ becomes infinite, and the curve is therefore discontinuous

(1) when $(x - 1) = 0$, and $x = 1$, and
(2) when $(x - 2) = 0$, and $x = 2$.

Similarly in $\frac{1}{\sqrt{2-x}}$ the denominator vanishes when $x = 2$, or more accurately, $\sqrt{2-x} \longrightarrow 0$ when $x \longrightarrow 2$. Consequently the function approaches infinity as $x \longrightarrow 2$.

All such cases must be examined to ascertain if a finite limit and therefore a definite value of the integral exists. For this purpose the property of an integral as stated in § 145, No. 2, can often be employed. In using this theorem the integral to be tested is expressed as the sum of two integrals in which the value of the variable for which the function becomes infinite is used as an end limit. Each of these must have a finite value if the original integral is finite and its value is given by that sum.

An example of this was given above, when it was pointed out that $\int_{-2}^{+2}\frac{dx}{x^2}$, when expressed as the sum of $\int_{0}^{2}\frac{dx}{x^2}$, and $\int_{-2}^{0}\frac{dx}{x^2}$ must be infinite, *i.e.*, it has no meaning, since each of the two component integrals had been shown to be infinite. A further example is given below in which a method is employed for determining whether a given definite integral is finite or not.

Example. *Determine if the definite integral* $\int_{0}^{3}\frac{dx}{\sqrt[3]{x-2}}$ *has a finite value.*

The integral approaches infinity as $x \longrightarrow 2$.

Using the above theorem (§ 145), the integral can be expressed as follows:

$$\int_0^3 \frac{dx}{\sqrt[3]{x-2}} = \int_2^3 \frac{dx}{\sqrt[3]{x-2}} + \int_0^2 \frac{dx}{\sqrt[3]{x-2}}.$$

It is necessary, if the original integral is to have a finite value, that each of these integrals should be finite. We, therefore, test these separately.

In the first let the end limit " 2 " be replaced by $2 + \alpha$, where α is a small positive number.

(1) Then
$$\int_{2+\alpha}^3 \frac{dx}{\sqrt[3]{x-2}} = \frac{3}{2}\Big[(x-2)^{\frac{2}{3}}\Big]_{2+\alpha}^3$$
$$= \frac{3}{2}[\{(3-2)^{\frac{2}{3}}\} - \{(2+\alpha) - 2\}^{\frac{2}{3}}]$$
$$= \frac{3}{2}(1 - \alpha^{\frac{2}{3}})$$
$$= \frac{3}{2} - \frac{3}{2}\alpha^{\frac{2}{3}}.$$

As $\alpha \longrightarrow 0$ and $(2+\alpha) \longrightarrow 2$, the value of the integral approaches $\frac{3}{2}$.

$\therefore$ in the limit the value of the integral is $\frac{3}{2}$.

(2)
$$\int_0^{2-\alpha} \frac{dx}{\sqrt[3]{x-2}} = \frac{3}{2}\Big[(x-2)^{\frac{2}{3}}\Big]_0^{2-\alpha}$$
$$= \frac{3}{2}[\{(2-\alpha) - 2\}^{\frac{2}{3}} - (0-2)^{\frac{2}{3}}]$$
$$= \frac{3}{2}\{(-\alpha)^{\frac{2}{3}} - (-2)^{\frac{2}{3}}\}.$$

In the limit when $\alpha \longrightarrow 0$, $(-\alpha)^{\frac{2}{3}} \longrightarrow 0$ and the value of the integral becomes $-\frac{3}{2}(-2)^{\frac{2}{3}} = -\frac{3}{2}\sqrt[3]{4}$.

As each of the definite integrals has a finite value, the whole integral is finite and is equal to the sum of the two integrals.

$$\therefore \int_0^3 \frac{dx}{\sqrt[3]{x-2}} = \frac{3}{2} - \frac{3}{2}\sqrt[3]{4}$$
$$= \frac{3}{2}(1 - \sqrt[3]{4}).$$

Exercise 30.

When possible calculate the values of the following definite integrals:

1. $\int_a^\infty \frac{dx}{x}$.

2. $\int_2^\infty \frac{dx}{x^3}$.

3. $\int_0^{+\infty} \frac{dx}{x^2+1}$.

4. $\int_2^{+\infty} \frac{dx}{x^2-1}$.

5. $\int_1^{\infty} \frac{dx}{\sqrt{x}}$.

6. $\int_0^{+\infty} e^{-x}dx$.

7. $\int_0^{\infty} e^{-x} \cos x dx$.

8. $\int_0^{\infty} \frac{xdx}{1+x^2}$.

9. $\int_1^{\infty} \frac{dx}{x^2(1+x)}$.

10. $\int_1^{\infty} \frac{dx}{x(1+x)^2}$.

11. $\int_0^1 \frac{dx}{x}$.

12. $\int_0^1 \sqrt{\left(\frac{x}{1-x}\right)} dx$.

13. $\int_{-1}^1 \sqrt{\left(\frac{1-x}{1+x}\right)} dx$.

14. $\int_0^1 x \log x dx$.

15. $\int_{-1}^1 \frac{1+x}{1-x} dx$.

16. $\int_0^{\infty} x^2 e^{-x} dx$.

17. $\int_0^1 \log x dx$.

18. $\int_0^1 \frac{dx}{\sqrt{1-x}}$.

19. $\int_0^3 \frac{dx}{(x-1)^2}$.

20. $\int_0^2 \frac{dx}{\sqrt[3]{x-1}}$.

CHAPTER XIV

INTEGRATION AS A SUMMATION. AREAS

149. Approximatiom to an area by division into small elements.

In the preceding chapter it was seen how, with the aid of integration, we could find the area of a figure bounded in part by a regular curve whose equation is known. We now proceed to the consideration of another, and a more general, treatment of the problem.

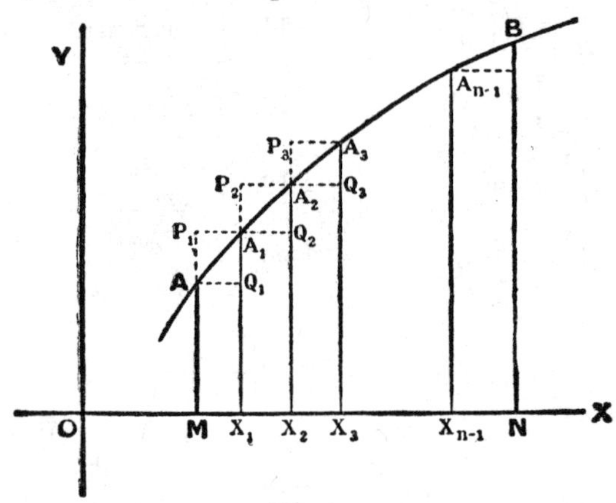

Fig. 34.

In Fig. 34 let AB be a portion of a curve whose equation is $y = \phi(x)$.

Let AM, BN be the ordinates of A and B, so that

$$OM = a, \quad ON = b.$$

$$\therefore \quad AM = \phi(a), \quad BN = \phi(b).$$

Let the co-ordinates of A be (x, y).

$ABNM$ is the figure whose area is required.

Let MN be divided into n equal parts at $X_1, X_2, X_3, \ldots$

Then MX_1 can be represented by δx. Hence each of the divisions X_1X_2, X_2X_3 . . . is equal to δx.

Let A_1X_1, A_2X_2, A_3X_3 . . . be ordinates corresponding to the points A_1, A_2, A_3 . . .

Complete the rectangles $AP_1A_1Q_1$, $A_1P_2A_2Q_2$, $A_2P_3A_3Q_3$ . . .

There are now two sets of rectangles corresponding to the divisions MX_1, X_1X_2, . . .

(1) $MP_1A_1X_1$, $X_1P_2A_2X_2$, $X_2P_3A_3X_3$, . . .
(2) MAQ_1X_1, $X_1A_1Q_2X_2$, $X_2A_2Q_3X_3$, . . .

The area beneath the curve, *i.e.*, the area of $MABN$, lies between the sums of the areas of the rectangles in sets (1) and (2).

If the number of divisions be increased the area of each of the two sets will approximate more nearly to the area of $MABN$.

If the number of divisions be increased indefinitely, δx will be decreased indefinitely and the area of each of the sets (1) and (2) approaches to equality with the area under the curve. It is therefore necessary to find expressions for the sums of these sets and then to obtain their limiting values when $\delta x \longrightarrow 0$.

The ordinates can be expressed thus:

$$\begin{aligned} AM &= \phi(a) \\ A_1X_1 &= \phi(a + \delta x) \\ A_2X_2 &= \phi(a + 2\delta x) \\ &\cdot \quad \cdot \quad \cdot \quad \cdot \quad \cdot \\ A_{n-1}X_{n-1} &= \phi\{a + (n-1)\delta x\} \\ \text{and} \qquad BN &= \phi(b) \end{aligned}$$

where n is the number of divisions.

$\therefore$ the areas of these rectangles in (2) are as follows:

$$\begin{aligned} \text{Area of } MAQ_1X_1 &= (AM \times MX_1) &&= \phi(a)\delta x \\ \text{,,} \quad X_1A_1Q_2X_2 &= (A_1X_1) \times (X_1X_2) &&= \phi(a + \delta x)\delta x \\ \text{,,} \quad X_2A_2Q_3X_3 &= (A_2X_2) \times (X_2X_3) &&= \phi(a + 2\delta x)\delta x \\ &\cdot \quad \cdot \quad \cdot \quad \cdot \quad \cdot \\ \text{,,} \quad X_{n-1}A_{n-1}Q_nN &= (A_{n-1}X_{n-1}) \times (X_{n-1}N) \\ &= \phi\{a + (n-1)\delta x\}\delta x. \end{aligned}$$

The sum of all these rectangles is

$$\delta x[\phi(a) + \phi(a + \delta x) + \ldots + \phi\{a + (n-1)\delta x\}] \quad \text{(A)}$$

Similarly the sum of all the rectangles in (1) is

$$\delta x[\phi(a + \delta x) + \phi(a + 2\delta x) \ldots + \phi\{a + (n - 1)\delta x\} + \phi(b)] \quad \text{(B)}$$

The area of the figure $AMNB$ lies between (A) and (B).

Then $\qquad$ (B) $-$ (A) $= \delta x\{\phi(b) - \phi(a)\}$.

In the limit when $\delta x \longrightarrow 0$ this difference vanishes. Thus each of the areas approaches the area of $AMNB$.

$\therefore$ **The area is the limit of the sum of either (A) or (B).**

The summation of such a series can be expressed concisely by the use of the symbol Σ (pronounced "sigma"), the Greek capital "S." Using this symbol the sum of the series may be written

$$\sum_{x=a}^{x=b} \phi(x)\delta x.$$

By this expression we mean, the sum of terms of the type $\phi(x)\delta x$, when we substitute for x the values

$$a,\ a + \delta x,\ a + 2\delta x,\ a + 3\delta x,\ \ldots$$

for all such possible values of it between $x = a$ and $x = b$.

The area of $AMNB$ is **the limit of this sum** when $\delta x \longrightarrow 0$, and this is written in the form

$$A = \underset{\delta x \to 0}{Lt} \sum_{x=a}^{x=b} \phi(x)\delta x.$$

But we have seen, (§ 142), that this area is given by the integral

$$\int_a^b \phi(x)dx.$$

$$\therefore \qquad \underset{\delta x \to 0}{Lt} \sum_{x=a}^{x=b} \phi(x)dx = \int_a^b \phi(x)dx.$$

150. The definite integral as the limit of a sum.

It is thus apparent that a definite integral can be regarded as a sum, or, more correctly,, the "limit of a sum," of the areas of an infinite number of rectangles, one side of each of which (dx in the above) is infinitesimally small.

The use of the term integral will now be clear, the word

integrate meaning " to give the total sum ". The first letter of the word sum appears in the sign $\int$, which is the old-fashioned elongated " s." It is also evident why the infinitesimal, dx, must necessarily appear as a factor in an integral.

The definite integral has been used in the illustration above to refer to the sum of areas. This, however, is used as a device for illustrating the process by a familiar geometrical example. Actually there was found the sum of an infinite number of algebraical products, one factor of which, in the limit, becomes infinitely small. The results, however, can be reached independently of any geometrical illustration.

Consequently, $\int_a^b \phi(x)dx$ **can be regarded as representing the sum of an infinite number of products, one factor of which is an infinitesimally small quantity.** The successive products must be of the nature of those appearing in the demonstration above, and must refer to successive values of the independent variable, x, between the limits $x = b$ and $x = a$.

This being the case, the method can be applied to the summation of any such series, subject to the conditions which have been stated.

This is of great practical importance, since it enables us to calculate not only areas, but also volumes, lengths of curves, centres of mass, moments of inertia, etc., such as are capable of being expressed in the form $\sum_{x=a}^{x=b} \phi(x)dx$. They can then be represented by the definite integral $\int_a^b \phi(x)dx$.

In the above demonstration $\phi(x)$ has been regarded as steadily and continuously increasing, but the arguments employed will apply equally when $\phi(x)$ is decreasing. It is essential, however, that the range of values of x between a and b can be divided into a definite number of parts, and that the corresponding values of $\phi(x)$ are continuously increasing or decreasing.

The practical applications of the above conclusions are very many, and some of them will be discussed in succeeding chapters.

The most obvious application, in view of the method followed in the demonstration, is to areas; so a beginning will be made by examining examples of them.

151. Examples on Areas.

Example 1. *Find the area between the curve of* $y = \frac{1}{2}x^2$, *the x-axis, and the ordinate of the curve corresponding to* $x = 2$.

The part of the curve involved is indicated in Fig. 35 by OQ, where the ordinate from Q corresponds to the point $x = 2$.

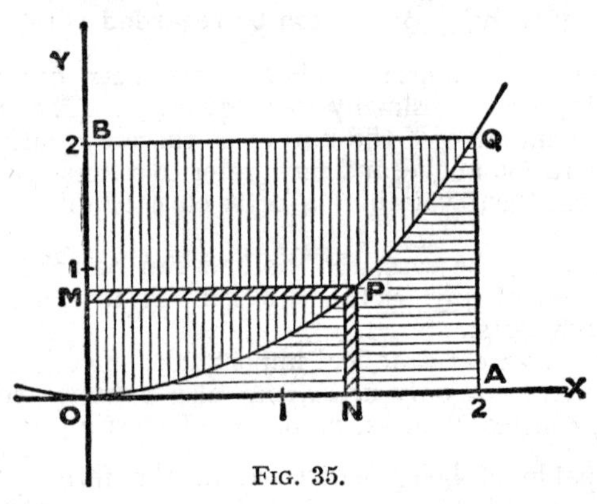

FIG. 35.

The area required is that of OAQ indicated by horizontal shading.

Let P, (x, y) be any point on the curve, so that $ON = x$. Let x be increased by δx, and drawing the corresponding ordinate there is enclosed what is approximately a small rectangle, as shown in the figure.

The area of this is approximately $y\delta x$.

When δx becomes indefinitely small, the sum of the area of all such rectangles throughout the range from $x = 0$ to $x = 2$ is equal to the required area.

The area of this very small rectangle is ydx.

This is called an **element of area,** and it is always necessary to obtain this element before proceeding to the solution.

The sum of all such areas is given by the definite integral

$$\int_0^2 y dx.$$

But $$y = \tfrac{1}{2}x^2.$$

$$\therefore \quad \text{Area} = \int_0^2 \tfrac{1}{2}x^2 dx = \tfrac{1}{2}\Big[\tfrac{1}{3}x^3\Big]_0^2$$

$$= \frac{4}{3} \text{ square units.}$$

Example 2. *Find the area between the curve of $y = \frac{1}{2}x^2$, the axis of y and the straight line $y = 2$.*

The curve is the same as in Example 1, and is shown in Fig. 35 with vertical shading; BQ is the line $y = 2$.

Take any point $P(x, y)$ on the curve; as before, $OM = y$, $ON = x$. PM represents a small element of area.

In this problem it is convenient to consider the area as being formed by the movement parallel to OX of PM, *i.e.,* y is regarded as being increased by δy to form the rectangle PM.

Then area of $PM = x\delta y$.

Then the rectangle becomes infinitely small, and when $\delta y \longrightarrow 0$, the element of area, is represented by xdy.

$\therefore$ Area of figure $OBQO$ is given by $\displaystyle\int_{y=0}^{y=2} xdy$.

Consequently there are two variables in the integral, and one of these must be expressed in terms of the other so that there remains one variable only.

Let us express x in terms of y, in which case the limits are unaltered.

Since $$y = \tfrac{1}{2}x^2$$
$$x = \sqrt{2y}.$$

Substituting, Area $$= \int_0^2 \sqrt{2y} \,.\, dy = \sqrt{2}\int_0^2 y^{\frac{1}{2}} dy$$

$$= \sqrt{2}\Big[\tfrac{2}{3}y^{\frac{3}{2}}\Big]_0^2 = \sqrt{2} \times \tfrac{2}{3}(\sqrt{2})^3.$$

$$= \frac{8}{3} \text{ square units.}$$

If dy had been expressed in terms of x, and it is seen that $dy = xdx$, then for the limits we must obtain the values of x corresponding to $y = 2$ and $y = 0$. In this case they are the same, since from $y = \frac{1}{2}x^2$, when $y = 2$, $x = 2$, and $y = 0$, $x = 0$.

Note.—Evidently the sum of this area and the preceding one in Example 1 must equal the area of the rectangle $OBQA$,

i.e., $$\frac{8}{3} + \frac{4}{3} = 4 \text{ square units.}$$

Example 3. *Area of a circle.*

(I) Area by rectangular co-ordinates.

The equation of a circle.

Before finding the area enclosed wholly or in part by a curve, it is necessary to know the equation of that curve. To help those students who have not studied co-ordinate geometry we will proceed to find the equation of the circumference of a circle in rectangular co-ordinates.

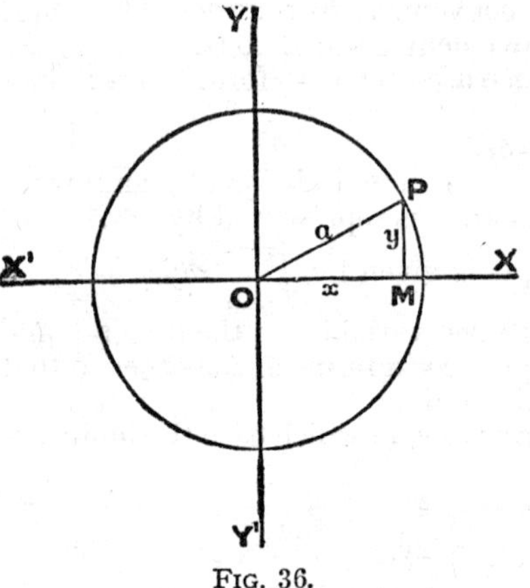

FIG. 36.

In any circle the centre can be taken as the origin of a system of co-ordinates, and two diameters at right angles to each other as the co-ordinate axes.

This is indicated in Fig 36.

Take any point P, (x, y) on the circumference and draw PM perpendicular to OX.

Let the radius of the circle be a.

Then $$OM = x, PM = y.$$

By the property of a right-angled triangle

$$OM^2 + MP^2 = OP^2$$

i.e.,
$$x^2 + y^2 = a^2.$$

This equation is true for any point on the circumference, and it states the relation which exists between the co-ordinates of any point and the constant which defines the circle, *i.e.*, the radius a.

$$\therefore \quad x^2 + y^2 = a^2$$

is the equation of a circle of radius a and the origin at its centre.

Area of the circle $x^2 + y^2 = a^2$.

Fig. 37 represents this circle.

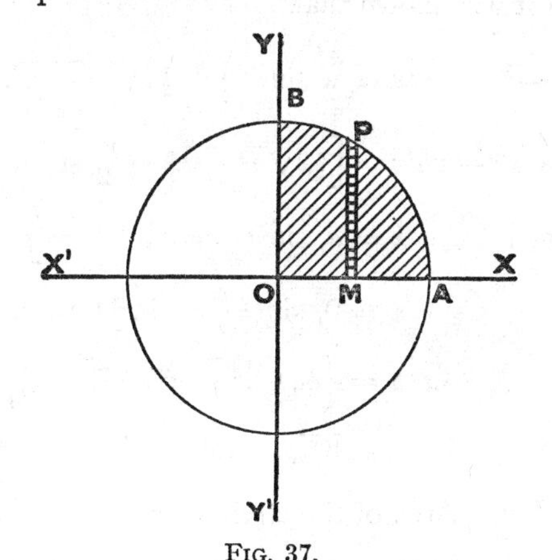

FIG. 37.

For reasons which will be apparent later it is better to find the area of the quadrant which is shaded; from this we get the area of the whole circle.

Let P (x, y) be any point on the circumference.

Then
$$OM = x, \; PM = y.$$

The element of area, as previously defined, can be represented by the small rectangle PM, and is given by $y\delta x$.

In the limit as $\delta x \longrightarrow 0$ the element of area is represented by ydx.

For the purposes of the definite integral which will give us the area, the limits of x for the quadrant are:

At O, $x = 0$.
At A, $x = a$.

$$\therefore \quad \text{Area} = \int_0^a ydx.$$

Since $x^2 + y^2 = a^2$

$$y = \sqrt{a^2 - x^2}.$$

$$\therefore \quad \text{Area} = \int_0^a \sqrt{a^2 - x^2}dx.$$

In § 117 it was shown that

$$\int \sqrt{a^2 - x^2}dx = \frac{a^2}{2}\sin^{-1}\frac{x}{a} + \tfrac{1}{2}x\sqrt{a^2 - x^2}.$$

$$\therefore \quad \int_0^a \sqrt{a^2 - x^2}dx = \left[\tfrac{1}{2}x\sqrt{a^2 - x^2} + \frac{a^2}{2}\sin^{-1}\frac{x}{a}\right]_0^a.$$

Now when $x = a,\ \sin^{-1}\frac{x}{a} = \sin^{-1}1 = \frac{\pi}{2}$.

When $x = 0,\ \sin^{-1}\frac{x}{a} = \sin^{-1}0 = 0$.

$$\therefore \quad \text{Area} = \left\{\tfrac{1}{2}a\sqrt{a^2 - a^2} + \frac{a^2}{2} \times \frac{\pi}{2}\right\} - 0$$

$$= \frac{\pi a^2}{4}.$$

$$\therefore \quad \text{Area of the circle} = \pi a^2.$$

(2) Alternative method.

The following method will be found useful in its applications.

The area of a circle can be conceived as the area of a plane figure which is traced out by a finite straight line as it rotates around one of its ends, and makes a complete rotation.

Thus in Fig. 38 if the straight line OP, length a units, starting from the fixed position OA on OX makes a complete rotation around a fixed point O, the point P describes the circumference of a circle, and the area marked out by OA is the area of the circle.

Let the point P have rotated from OA, so that it has described the angle, θ, AOP being consequently a sector of a circle.

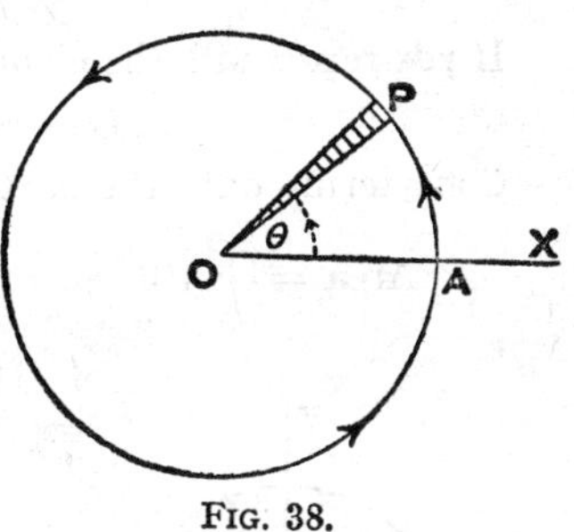

FIG. 38.

Now suppose OP to rotate further through an infinitesimally small angle $\delta\theta$. The infinitely small sector so described would be an element of area, and the sum of all such sectors when OP makes a complete rotation from OX, back again to its original position, will be the area of the circle.

The infinitely small arc subtended by $\delta\theta$ in the limit can be regarded as a straight line, and the infinitely small sector as a triangle.

The length of the arc is $a\delta\theta$ (*Trigonometry*, § 120).

The altitude of the triangle can be regarded, in the limit, as the radius of the circle, a.

$\therefore$ Using the formula for the area of a triangle

$$\text{Element of area} = \tfrac{1}{2} \times a\delta\theta \times a = \tfrac{1}{2}a^2\delta\theta.$$

And the angle corresponding to a complete rotation is 2π radians.

$$\begin{aligned}\therefore \text{Area} = \int_0^{2\pi} \tfrac{1}{2}a^2\delta\theta &= \Big[\tfrac{1}{2}a^2\theta\Big]_0^{2\pi} \\ &= \tfrac{1}{2}a^2 \times 2\pi \\ &= \pi a^2 \text{ square units.}\end{aligned}$$

Example 4. *Area of part of a circle between two parallel chords.*

In the circle $x^2 + y^2 = 9$ *find the area contained between the lines* $x = 1$, $x = 2$.

The radius of this circle is 3 and the centre is at the

origin. The area which it is required to find is shown in Fig. 39.

Since
$$x^2 + y^2 = 9$$
$$y = \sqrt{9 - x^2}.$$

If ydx represents the element of area, then
$$ydx = \sqrt{9 - x^2}dx.$$

Considering only the part of the area above OX, then
$$\text{Area} = \int_1^2 \sqrt{9 - x^2}\,.\,dx.$$

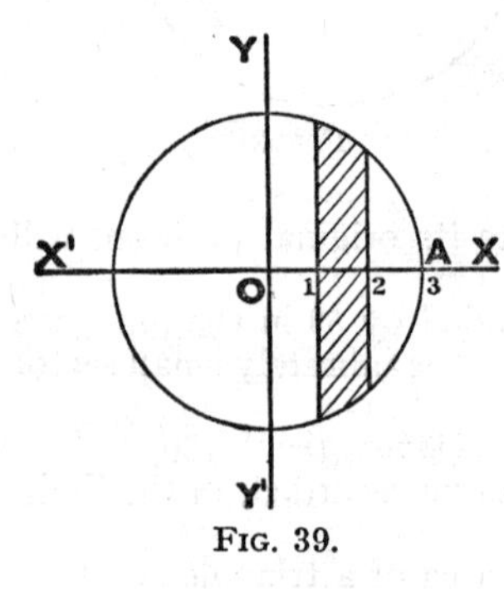

FIG. 39.

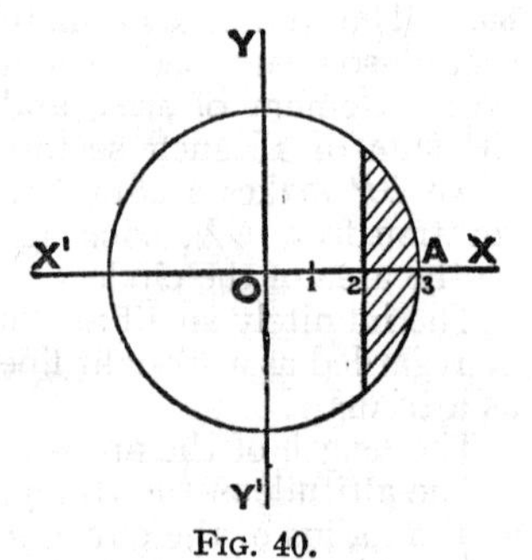

FIG. 40.

Using the integral
$$\int \sqrt{a^2 - x^2dx} = \tfrac{1}{2}x\sqrt{a^2 - x^2} + \tfrac{1}{2}a^2 \sin^{-1}\frac{x}{a} \qquad (\S\ 117)$$
$$\int_1^2 \sqrt{9 - x^2}dx = \left[\tfrac{1}{2}x\sqrt{9 - x^2} + \tfrac{9}{2}\sin^{-1}\frac{x}{3}\right]_1^2$$
$$= \{\tfrac{1}{2}2\sqrt{9-4} + \tfrac{9}{2}\sin^{-1}\tfrac{2}{3}\} - \{\tfrac{1}{2}\sqrt{9-1} + \tfrac{9}{2}\sin^{-1}\tfrac{1}{3}\}$$
$$= (\sqrt{5} + \tfrac{9}{2}\sin^{-1}\tfrac{2}{3}) - (\tfrac{1}{2}\sqrt{8} + \tfrac{9}{2}\sin^{-1}\tfrac{1}{3})$$
$$= (\sqrt{5} - \sqrt{2}) + \tfrac{9}{2}(\sin^{-1}\tfrac{2}{3} - \sin^{-1}\tfrac{1}{3})$$

Now
$$\sin^{-1}\tfrac{2}{3} = 41^\circ\ 48' = 0{\cdot}730 \text{ radians (approx.)}$$
and $\sin^{-1}\tfrac{1}{3} = 19^\circ\ 30' = 0{\cdot}340$ " "
$$\therefore \quad \text{Area} = 0{\cdot}822 + \tfrac{9}{2}(0{\cdot}730 - 0{\cdot}340)$$
$$= 2{\cdot}582 \text{ (approx.)}.$$

$\therefore$ Area of the whole

$$= 2{\cdot}582 \times 2$$
$$= 5{\cdot}164 \text{ square units (approx.)}.$$

Example 5. *Area of a segment of a circle.*

Find the area of the segment cut off from the circle $x^2 + y^2 = 9$ by the line $x = 2$.

This is the same circle as in the previous example, and the area required is that which is shaded in Fig. 40. Considering only the area of that part lying above OX, we have:

$$\text{Area} = \int_2^3 y dx = \int_2^3 \sqrt{9 - x^2} dx.$$

Using the result obtained in the previous example

$$\int_2^3 \sqrt{a^2 - x^2} dx = \left[\tfrac{1}{2}x\sqrt{9 - x^2} + \tfrac{9}{2}\sin^{-1}\frac{x}{3}\right]_2^3$$
$$= \{0 + \tfrac{9}{2}\sin^{-1}\tfrac{3}{3}\} - \{\sqrt{5} + \tfrac{9}{2}\sin^{-1}\tfrac{2}{3}\}$$
$$= \tfrac{9}{2} \times \frac{\pi}{2} - \{2{\cdot}236 + \tfrac{9}{2} \times 0{\cdot}730\} \text{ (from above)}$$
$$= \frac{9\pi}{4} - (2{\cdot}236 + 3{\cdot}29)$$
$$= 1{\cdot}543$$

$\therefore$ total area $= 3{\cdot}086 = 3{\cdot}09$ square units (approx.).

Note.—As a check, the student should find the area of the segment cut off by the line $x = 1$. It should be the sum of those above.

Example 6. *The area of an ellipse.*

Fig. 41 represents an ellipse in which the origin is the centre, *i.e.*, the point of intersection of the major axis AA^1 and the minor axis BB^1.

Let the length of AA^1 be $2a$.

" " BB^1 " $2b$.

Then $OA = a$, and $OB = b$.

It is shown in co-ordinate geometry that the equation of such an ellipse is

$$\frac{x^2}{a^2} + \frac{y^2}{b^2} = 1$$

whence
$$y = \frac{b}{a}\sqrt{a^2 - x^2}.$$

The element of area, ydx, is $\frac{b}{a}\sqrt{a^2 - x^2}dx$.

Considering the area of one quadrant of the ellipse, such as that which is shaded in Fig. 41.

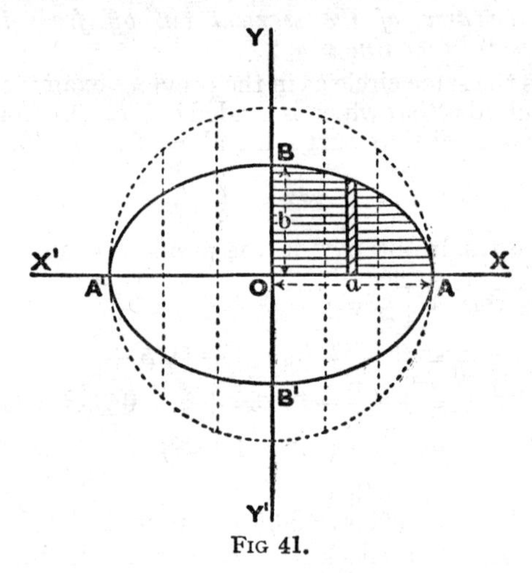

FIG 41.

$$\text{Area of quadrant} = \int_0^a \frac{b}{a}\sqrt{a^2 - x^2}dx$$
$$= \frac{b}{a}\int_0^a \sqrt{a^2 - x^2}dx$$
$$= \frac{b}{a}\left[\tfrac{1}{2}x\sqrt{a^2 - x^2} + \tfrac{1}{2}a^2 \sin^{-1}\frac{x}{a}\right]_0^a.$$

The total area is four times this.

Comparing this with the area of the circle, radius a, in Example 3, it is seen that the ratio of the area of the quadrant of the ellipse to that of the corresponding area of the circle of radius a is $\frac{b}{a}$, *i.e.*, the ratio of the minor axis to the major. This is also the ratio of corresponding ordinates of the two curves.

Example 7. *Area of a segment of a hyperbola.*

It is not possible within the limits of this book to give any satisfactory account of the geometry of a hyperbola, or the method of arriving at its equation. For this the student is referred to a book on co-ordinate geometry.

The curve of $y = \frac{a}{x}$, which has been discussed previously, is an example of a hyperbola (see § 14). In this form of the equation the co-ordinate axes are the asymptotes of the curve. There are two branches of the curve, and in each the curve proceeds to infinity as x becomes infinite.

In the general form of the equation to the curve, the axis of symmetry of the curve is taken as the x-axis and the curve appears as represented in Fig. 42.

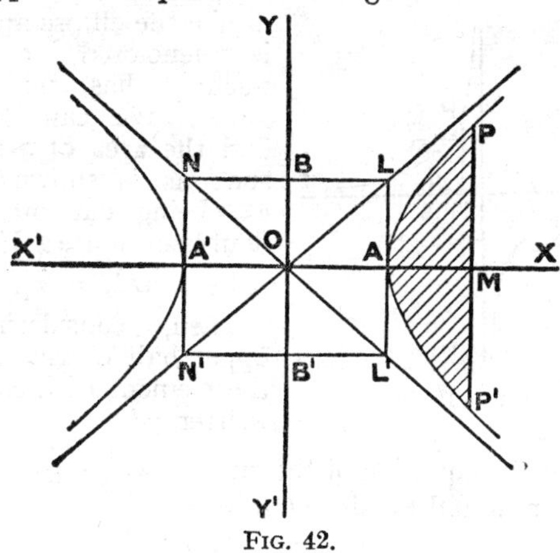

FIG. 42.

AA^1, the line joining the apexes of the two curves, is called the **transverse axis.**

Let its length be $2a$, so that $OA = a$. Draw tangents to the curve at A and A'. On them take AL, $A'N$ each equal to b. Then $\tan LOA = \frac{b}{a}$.

Note.—The relation between a and b cannot be discussed here.

The straight lines $N'OL$, NOL' are asymptotes to the curve. It is shown in co-ordinate geometry that the equation of the hyperbola is

$$\frac{x^2}{a^2} - \frac{y^2}{b^2} = 1.$$

The similarity to the equation of the ellipse will be noticed.

If $b = a$, *i.e.*, $AO = AL$, $\angle AOL = 45°$.

Thus $\angle LOL^1$, between the asymptotes, is a right angle, and the equation of the curve can be written

$$x^2 - y^2 = a^2.$$

This form of the curve is called a **rectangular hyperbola.**

The area of the hyperbola, unlike the ellipse and circle, is unenclosed, and consequently has no definite value. We can, however, find the **area of a segment** such as is shown in Fig. 42, being cut off by the double ordinate PMP^1.

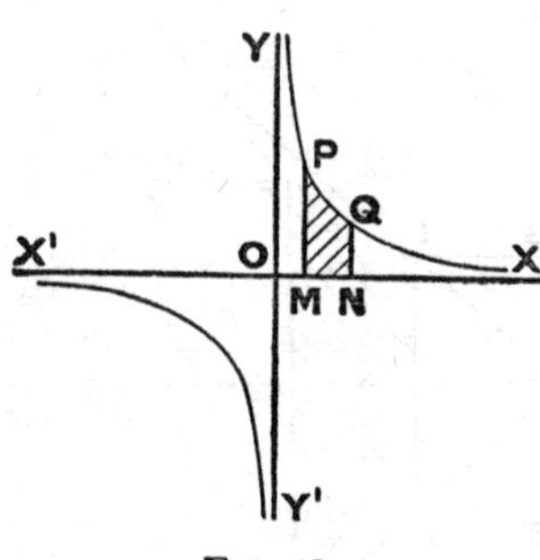

FIG. 43.

Let $\quad OM = x_1$.

Then, considering the upper half of the segment, the element of area can be written ydx.

But from the equation of the curve $y = \frac{b}{a}\sqrt{x^2 - a^2}$, and the limits are a and x_1, since $OA = a$.

$$\therefore \text{Area of whole segment} = 2\int_a^{x_1} \frac{b}{a}\sqrt{x^2 - a^2dx}$$

$$= \frac{2b}{a}\left[\tfrac{1}{2}x\sqrt{x^2 - a^2} - \frac{a^2}{2}\log\left(x + \sqrt{x^2 - a^2}\right)\right]_a^{x_1} \quad (\S\ 117).$$

Equation of a hyperbola referred to its asymptotes as axes.

This form has been mentioned above.

The curve is represented in Fig. 43.

The general form of the equation is shown in co-ordinate geometry to be

$$xy = c^2.$$

The area required to be found is usually that under a portion of the curve as shown by the shaded portion of the figure. This can be found in the usual way. A modified form is worked out in the next example.

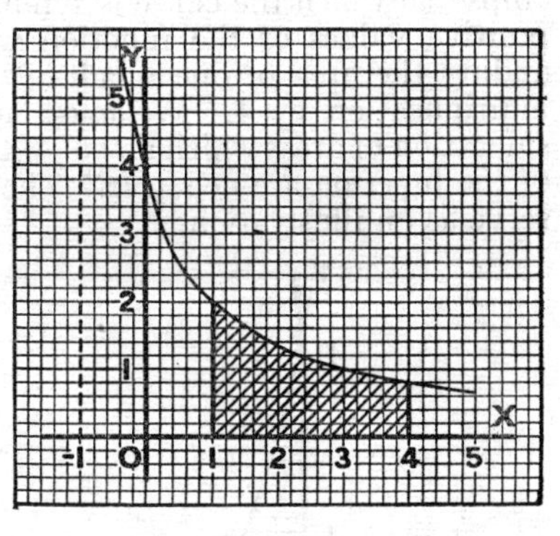

FIG. 44.

Example 8. *Find the area enclosed between the curve of* $y = \frac{4}{x+1}$, *the axis of* x, *and the ordinates* $x = 1$, $x = 4$.

The curve of this function is a hyperbola (see Fig. 44).

When $x \longrightarrow -1, y \longrightarrow \infty$.

$\therefore$ the ordinate $x = -1$ (dotted in the figure) is an asymptote to the curve. So also is the x-axis.

The area which it is required to find is that which is shaded.

Taking the element of area as ydx and substituting

$y = \frac{4}{x+1}$, we have

$$\begin{aligned}
\text{Area} &= \int_1^4 \frac{4}{x+1}\,dx \\
&= 4\Big[\log(x+1)\Big]_1^4 \\
&= 4(\log 5 - \log 2) \\
&= 4(\log \tfrac{5}{2}) \\
&= 4(0{\cdot}9163) \text{ (remembering the logs are hyperbolic)} \\
&= 3{\cdot}665 \text{ (approx.)}
\end{aligned}$$

152. Sign of an area.

It will be seen that in the foregoing examples of the determination of areas, these were, in most cases, lying above the axis of x, and the values of the function were consequently positive. In the examples of the circle and ellipse, in which the curve is symmetrical about both axes, positive values of the function were still adhered to by finding the area of one quadrant and then the whole by multiplication by 4. We must now proceed to the consideration of areas which lie below the x axes, and the values of the function are negative. The following examples will serve as an illustration:

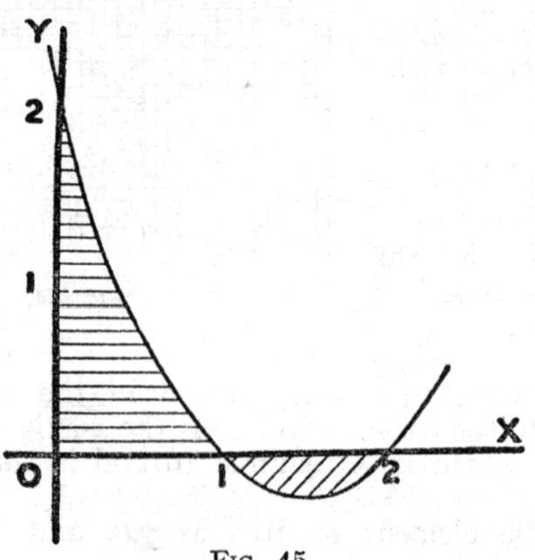

FIG. 45.

Example 1. *Find the area enclosed between the curve* $y = x^2 - 3x + 2$ *and the axis of* x.

Since $x^2 - 3x + 2 = (x - 1)(x - 2)$.
The curve cuts OX at $x = 1$, $x = 2$.

Also $\dfrac{dy}{dx} = 2x - 3$. $\therefore$ there is a turning-point when $x = \frac{3}{2}$.

Since $\dfrac{d^2y}{dx^2} = 2$ and is always positive, this point is a minimum.

The curve is represented in Fig. 45, and the area required lies entirely below OX.

Let A represent this area.

Then
$$A = \int_1^2 (x^2 - 3x + 2)dx$$
$$= \left[\frac{x^3}{3} - \tfrac{3}{2}x^2 + 2x\right]_1^2$$
$$= (\tfrac{8}{3} - 6 + 4) - (\tfrac{1}{3} - \tfrac{3}{2} + 2)$$
$$= -\tfrac{1}{6}.$$

The result is a negative area. But an area, fundamentally, is signless. How, then, is this result to be interpreted? It will probably not come as a surprise to the student because he will have seen that the definite integral $\int_a^b ydx$ **represents the sum** of an infinite number of products which are themselves infinitely small. When the area lies below OX, all values of the function, *i.e.*, y, are negative, and since dx, being the limit of δx and representing an **increase**, is positive, all the products must be negative. Hence the sum is negative. It has been pointed out (§ 150) that the summation is general for all such products, and the representation of an area by it is but one of the applications. Hence if we are finding an actual area by the integration, the negative sign must be disregarded, Since by the convention of signs used in the graphical representation of a function ordinates below the axis are negative, the corresponding areas are also negative. Hence as a matter of convention, areas above the x-axis are considered positive and below the axis are negative.

The student may note the following in connection with the above examples:

(1) The area below the curve between $x = 0$ and $x = 1$, *i.e.*, the area with horizontal shading in Fig. 45, is given by

$$\int_0^1 (x^2 - 3x + 2)dx = \tfrac{5}{6}.$$

(2) Consequently the total actual area, *i.e.*, disregarding the negative sign of the above integral, is $\tfrac{5}{6} + \tfrac{1}{6} = 1$.

(3) The total area as given by the integral

$$\int_0^2 (x - 3x + 2)dx = \tfrac{2}{3},\ i.e.,\ \tfrac{5}{6} - \tfrac{1}{6}.$$

Example 2. *Find the area enclosed between the curve of* $y = 4x(x - 1)(x - 2)$ *and the axis of* x.

The function $4x(x - 1)(x - 2)$ vanishes when $x = 0$, 1 and 2. Consequently its curve cuts the x-axis for these values of x. Proceeding as shown in § 57, there are found to be two turning points, as follows:

(1) a maximum value 1·55 when $x = 0{\cdot}45$;
(2) a minimum value $-$ 1·55 when $x = 1{\cdot}55$.

That part of the curve with which we are concerned is shown in Fig. 46. The areas required are shaded.

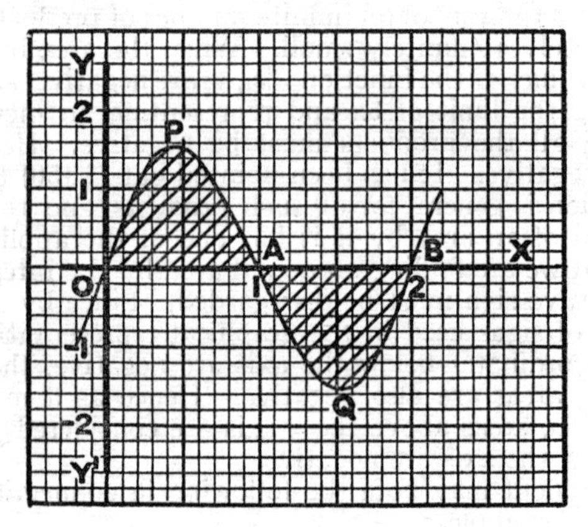

Fig. 46.

$$\begin{aligned}
\text{(1) Area of } OPA &= \int_0^1 4x(x - 1)(x - 2)dx \\
&= \int_0^1 (4x^3 - 12x^2 + 8x)dx \\
&= \Big[x^4 - 4x^3 + 4x^2\Big]_0^1 \\
&= (1 - 4 + 4) - 0 \\
&= 1 \text{ square unit.}
\end{aligned}$$

(2) Area of $AQB = \int_1^2 4x(x-1)(x-2)dx$

$$= \left[x^4 - 4x^3 + 4x^2\right]_1^2$$

$$= -1 \text{ square unit.}$$

Hence, disregarding the negative sign of the lower area, the total actual area of the shaded portions is 2 square units.

If we integrated, for the whole area between the limits 0 and 2 we get:

$$\text{Area} = \int_0^2 4x(x-1)(x-2)dx$$

$$= \left[x^4 - 4x^3 + 4x^2\right]_0^2$$

$$= 16 - 32 + 16$$

$$= 0.$$

This agrees with the **algebraical** sum of the two areas found separately.

From these examples we conclude that when finding the total area enclosed by a curve and the axis of x when it crosses the axis, we must find separately the areas above and below the axis. The sum of these, disregarding the signs, will be the actual area required.

Other examples follow.

Example 3. *Find the area enclosed between the axis of x and the curve of $y = \cos x$, between the limits*

(1) 0 and $\frac{\pi}{2}$.

(2) $\frac{\pi}{2}$ and π.

(3) 0 and π.

(1) The first area is shown in Fig. 47, in which it is the area above OX with shading.

$$\text{Area} = \int_0^{\frac{\pi}{2}} \cos\theta d\theta = \left[\sin\theta\right]_0^{\frac{\pi}{2}}$$

$$= \sin\frac{\pi}{2} - \sin 0$$

$$= 1.$$

(2) The second area is shown with shading below OX.

$$\text{Area} = \int_{\frac{\pi}{2}}^{\pi} \cos \theta d\theta = \Big[\sin \theta\Big]_{\frac{\pi}{2}}^{\pi}$$
$$= \sin \pi - \sin \frac{\pi}{2}$$
$$= 0 - 1$$
$$= -1.$$

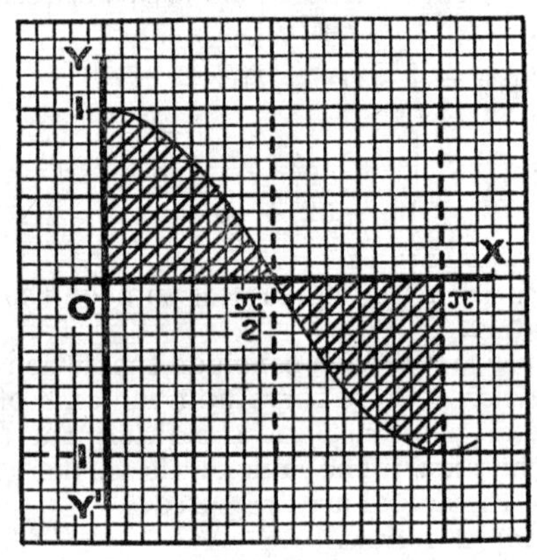

Fig. 47.

(3) The third area is composed of both (1) and (2).

$$\int_{0}^{\pi} \cos \theta d\theta = \Big[\sin \theta\Big]_{0}^{\pi}$$
$$= \sin \pi - \sin 0.$$
$$= 0.$$

These results agree algebraically, but if we require to know the actual area between 0 and π, the negative sign of the second area must be disregarded, and consequently the area of the two parts is **2 square units.**

Example 4. *Find the area enclosed between the axes of* **x** *and the curve of* $y = \sin x$, *for values of* **x** *between*

(1) 0 and π.
(2) π and 2π.

It is evident from the part of the graph of $y = \sin x$ in Fig. 48, that the area enclosed between the curve and the

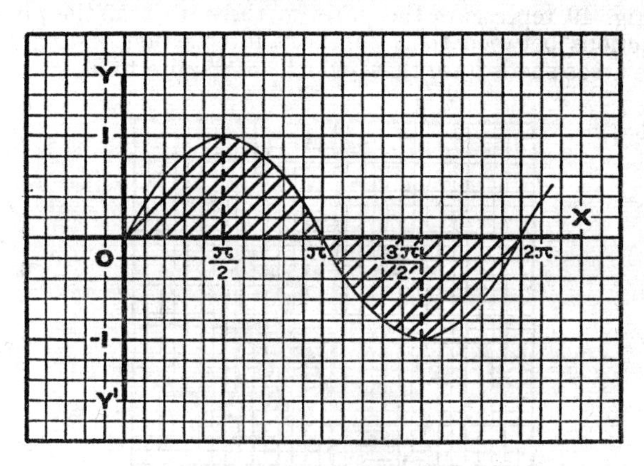

FIG. 48.

x-axis consists of a series of loops of equal area, each corresponding to a range of π radians, and lying alternately above and below the x-axis; consequently they are alternately positive and negative.

(1) Area of first loop

$$= \int_0^{\pi} \sin x dx = \Big[-\cos x\Big]_0^{\pi} = -(\cos \pi - \cos 0)$$
$$= -(-1 - 1) = 2$$

(2) Area of second loop

$$= \int_{\pi}^{2\pi} \sin x dx = \Big[-\cos x\Big]_{\pi}^{2\pi} = -(\cos 2\pi - \cos \pi)$$
$$= -\{+1 - (-1)\} = -2.$$

It is evident that if there are n loops, when n is an odd number, the total area, regard being paid to the negative signs, is 2, but if n is even, the area thus calculated is zero.

The **actual area** of n loops is $2n$.

Example 5. *Find the area contained between the curve of $y = x^3$ and the straight line $y = 2x$.*

Fig. 49 represents the parts of the curves of the given functions between their points of intersection, A and A^1. The areas shaded are those which we require to find.

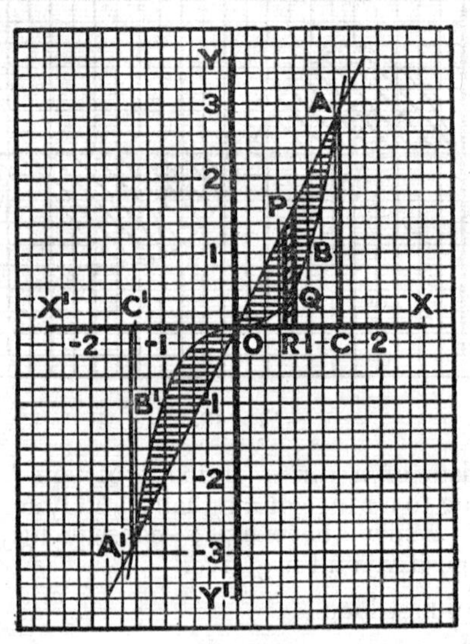

Fig. 49.

From symmetry it is evident that the parts above and below the x-axis will be equal in magnitude but of opposite signs.

We therefore proceed to find the area of $OABO$ (the shaded area). This is the difference between (1) the triangle OAC, and (2) the area beneath the curve of $y = x^3$, viz. $OBAC$.

We first find as usual an expression for the element of area.

From any point P on the line $y = 2x$ draw the ordinate PR, cutting the curve of $y = x^3$ in Q.

As before, construct a small rectangle represented by PR.

This represents the element of area for the triangle, while QR represents the element of area for $OBAC$.

$\therefore$ their difference PQ represents the element of area for the shaded part.

Let $PR = y_1$, $QR = y_2$.

Then element of area represented by PR is equal to $y_1 dx$, in the limit.

Then element of area represented by QR is equal to $y_2 dx$, in the limit.

$\therefore$ the element of area PQ is represented by $(y_1 - y_2)dx$, in the limit. Before we can integrate, the limits of the integral must be found. These will be the value of x at O and A, the points of intersection.

To find the value of x at these points we solve simultaneously

$$y = 2x$$
$$y = x^3.$$

Then
$$x^3 = 2x$$

and the roots are $0, +\sqrt{2}, -\sqrt{2}$.

These are values of x at O, A and A^1 respectively.

$\therefore$ for the positive area $OABO$ the limits are

$$x = 0 \quad \text{and} \quad x = +\sqrt{2}.$$

$$\therefore \text{the area required} = \int_0^{\sqrt{2}} (y_1 - y_2)dx$$
$$= \int_0^{\sqrt{2}} (2x - x^3)dx = \left[x^2 - \frac{x^4}{4}\right]_0^{\sqrt{2}}$$
$$= (\sqrt{2})^2 - \frac{(\sqrt{2})^4}{4}$$
$$= 1 \text{ square unit.}$$

From symmetry and previous considerations we conclude that the area below the x-axis is -1 square unit. This can be verified as follows:

$$A = \int_{-\sqrt{2}}^{0} (2x - x^3)dx = \left[x^2 - \frac{x^4}{4}\right]_{-\sqrt{2}}^{0}$$
$$= 0 - \left[(\sqrt{2})^2 - \frac{(-\sqrt{2})^4}{4}\right]$$
$$= 0 - 1 = -1.$$

Disregarding the negative sign the actual area of the two loops is **2 square units.**

The student, as an exercise should verify by finding the area of the two loops by the integral

$$\int_{-\sqrt{2}}^{+\sqrt{2}} (2x - x^3)dx.$$

Exercise 31.

Note.—The student is recommended to draw the figure which represents each problem, even though the drawing might be rough.

1. Find the area bounded by the curve of $y = x^3$, the x-axis and the ordinates $x = 2$, $x = 5$.
2. Find the area bounded by the straight line $2y = 5x + 7$, the x-axis and the ordinates $x = 2$, $x = 5$.
3. Find the area between the curve of $y = \log x$, the x-axis and the ordinates $x = 1$, $x = 5$.
4. Find the area enclosed by the curve of $y = 4x^2$, the y-axis and the straight lines $y = 1$, $y = 4$.
5. Find the area between the curve of $y^2 = 4x$, the x-axis and the ordinates $x = 4$, $x = 9$.
6. Find by the method of integration the area of the circle $x^2 + y^2 = 4$.
7. In the circle $x^2 + y^2 = 16$ find the area included between the parallel chords whose perpendicular distances from the centre are 2 and 3 units.

   Find also the area of the segment cut off from the circle $x^2 + y^2 = 16$ by the chord whose distance from the centre is 3 units.
8. Find by integration the area of the ellipse $\frac{x^2}{16} + \frac{y^2}{9} = 1$.
9. Find the area of the segment cut off from the hyperbola $\frac{x^2}{9} - \frac{y^2}{4} = 1$ by the chord $x = 4$.
10. Find the area between the hyperbola $xy = 4$, the x-axis and the ordinates $x = 2$, $x = 4$.
11. Find the area included between the curve of $y = 2x - 3x^2$ and the x axis.
12. Find the area bounded by $y = e^x$ and the x-axis between the ordinates $x = 0$ and $x = 3$.

13. Find the area cut off by the x-axis from the curve of $y = x^2 - x - 2$.

14. Find the whole area included between the curve of $y^2 = x^3$ and the line $x = 4$.

15. Find the area of the segment cut off from the curve of $xy = 2$ by the straight line $x + y = 3$.

16. Find the total area of the segments enclosed between the x-axis and the curve of $y = x(x - 3)(x + 2)$.

17. Find the area between the curves of $y = 8x^2$ and $y = x^3$.

18. Find the area which is common to the two curves $y = x^2$ and $y^2 = x$.

19. Find the area between the catenary, $y = \cosh \frac{x}{2}$ (see § 91), the x-axis and the ordinates $x = 0$, $x = 2$.

20. Find the actual area between the curve of $y = x^2 - 8x + 12$, the x-axis and the ordinates $x = 1$, $x = 9$.

21. Find the actual area between the curve of $y = x^3$ and the straight line $y = \frac{x}{4}$.

153. Polar co-ordinates.

The equations of curves are frequently more simple and the determination of areas easier when polar co-ordinates are employed, instead of rectangular. For the benefit of those students who have had no previous acquaintance with them, a very brief account is accordingly given below. For a full treatment the student should consult a text-book on Co-ordinate Geometry.

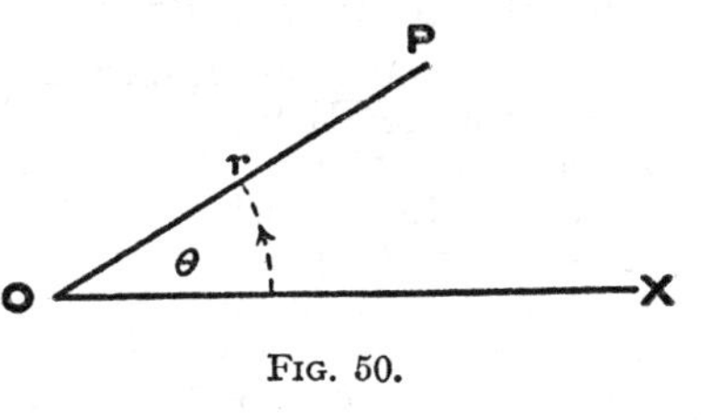

FIG. 50.

(*a*) **Definitions.** Let OX (Fig. 50) be a fixed straight line and O a fixed point on it.

Then the position of any point P is defined with reference to these when we know

(1) its distance from O,
(2) the angle made by OP with OX.

Let r be this distance.

Let θ be the angle made by OP with OX.

Then (r, θ) are called the **polar co-ordinates** of P.

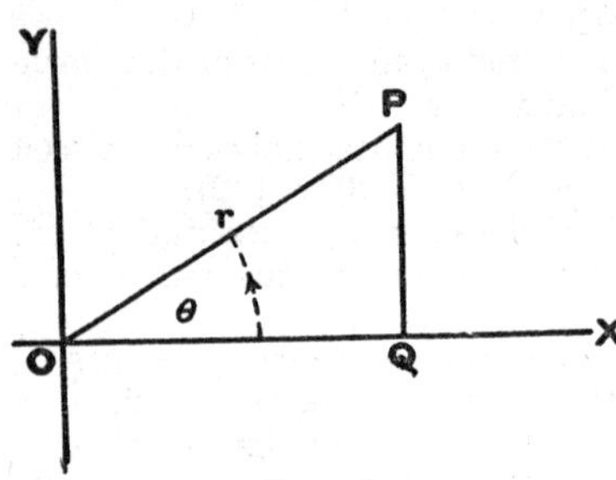

FIG. 51.

O, the fixed point, is called the **pole**, OP is called the **radius vector**, θ the **vectorial angle**, and OX the **initial line**.

θ is the angle which would be described by the radius vector, in rotating in a positive direction from OX.

(b) Connection between rectangular co-ordinates of a point and the polar co-ordinates.

Let P be a point (Fig. 51) whose polar co-ordinates are (r, θ), and rectangular co-ordinates (x, y), viz. OQ and PQ.

Then it is evident that
$$x = r \cos \theta$$
$$y = r \sin \theta$$
$$x^2 + y^2 = r^2.$$

(c) Polar equation of a curve.

If a point moves along a curve, as θ changes, r in general will also change. Hence

***r* is a function of θ.**

The equation which states the relation between r and θ for a given curve is called the **polar equation of the curve.**

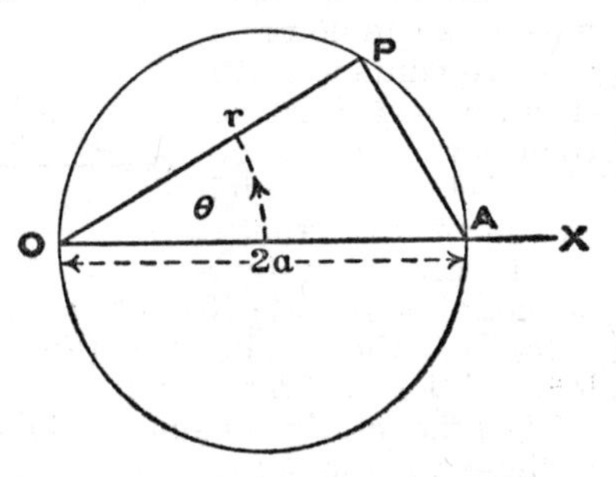

FIG. 52.

(d) Example of a polar equation.

Let a point P move along the circumference of a circle (Fig. 52).

Let O be a fixed point at the extremity of a fixed diameter.

Let $2a$ = the diameter of the circle.

Then, for any position of P with reference to O and OA, the polar co-ordinates are:

$$OP = r$$
$$\angle AOP = \theta.$$

From geometry it is known that $\angle OPA$ is a right angle.

$$\therefore \quad r = 2a \cos \theta.$$

This is the polar equation of the circle with the above conditions. It may be noted that if the centre of the circle were taken as the pole, r is always equal to a; *i.e.*, the polar equation is then

$$r = a.$$

In such a case r is a constant, being the radius of the circle, and has no functional relation to θ.

The equation of the circle may take other forms.

154. Plotting curves from their equations in polar co-ordinates.

Many curves are easily drawn from their polar equations, though the plotting of points may be difficult when using the equations of the curves in rectangular co-ordinates. The following example is given as typical of the method employed.

Example. *Draw the curve whose polar equation is*

$$r = a(1 + \cos \theta)$$
$$= a + a \cos \theta.$$

The general method is to select values of θ, find the corresponding values of r; then plot the points obtained.

As has been shown above, $r = a \cos \theta$ is the equation of a circle of diameter a, when the pole is on the circumference. It is evident therefore that if for any value of θ, the value of r for the circle is increased by a, the result is the value of r for the required curve.

Draw a circle of radius $\frac{a}{2}$ (Fig. 53).

Take a point O at the end of a diameter OA. O will be the pole for the curve.

Since cos θ is a maximum, viz., unity, when $\theta = 0$, the maximum value of r for the curve will be at the point B, where $AB = a$.

Thus the **maximum value of r is $2a$.**

When $\theta = \frac{\pi}{2}$ and $\frac{3\pi}{2}$, $\cos\theta = 0$. $\therefore$ $r = a$. Hence we get the points C and D.

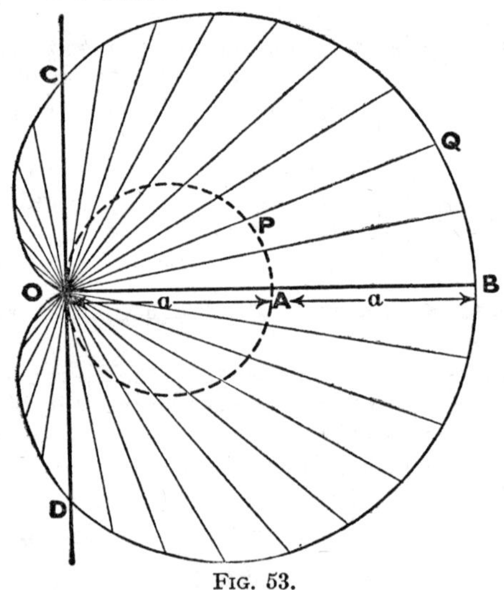

Fig. 53.

In the 2nd quadrant, $\cos\theta$ is decreasing to -1, at π, then $r = a - a = O$.

Similarly the general path of the curve may be found for the third and fourth quadrants.

Finally, when $\theta = 2\pi$, $\cos\theta = 1$.

$\therefore$ the curve is closed at B.

To get other points on the curve between the special points considered above, draw a series of chords of the circle, for increasing values of θ. If OP be one of these, produce it and mark off PQ equal to a. Then Q is a point on the curve. The complete curve is as shown in Fig. 53.

It is known as the **cardioid**, from its heart-like shape. It is of importance in optics.

Other examples of curves which are readily drawn from their polar equations are

(1) The **lemniscate**, $r^2 = a^2 \cos 2\theta$.
(2) The **limacon**, $r = b - a \cos \theta$.
(3) The **spiral of Archimedes**, $r = a\theta$.
(4) The **logarithmic** or **equiangular spiral**, $\log r = a\theta$.
(5) The **hyperbolic spiral**, $r\theta = a$.

155. Areas in Polar Co-ordinates.

Let AB, Fig. 54, be part of a curve whose equation is known in polar co-ordinates.

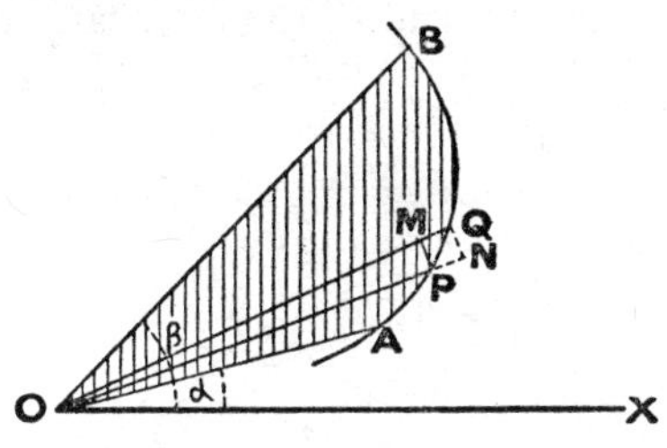

FIG. 54.

Suppose it is required to find the area of the sector OAB, contained between the curve and the two radii OA, OB, the angles made by them with the fixed line OX being

$$\angle AOX = \alpha$$
$$\angle BOX = \beta.$$

Let P be any point on the curve, and its polar co-ordinates (r, θ).

$\therefore \angle POX = \theta$.

Let θ receive an increment $\delta\theta$, and r, in consequence be increased by δr. The polar co-ordinates of Q, the new position on the curve, are

$$((r + \delta r), (\theta + \delta\theta))$$

Then, with the construction shown in the figure, the area of the sector OPQ lies between the areas of the Δs OPM, ONQ, the areas of which are

$$\Delta OPM = \tfrac{1}{2}r^2\delta\theta$$
$$\Delta ONQ = \tfrac{1}{2}(r + \delta r)^2\delta\theta.$$

If the angle $\delta\theta$ be now decreased indefinitely, then as

$$\delta\theta \longrightarrow O, \ (r + \delta r) \longrightarrow r$$

and the area of the infinitely small sector approaches $\frac{1}{2}r^2d\theta$.

This is, therefore, the element of area, and the sum of all such sectors between the limits $\theta = \alpha$ and $\theta = \beta$ will be the area of the sector OAB.

Expressing this as an integral, as before,

$$\text{Area of sector } OAB = \int_{\alpha}^{\beta} \tfrac{1}{2}r^2d\theta.$$

When the polar equation of the curve is known, r can be expressed in terms of θ and the integral can be evaluated.

Example. *Find the area of the circle whose polar equation is* $r = 2a \cos \theta$ (§ 153, *d*).

If P be a point moving round the curve, the radius vector describes the area of the circle.

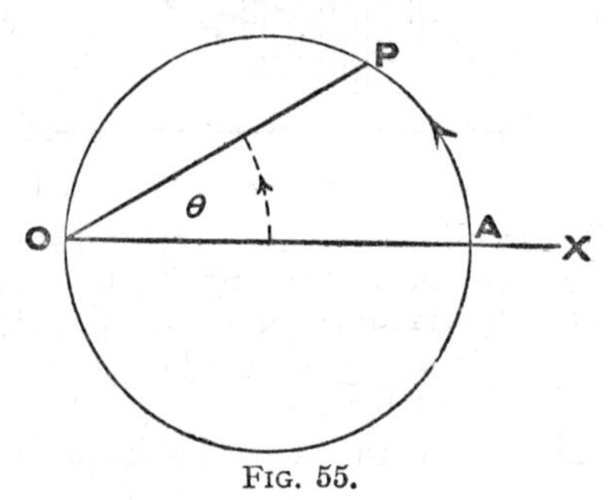

FIG. 55.

When P is at A, $\theta = 0$.

When P is at O, $\theta = \frac{\pi}{2}$.

$\therefore$ as P moves from A to O, and the vectorial angle θ changes from O to $\frac{\pi}{2}$, the area described is a semi-circle. Using the formula obtained above

$$\text{Area of semi-circle} = \int_{0}^{\frac{\pi}{2}} \tfrac{1}{2}r^2d\theta$$

The area of the circle is twice this.

$$\therefore \text{ Area of circle} = \int_0^{\frac{\pi}{2}} r^2 d\theta.$$

But $r = 2a \cos \theta$.

$$\begin{aligned}
\therefore \quad \text{Area} &= \int_0^{\frac{\pi}{2}} 4a^2 \cos^2 \theta d\theta \\
&= 4a^2 \int_0^{\frac{\pi}{2}} \cos^2 \theta d\theta \\
&= 4a^2 \int_0^{\frac{\pi}{2}} \tfrac{1}{2}(1 + \cos 2\theta) d\theta \quad (\S\, 114) \\
&= 2a^2 \Big[\theta + \tfrac{1}{2} \sin 2\theta\Big]_0^{\frac{\pi}{2}} \\
&= 2a^2 \Big[\frac{\pi}{2} + \tfrac{1}{2} \sin \pi\Big] \\
&= 2a^2 \times \frac{\pi}{2} \\
&= \pi a^2.
\end{aligned}$$

Exercise 32.

1. Find the area of the cardioid whose equation is $r = a(1 + \cos \theta)$, the limits of θ being 2π and 0.

2. Find the area of one loop of the curve $r = a \sin 2\theta$, *i.e.*, between the limits 0 and $\frac{\pi}{2}$. How many loops are there between 0 and 2π.

   Note.—$a \sin 2\theta$ vanishes when $\theta = 0$ and $\theta = \frac{\pi}{2}$. As the function is continuous between these values, the curve must form a loop between them. The student should draw roughly the whole curve.

3. Find the area of one loop of the lemniscate

$$r^2 = a^2 \cos 2\theta.$$

How many loops are there in the complete curve?

4. If the radius vector of the function $r = a\theta$ makes one complete rotation from 0 to 2π, find the area thus passed over.

5. Find the area which is described in the curve $r = a \sec^2 \frac{\theta}{2}$ from $\theta = 0$ to $\theta = \frac{\pi}{2}$.

6. Find the area enclosed by the curve $r = 3 \cos \theta + 5$ between $\theta = 2\pi$ and $\theta = 0$.

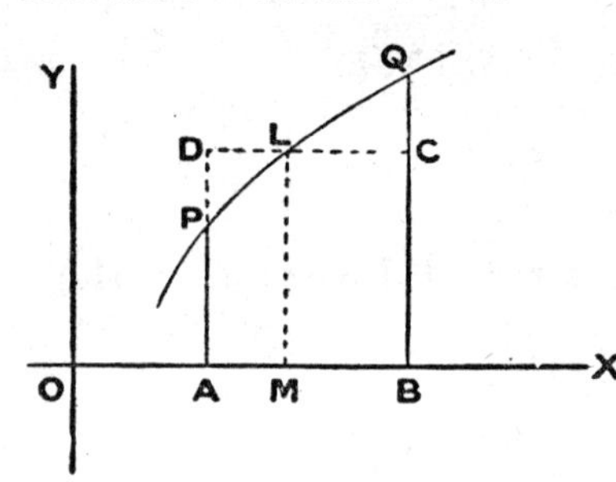

FIG. 55A.

156. Mean value.

Let PQ (Fig. 55A) represent part of the curve of a continuous function

$$y = f(x).$$

Let PA, QB be the ordinates at P and Q, where $OA = a$, $OB = b$.

Then from previous work we know that

$$\text{Area of } APQB = \int_a^b f(x)dx.$$

Let $ABCD$ be a rectangle whose area is equal to that of $APQB$, *i.e.*, to $\int_a^b f(x)dx$.

Draw LM parallel to OY, from L, the intersection of the curve, and DC parallel to OX.

$$\text{Area} \quad ABCD = AB \times LM$$

$$\therefore \quad LM = \frac{\text{Area of } ABCD}{AB}$$

$$= \frac{\text{Area of } APQB}{AB}$$

$$= \frac{\int_a^b f(x)dx}{AB}$$

$$= \frac{\int_a^b f(x)dx}{b - a}.$$

LM is said to be the mean value of ordinates of the curve for the range of values $x = a$ to $x = b$.

$$\therefore \text{ Mean value of } f(x) \text{ from } a \text{ to } b = \frac{\int_a^b f(x)dx}{b - a}.$$

Example. *Find the mean value of* $2 \cos t - \sin 3t$ *between the values* $t = 0$ and $t = \frac{\pi}{6}$.

From the above, mean value

$$= \frac{\int_0^{\frac{\pi}{6}} (2 \cos t - \sin 3t)dt}{\frac{\pi}{6} - 0}$$

$$= \frac{\left[2 \sin t + \frac{1}{3} \cos 3t\right]_0^{\frac{\pi}{6}}}{\frac{\pi}{6}}$$

$$= \frac{\left\{2 \sin \frac{\pi}{6} + \frac{1}{3} \cos \frac{\pi}{2}\right\} - \{2 \sin 0 + \frac{1}{3} \cos 0\}}{\frac{\pi}{6}}$$

$$= \frac{1 - \frac{1}{3}}{\frac{\pi}{6}}$$

$$= \frac{4}{\pi}.$$

Exercise 33.

1. Find the mean value of the function $\sin x$ over the range of values $x = 0$ to $x = \pi$.
2. Find the mean value of the function $\sin^2 x$ over the range of values $x = 0$ to $x = \pi$.
3. Find the mean value of $y = \frac{1}{x}$ for the range of values $x = 1$ to $x = 10$.
4. Find the mean value of $y^2 = 4x$ between $x = 4$ and $x = 0$.
5. The equation of a curve is $y = b \sin^2 \frac{\pi x}{a}$. Find the

mean height of the portion for which x lies between b and a.

6. Find the mean value of $\cos x$ between $x = 0$ and $x = \frac{\pi}{4}$.

7. Find the mean value of the function $y = a \sin bx$ between the values $x = 0$ and $x = \frac{\pi}{b}$.

8. The range of a projectile fired with initial velocity v_0 and an elevation θ is $\frac{v_0^2}{g} \sin 2\theta$. Find the mean range as θ varies from 0 to $\frac{\pi}{2}$.

157. Irregular areas.

The determination of irregular areas, *i.e.*, areas the boundaries of which cannot be expressed by formal equations, is often a matter of great practical importance. There are certain practical methods, such as using squared paper and counting squares, which yield rough approximate results, but there are also methods of calculation by which the area can be determined with greater accuracy, though still approximate. The first of these is the trapezoidal rule which is as follows:

158. The trapezoidal rule.

Let the area which it is required to determine be that enclosed by the irregular curve PV (Fig. 56), the x-axis and the ordinates PA and VG. Divide AG into any number of equal parts, at B, C, D . . ., each of length l, and draw the corresponding ordinate PA, QB, RC . . .

Join PQ, QR, RS . . . UV.

Let the lengths of the ordinates be y_1, y_2, y_3 . . .

Then each of the figures formed by these constructions, such as $APQB$, is a trapezium, and their areas are

$$\tfrac{1}{2}l(y_1 + y_2) + \tfrac{1}{2}l(y_2 + y_3) + \ldots + \tfrac{1}{2}l(y_6 + y_7).$$

The areas of the trapeziums approximate to the areas of those figures in which the straight line PQ is replaced by the curve PQ, and so for the others. Consequently the sum of all these approximates to the area of the whole figure which is required, and the greater the number, the closer will be the approximation.

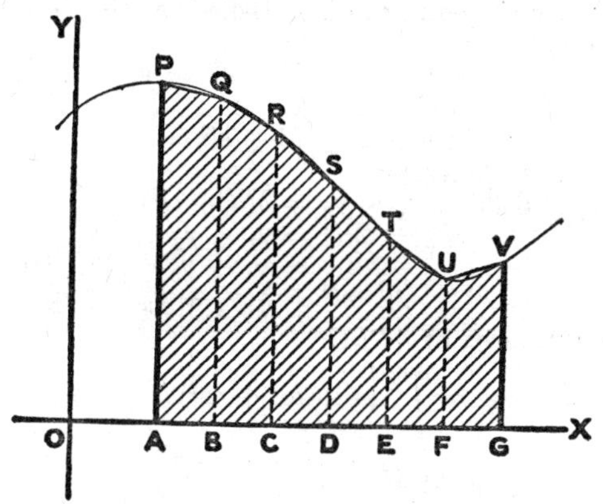

FIG. 56.

∴ the area is approximately equal to

$\frac{1}{2}l\{(y_1 + y_2) + (y_2 + y_3) + (y_3 + y_4) \ldots + (y_6 + y_7)\}$
$= \frac{1}{2}l\{(y_1 + y_7) + 2(y_2 + y_3 + y_4 \ldots + y_6)\}$
= (half the distance between the strips) × {(sum of first and last ordinates) + (twice the sum of other ordinates)}.

159. Simpson's rule for area.

Considering again the irregular curve of the previous section, it is evident that if the chords PQ, QR, RS . . . were to be replaced by the arcs of suitable regular curves, and the areas so obtained be found by previous methods, the approximation to the area would be closer than that found by the trapezoidal rule.

Accordingly we assume that the part of the curve joining three consecutive points, such as P, Q, R, is the arc of a parabola.

Assume the origin for this parabola to be at B, so that the co-ordinates of A, B and C are $-l$, 0, $+l$, then the area of $APRC$ can be found by integration.

Let the equation of the parabola, of which PQR is an arc, be

$$y = a + bx + cx^2.$$

Then, since the equation is satisfied by the co-ordinates of A, B, C

$$AP = a - bl + cl^2 = y_1 \quad . \quad . \quad . \quad (1)$$
$$BQ = a = y_2 \quad . \quad . \quad . \quad . \quad . \quad . \quad (2)$$
$$CR = a + bl + cl^2 = y_3 \quad . \quad . \quad . \quad (3)$$

$$\therefore \quad y_2 = a$$

adding (1) and (3) $\quad y_1 + y_3 = 2(a + cl^2)$

whence $\quad 2cl^2 = y_1 + y_3 - 2a$

$$= y_1 + y_3 - 2y_2 \text{ from (2)}$$

$$\therefore \quad cl^2 = \tfrac{1}{2}(y_1 + y_3 - 2y_2) \quad . \quad (A)$$

Integrating area of $APRC$

$$= \int_{-l}^{l} (a + bx + cx^2)dx$$
$$= \Big[ax + \tfrac{1}{2}bx^2 + \tfrac{1}{3}cx^3\Big]_{-l}^{l}$$
$$= 2al + \tfrac{2}{3}cl^3$$
$$= 2l(a + \tfrac{1}{3}cl^2)$$
$$= 2l\{y_2 + \tfrac{1}{6}(y_1 + y_3 - 2y_2)\} \quad . \quad \text{(A)}$$
$$= 2l\left(\frac{4y_2 + y_1 + y_3}{6}\right)$$
$$= l\left(\frac{y_1 + 4y_2 + y_3}{3}\right)$$

Similarly, area of $RCET$

$$= l\left(\frac{y_3 + 4y_4 + y_5}{3}\right)$$

and area of $TEGV$

$$= l\left(\frac{y_5 + 4y_6 + y_7}{3}\right).$$

$\therefore$ area of whole

$$= \frac{l}{3}\{(y_1 + 4y_2 + y_3) + (y_3 + 4y_4 + y_5) + (y_5 + 4y_6 + y_7)\}$$
$$= \frac{l}{3}\{(y_1 + y_7) + 2(y_3 + y_5) + 4(y_2 + y_4 + y_6)\}.$$

Clearly, this process can be applied to any **even** number of intervals, which involves an **odd** number of ordinates.

Thus, if there be $2n$ intervals, there will be $2n + 1$ ordinates. From the consideration of these results we may deduce:

Simpson's rule for areas.

If the area be divided into an even number of strips by equidistant ordinates, then

$$\text{Area} = \frac{\text{width of strip}}{3}\{(\text{sum of first and last ordinates})$$
$$+ 2(\text{sum of odd ordinates}) + 4(\text{sum of even ordinates})\}.$$

It will readily be undersood that the greater number of strips which are taken, the greater will be the accuracy of the approximation to the area.

160. Worked example. *Find the area of a quadrant of a circle of 2 cm radius.*

In this example the result as obtained by Simpson's Rule can be compared with the calculated area of the quadrant.

Fig. 57 represents the quadrant.

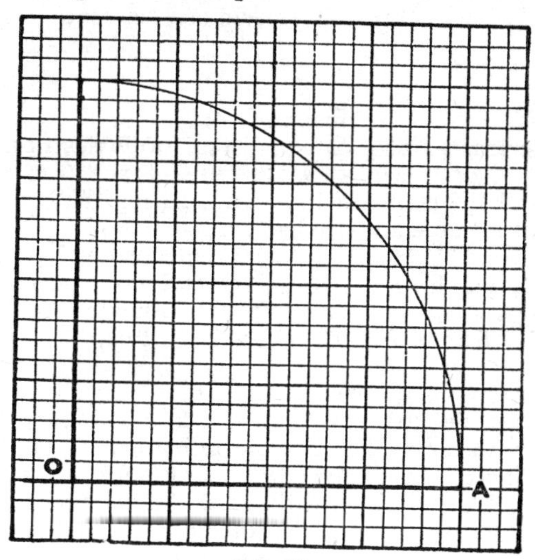

FIG. 57.

Divide the radius OA into 10 equal divisions each of 2 mm.

Then the ordinates will be represented by $y_1, y_2, y_3 \ldots y_{11}$.

Measuring these, the working is arranged as follows:

(1) First and last.	(2) Odd ordinates.	(3) Even ordinates.
$y_1 = 2$	$y_3 = 1{\cdot}96$	$y_2 = 1{\cdot}99$
$y_{11} = 0$	$y_5 = 1{\cdot}83$	$y_4 = 1{\cdot}91$
	$y_7 = 1{\cdot}6$	$y_6 = 1{\cdot}73$
sum 2	$y_9 = 1{\cdot}2$	$y_8 = 1{\cdot}42$
		$y_{10} = 0{\cdot}86$
	sum 6·59	
		sum 7·91

∴ By Simpson's rule

$$\text{Area} = \frac{0{\cdot}2}{3}\{2 + (2 \times 6{\cdot}59) + (4 \times 7{\cdot}91)\}$$

$$= \frac{0{\cdot}2}{3} \times 46{\cdot}82 = 3{\cdot}12 \text{ cm}^2.$$

Calculated areas $= \frac{1}{4}\pi r^2 = \frac{1}{4} \times \pi \times 4 = 3{\cdot}14 \text{ cm}^2$.

The error 0·2 in 3·14 is less than one per cent.

Exercise 34.

1. The lengths of nine equidistant ordinates of a curve are 8, 10·5, 12·3, 11·6, 12·9, 13·8, 10·2, 8 and 6 m respectively, and the length of the base is 24 m. Find the area between the curve and the base.

2. An area is divided into ten equal parts by parallel ordinates, 0·2 m apart, the first and last touching the bounding curve. The lengths of the ordinates are 0, 1·24, 2·37, 4·10, 5·28, 4·76, 4·60, 4·36, 2·45, 1·62, 0. Find the area.

3. The lengths of the ordinates of a curve in mm are 2·3, 3·8, 4·4, 6·0, 7·1, 8·3, 8·2, 7·9, 6·2, 5·0, 3·9. Find the area under the curve if each of the ordinates are 1 mm apart.

4. Ordinates at a common distance of 10 m are of length in m, 6·5, 9, 13, 18·5, 22, 23, 22, 18·5, 14·5. Find the area bounded by the curve, the axis of x, and the end ordinates.

5. Find the area under the curve shown in Fig. 58, the ordinates being drawn at the points marked 1 to 12, each division representing one m.

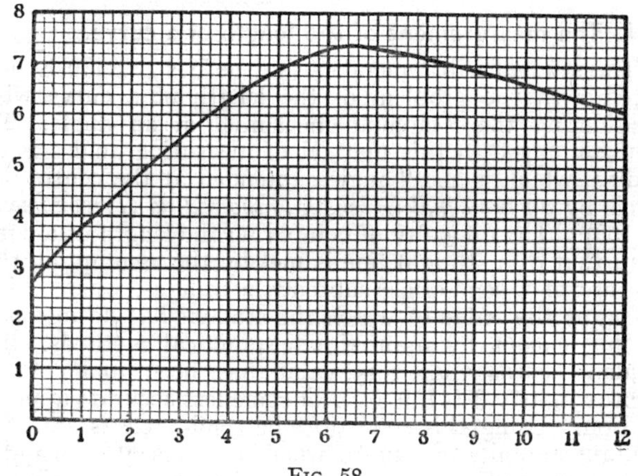

FIG. 58.

CHAPTER XV

THE LENGTHS OF CURVES

161. The measurement of the length of a curve.

The student will remember that he has previously been faced with the problem of the length of a curve when considering the "circular measure" of an angle. The unit employed in this method of measuring angles is the **radian**, which is the "angle subtended at the centre of a circle by *an arc equal in length to the radius* (*Trigonometry*, p. 150). The difficulty of comparing the length of a curve with that of a straight line is met by the assumption that the arc of a semi-circle subtends π radians, where π is a constant the value of which the student has no means of finding except by approximate practical methods. The student learns that this value has been found to be approximately 3·14159 . . . or some less accurate approximation. Using this constant, the semi-circle is stated to contain πr units of length, where r represents the radius and that the length of the circumference of the circle is $2\pi r$ units.

It will be observed that this "formula" for the circumference of a circle is, in reality, merely a statement that the ratio of the length of the circumference of a circle to its diameter is represented by the Greek letter π, where the value of π is undetermined. The determination of its value occupied mathematicians through the centuries, and by various ingenious devices, with which we are not concerned here, approximations were found.

Modern mathematics, however, with the help of the calculus, as the student will see later, has solved the problem, and it can now be proved that the ratio is incommensurable, but that its value to any required degree of accuracy can be calculated with certainty.

Since no part of a curve, however small, can be superimposed on any portion of a straight line, so that it coincides with it, its length cannot thus be found by comparison with a straight line of known length. Integration, however, supplies a method of determining the length of any **regular** curve. This method, as the student has probably anticipated, is similar to that used for areas. An expression is found for "*an element of length*" of the curve and the sum of all such elements is obtained by integration.

This process is called "**the rectification of a curve.**"

162. General formula for the length of a curve in cartesian co-ordinates.

Let AB (Fig. 59) represent a portion of the curve of a function $y = f(x)$ between the points A, where $x = a$, and B, where $x = b$.

Let P, Q be two points on the curve, and PQ the chord of the curve through them.

Let P be (x, y).

Let s be the length of the arc from A to B.

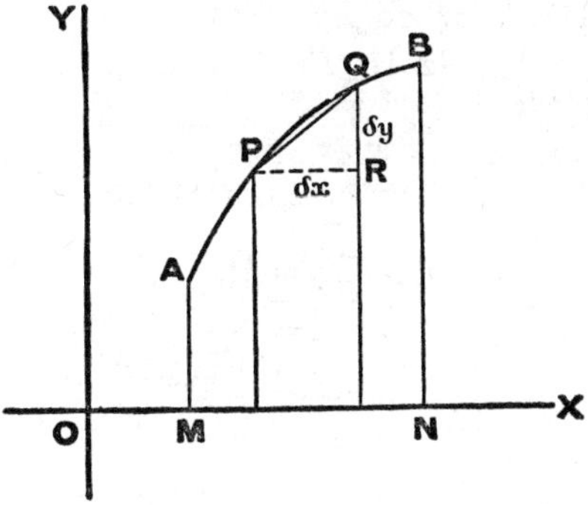

FIG. 59.

When x is increased by δx
y ,, δy
then s ,, δs.

i.e., δs represents the length of the arc PQ.

Then by geometry the chord $PQ = \sqrt{(\delta x)^2 + (\delta y)^2}$.

If Q be taken close to P, *i.e.*, δx becomes small, the length of the chord is nearly equal to the length of the arc.

If Q is indefinitely close to P, in the limit when $\delta x \longrightarrow 0$, the chord approaches to coincidence with the curve and the sum of these chords is equal to the length of the arc.

$$\text{Then} \quad ds = \sqrt{(dx)^2 + (dy)^2}$$

$$= \sqrt{1 + \left(\frac{dy}{dx}\right)^2} \cdot dx \quad \text{or} \quad \sqrt{\left(\frac{dx}{dy}\right)^2 + 1} \cdot dy.$$

$\therefore$ Integrating

$$s = \int_a^b \sqrt{1 + \left(\frac{dy}{dx}\right)^2} \cdot dx.$$

If the integration is more conveniently performed with respect to values of y, then $s = \int_d^c \sqrt{\left(\frac{dx}{dy}\right)^2 + 1} \, . \, dy$, where c and d are the limits of y.

In many cases the evaluation of the integral is difficult and requires a more advanced knowledge of the subject than is contained in this volume.

163. Worked examples.

Example 1. *Find the length of the circumference of the circle $x^2 + y^2 = a^2$.*

Since $x^2 + y^2 = a^2$

$$y = \sqrt{a^2 - x^2} = (a^2 - x^2)^{\frac{1}{2}}$$

$$\therefore \quad \frac{dy}{dx} = \tfrac{1}{2}(a^2 - x^2)^{-\frac{1}{2}} \times (-2x)$$

$$= \frac{-x}{\sqrt{a^2 - x^2}}.$$

$$\therefore \quad \left(\frac{dy}{dx}\right)^2 = \frac{x^2}{a^2 - x^2}.$$

Considering the area of a quadrant the limits will be a and 0.

Using the formula above, viz.

$$s = \int_a^b \sqrt{1 + \left(\frac{dy}{dx}\right)^2} \, . \, dx$$

then

$$s = \int_0^a \sqrt{1 + \frac{x^2}{a^2 - x^2}} \, . \, dx$$

$$= \int_0^a \sqrt{\frac{a^2}{a^2 - x^2}} \, . \, dx$$

$$= a \int_0^a \frac{dx}{\sqrt{a^2 - x^2}}$$

$$= a \times \left[\sin^{-1} \frac{x}{a}\right]_0^a$$

$$= a \times \left(\frac{\pi}{2} - 0\right)$$

$$= \frac{\pi a}{2}.$$

$\therefore$ circumference of the circle

$$= 4 \times \frac{\pi a}{2}$$
$$= 2\pi a.$$

Note.—The use of π is necessitated in the evaluation of the definite integral, and it is there employed in the same way as referred to in § 161.

Example 2. *Find the length of the arc of the parabola* $x^2 = 4y$ *from the vertex to the point where* $x = 2$.

The equation can be written in the form:

$$y = \frac{x^2}{4}$$

whence
$$\frac{dy}{dx} = \frac{x}{2}.$$

A sketch of the curve is shown in Fig. 60, where OQ

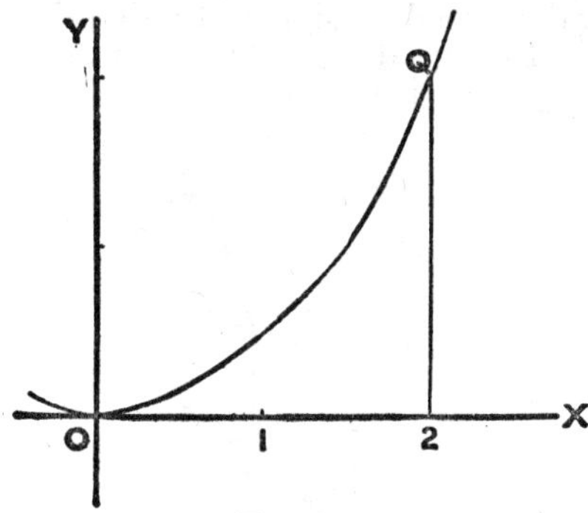

FIG. 60.

represents the part of the curve of which the length is required. The limits of x are clearly 0 and 2.

Using

$$s = \int_a^b \sqrt{1 + \left(\frac{dy}{dx}\right)^2} \, . \, dx$$

on substitution

$$s = \int_0^2 \sqrt{1 + \frac{x^2}{4}} \, . \, dx$$

$$= \frac{1}{2} \int_0^2 \sqrt{x^2 + 4} \, . \, dx$$

$\therefore$ by the formula of § 117, page 177,

$$s = \frac{1}{2}\left[\frac{1}{2}x\sqrt{x^2 + 4} + \frac{4}{2}\log \frac{x + \sqrt{x^2 + 4}}{2}\right]_0^2$$

$$= \frac{1}{2}[\frac{1}{2} \times 2\sqrt{8} + 2\{\log (2 + \sqrt{8}) - \log 2\}]$$

$$= \sqrt{2} + \log \frac{2 + 2\sqrt{2}}{2}$$

$$= \sqrt{2} + \log (1 + \sqrt{2})$$

$$= 2{\cdot}295 \text{ (approx.)}.$$

(The logs being to base **e**.)

164. Equation for the length of a curve in polar co-ordinates.

The general method is similar to that in rectangular co-ordinates. In Fig. 61 let AB represent part of a curve whose polar equation is known.

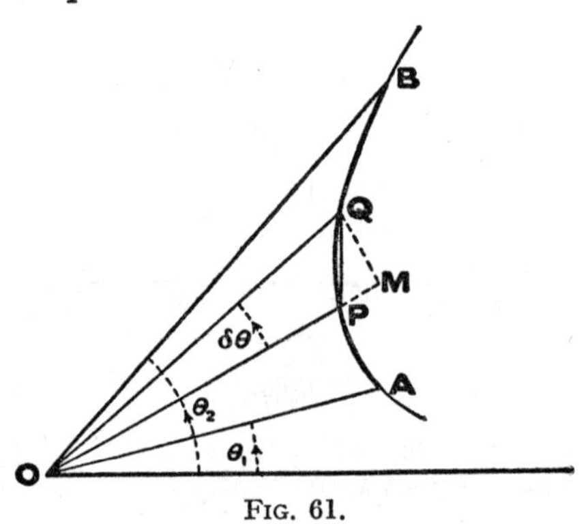

Fig. 61.

Let the angles made by OA and OB with OX be θ_1 and θ_2. Let s be the length of AB.

Let P be any point (r, θ) on the curve.

Let Q be a point on the curve near to P, so that $\angle QOM$, the increase in θ is $\delta\theta$ and PM is the increase in r, *i.e.*, δr.

Whence Q is the point $(r + \delta r, \theta + \delta\theta)$.

Let PQ be the chord joining P to Q.

Then $QM = r\delta\theta$ and the arc PQ represents δs.

With the construction shown

$$PM = \delta r.$$

Then
$$PQ^2 = (r\delta\theta)^2 + (\delta r)^2$$

When Q is taken indefinitely close to P, *i.e.*, $\delta\theta \longrightarrow 0$, in the limit

$$(ds)^2 = (rd\theta)^2 + (dr)^2$$

$$\therefore \quad ds = \sqrt{r^2(d\theta)^2 + (dr)^2}$$

$$= \sqrt{r^2 + \left(\frac{dr}{d\theta}\right)^2} \,.\, d\theta \quad . \quad . \quad . \quad \text{(A)}$$

The limits of the integral are θ_1 and θ_2.

$$\therefore \text{ Integrating,} \quad S = \int_{\theta_1}^{\theta_2} \sqrt{r^2 + \left(\frac{dr}{d\theta}\right)^2} \,.\, d\theta \quad . \quad \text{(I)}$$

We may also write (A) in the form

$$ds = \left\{\sqrt{1 + r^2\left(\frac{d\theta}{dr}\right)^2} \,.\, dr,\right.$$

i.e., we regard θ as a function of **r**; hence if the limits of **r** are r_1, r_2

$$s = \int_{r_1}^{r_2} \sqrt{1 + r^2\left(\frac{d\theta}{dr}\right)^2} \,.\, dr \,. \quad . \quad \text{(II)}$$

165. Worked example.

Find the complete length of the cardioid whose equation is $r = a(1 - \cos\theta)$.

As was seen in § 154, the construction of a complete cardioid involves a complete rotation of the radius vector, so that θ increases from 0 to 2π.

Since
$$r = a(1 - \cos\theta)$$

$$\frac{dr}{d\theta} = a \sin\theta.$$

Using formula (1) above

$$s = \int_0^{2\pi} \sqrt{\{a(1 - \cos\theta)\}^2 + (a\sin\theta)^2}d\theta$$
$$= \int_0^{2\pi} \sqrt{\{a^2(1 - 2\cos\theta + \cos^2\theta) + a^2\sin^2\theta\}}d\theta$$
$$= a\sqrt{2}\int_0^{2\pi} \sqrt{1 - \cos\theta}d\theta \text{ (on simplification)}$$
$$= a\sqrt{2}\int_0^{2\pi} \sqrt{2\sin^2\frac{\theta}{2}}\,d\theta$$
$$= 2a\int_0^{2\pi} \sin\frac{\theta}{2}\,d\theta$$
$$= 2a\Big[-2\cos\frac{\theta}{2}\Big]_0^{2\pi}$$
$$= 4a[-\cos\pi + \cos 0]$$
$$= 8a.$$

Exercise 35.

1. Find the length of the arc of the parabola $y = \frac{1}{2}x^2$ between the origin and the ordinate $x = 2$.

2. Find the length of the arc of the parabola $y^2 = 4x$ from $x = 0$ to $x = 4$.

3. Find the length of the arc of the curve $y^2 = x^3$ from $x = 0$ to $x = 5$.

4. Find the length of the arc of the catenary $y = \cosh x$ from the vertex to the point where $x = 1$.

5. Find the length of the arc of the curve $y = \log_e x$ between the points where $x = 1$ and $x = 2$. (For the integral see Ex. 28, No. 11.)

6. Find the length of the part of the curve of $y = \log\sec x$ between the values $x = 0$ and $x = \frac{\pi}{3}$.

7. Find the length of the circumference of the circle whose equation is $r = 2a\cos\theta$.

8. Find the length of the arc of the spiral of Archimedes, $r = a\theta$, between the points where $\theta = 0$ and $\theta = \pi$.

(*Note.*—The student should draw the curve.)

9. Find the length of the curve of the hyperbolic spiral $r\theta = a$ from $\theta = \frac{1}{2}$ to $\theta = 1$. (For the integral see Ex. 28, No. 13.)

10. Find the whole length of the curve of $r = a\sin^3\frac{\theta}{3}$.

CHAPTER XVI

SOLIDS OF REVOLUTION. VOLUMES AND AREAS OF SURFACES

166. Solids of revolution.

It is obvious that the methods of integration which enabled us to find areas of plane figures may be extended to the determination of the volumes of regular solids.

The solids with which we shall chiefly be concerned are those which are marked out in space when a regular curve or area is rotated about some axis. These are termed **Solids of Revolution.** For example, if a semi-circle is rotated about its diameter it will generate a sphere. Similarly, a rectangle rotated about one side will describe a cylinder in a complete rotation.

167. Volume of a cone.

The method employed for the determination of the volumes of solids or revolution can be illustrated by the example of a cone. If a right-angled triangle rotates completely about one of the sides containing the right angle as an axis, the solid generated is a **cone.**

Or, if a straight line, equation $y = mx$, is rotated about the x-axis (or y-axis) so that it makes a constant angle with the axis, it will generate a cone. Since the straight line passes through the origin, and is of undetermined length,

(1) The volume will be undetermined.
(2) The complete solid will be a double cone with the origin as a common apex.

Incidentally, if the complete cone be cut by a plane parallel to the x-axis, the section will be a hyperbola. Hence it is that the curve as stated in § 151, Example 7, has two symmetrical branches.

The volume becomes definite if an ordinate from a point on $y = mx$ is also rotated to enclose a definite portion of the cone. It is the volume of such a cone that we will proceed to determine.

In Fig. 62 let OA be the straight line $y = mx$, A being any point on it.

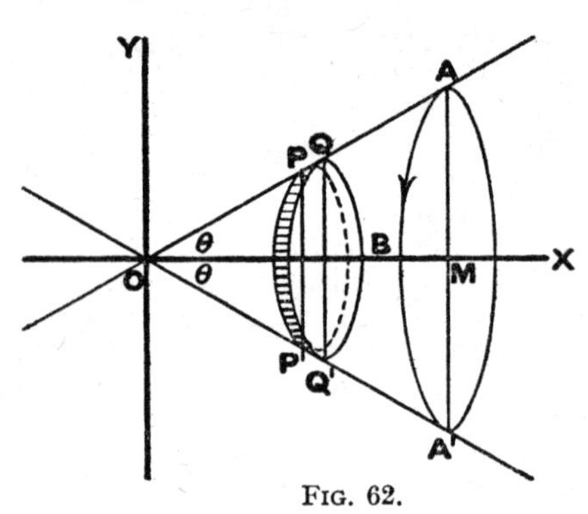

Fig. 62.

Let θ be the angle made with OX.

$\therefore \quad \tan\theta = m.$

Let OA rotate around OX so that the angle made with OX is always θ.

Let OA^1 be the position after half a complete rotation. Then A, and every other point on OA after a complete rotation, will describe a circle, and a cone will be generated with apex at 0.

Let AMA^1 be the double ordinate joining A and A^1. It is also a diameter of the circle formed by the rotation of A—viz., ABA^1.

Let V be the volume of the cone of which O is the vertex and the circle ABA^1 the base.

OM represents the height of the cone. Let this be h.

Let P be any point on OA and its co-ordinates (x, y).

Let x be increased by δx so that the corresponding point Q on OA has co-ordinates $(x + \delta x, y + \delta y)$.

PQ, on rotating, describes a small slice of the cone of which the ends are the circles described by P and Q.

The thickness of the slab is δx.

Its volume lies between the cylinders whose volumes are

$$\pi y^2 \delta x \quad \text{and} \quad \pi(y + \delta y)^2 \delta x.$$

Let Q become infinitely close to P, so that δx tends to become infinitely small and in limit is represented by dx.

Thus as $\delta x \longrightarrow 0$ the volume of the slice $\longrightarrow \pi y^2 dx$.

$\pi y^2 dx$ is therefore the element of volume.

$\therefore$ The volume of the cone is the sum of all such elements between the values $x = 0$ and $x = h$.

$$\therefore \quad V = \int_0^h \pi y^2 dx$$
$$= \pi \int_0^h (mx)^2 dx$$
$$= \pi m^2 \left[\frac{x^3}{3}\right]_0^h$$
$$= \tfrac{1}{3}\pi m^2 h^3.$$

But $$m = \tan\theta = \frac{AM}{h}$$

$$\therefore \quad V = \tfrac{1}{3}\pi \cdot \frac{AM^2}{h^2} h^3$$
$$= \tfrac{1}{3}\pi AM^2 h$$

or, if $AM = y_1$, the radius of the base

$$V = \tfrac{1}{3}\pi y_1^2 h$$

or volume of cone $= \frac{1}{3}$ (area of base $\times$ height).

168. General formula for volumes of solids of revolution.

(A) Rotation around the x-axis.

Let AB (Fig. 63) be part of a curve whose equation is $y = f(x)$.

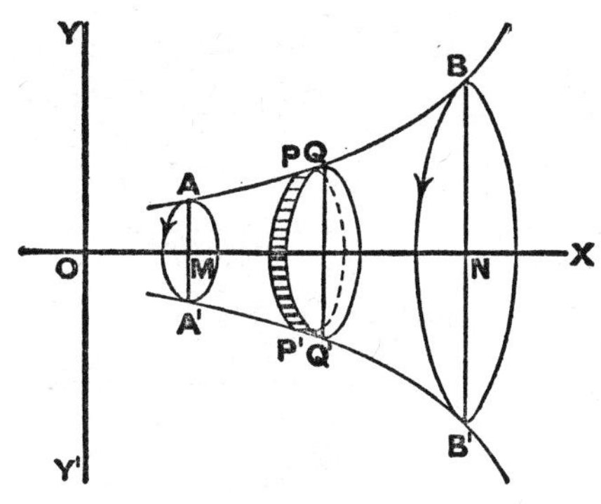

FIG. 63.

Let it rotate around OX, generating a solid which is depicted in the figure.

Let AMA^1, BNB^1 be double ordinates so that $OM = a$, $ON = b$.

Let $P\ (x, y)$ be any point on the curve.

Let x be increased by δx, so that Q, the corresponding point on the curve, is $(x + \delta x, y + \delta y)$.

Then, the volume of the slab described by PQ on rotation lies between $\pi y^2 \delta x$ and $\pi(y + \delta y)^2 \delta x$.

In the limit when Q is infinitely close to P,

$$\text{as } \delta x \longrightarrow 0, \text{ and } \delta y \longrightarrow 0, \text{ volume} \longrightarrow \pi y^2 dx.$$

The volume of the whole solid is the sum of all such slabs between the limits $x = a$ and $x = b$. Let V be this volume.

$$\therefore \quad V = \int_a^b \pi y^2 dx \quad . \quad . \quad . \quad . \quad (1)$$

Since $y = f(x)$ we can substitute for y in terms of x and integrate.

B. Rotation around the y axis.

Let AB (Fig. 64) be a portion of the curve of $y = f(x)$.

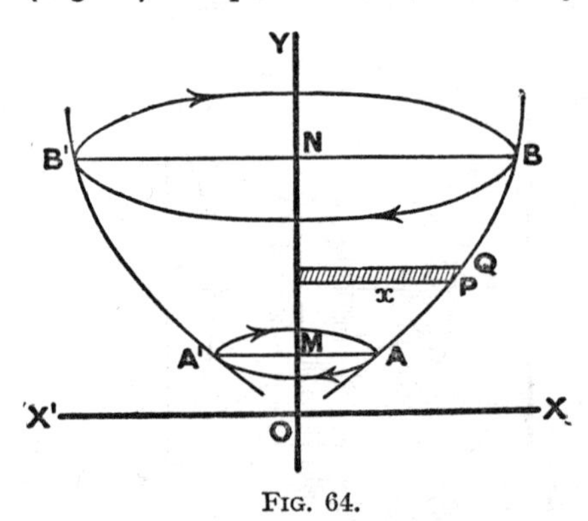

FIG. 64.

Let it rotate around OY so that A and B describe circles as indicated, centres M and N.

Let $OM = a$, $ON = b$.

Let P (x, y) be any point on the curve and Q another point with co-ordinates $(x + \delta x, y + \delta y)$.

Then, using the method of the previous example, the slab generated by PQ becomes, in the limit,

$$\pi x^2 dy.$$

$\therefore$ the volume of the whole solid is the sum of all such slabs between the limits $y = a$, $y = b$.

$$\therefore \quad V = \int_a^b \pi x^2 dy \quad . \quad . \quad . \quad . \quad . \quad (2)$$

From the equation $y = f(x)$, x can be found in terms of y and substituted in the integral.

169. Volume of a sphere.

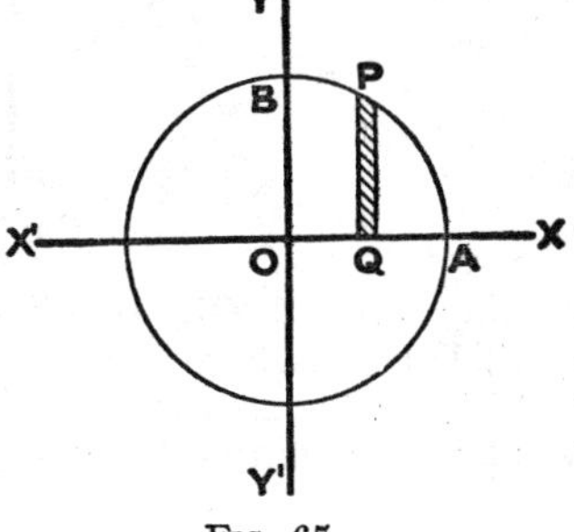

FIG. 65.

Let the equation of the circle in Fig. 65 be

$$x^2 + y^2 = a^2.$$

The centre is at the origin and radius $OA = a$.

Let the quadrant OAB be rotated about OX. The volume described will be that of a hemi-sphere.

Using formula (1) of the preceding section, and representing the volume of the sphere by V, we have:

$$V = 2 \times \int \pi y^2 dx$$

$$= 2\int_0^a \pi(a^2 - x^2)\,dx \quad . \quad . \quad . \quad . \quad . \quad (A)$$

$$= 2\pi\left[a^2x - \tfrac{1}{3}x^3\right]_0^a$$

$$= 2\pi(a^3 - \tfrac{1}{3}a^3) = 2\pi \times \tfrac{2}{3}a^3$$

$$= \frac{4}{3}\pi a^3$$

170. Volume of part of a sphere between two parallel planes.

In Fig. 66 let the quadrant OCD of the circle $x^2 + y^2 = r^2$

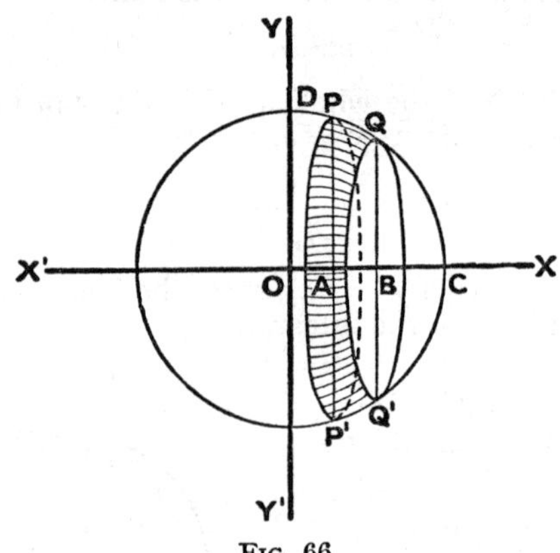

FIG. 66.

rotating around OX describe a hemi-sphere. Let two parallel planes whose distances from O are given by $OA = a$, $OB = b$, mark out the segment whose volume (V) is required. We may use equation (A) in the example above to express V.

$$\text{Then} \qquad V = \int_a^b \pi(r^2 - x^2)\,dx$$
$$= \pi\Big[r^2x - \tfrac{1}{3}x^3\Big]_a^b$$
$$= \pi[(r^2b - \tfrac{1}{3}b^3) - (r^2a - \tfrac{1}{3}a^3)]$$
$$= \pi\{r^2(b - a) - \tfrac{1}{3}(b^3 - a^3)\}$$
$$= \pi(b - a)\{r^2 - \tfrac{1}{3}(b^2 + ab + b^2)\}$$

If $b = r$ the part of the sphere becomes = **spherical cap.**

$$\text{Then} \qquad V = \pi(r - a)\{r^2 - \tfrac{1}{3}(r^2 + ar + a^2)\}.$$

Note.—When in this result $a = 0$, the spherical cap becomes a hemi-sphere, and the result is one-half of the volume of the sphere found above.

171. Volume of an ellipsoid of revolution.

This is the solid formed by the rotation of an ellipse

(1) about its major axis,

or (2) about its minor axis.

(1) Rotation about the major axis.

The rotation as shown in Fig. 67 is supposed to be about AA^1, *i.e.*, OX.

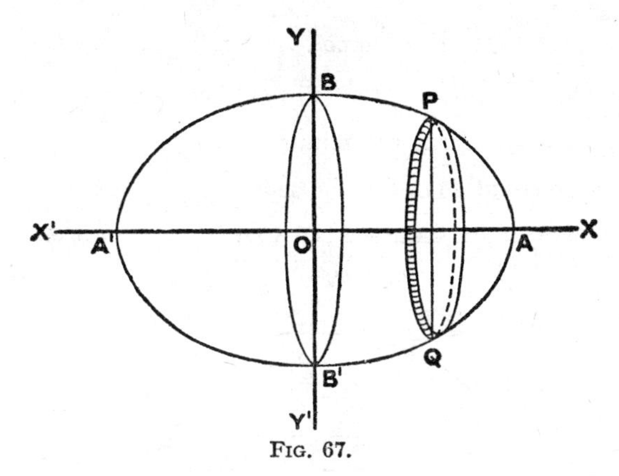

FIG. 67.

Consequently any section perpendicular to OX is a circle.
Let the equation of the ellipse be

$$\frac{x^2}{a^2} + \frac{y^2}{b^2} = 1.$$

$$\therefore \quad y^2 = \frac{b^2}{a^2}(a^2 - x^2).$$

Let V be the volume of the ellipsoid.

Consider the volume marked out by the rotation of the quadrant OAB, the limits being o and a.
Then, using formula (1) of § 168

This volume $\quad = \int_0^a \pi y^2 dx.$

$$\therefore \quad V = 2\int_0^a \pi\left\{\frac{b^2}{a^2}(a^2 - x^2)dx\right.$$

$$= \frac{2\pi b^2}{a^2}\int_0^a (a^2 - x^2)dx$$

$$= \frac{2\pi b^2}{a^2}\left[a^2x - \tfrac{1}{3}x^3\right]_0^a$$

$$= \frac{2\pi b^2}{a^2}(a^3 - \tfrac{1}{3}a^3)$$

$$\therefore \quad V = \frac{4}{3}\pi ab^2.$$

Note.—If $b = a$, the ellipsoid becomes a sphere.

(2) Rotation about the minor axis.

Let the equation of the ellipse be $\frac{x^3}{a^2} + \frac{y^3}{b^2} = 1.$

In this case, as indicated in Fig. 68, the rotation being

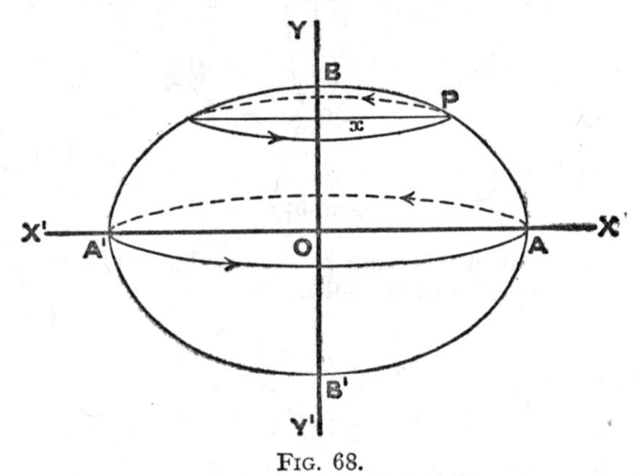

Fig. 68.

about OY, any point P (x, y) on the circumference of the ellipse will describe a circle radius x and centre on OY.

The area of such a circle is πx^2.

$\therefore$ Volume of slab between two such circles infinitely close together is

$$\pi x^2 dy.$$

∴ Using formula (2) of § 168 and considering the half of the ellipsoid above OX, we have, the limits of y being b and o:

Volume of half the ellipsoid

$$= \int_0^b \pi x^2 dy$$

∴ Volume of whole ellipsoid

$$= 2\pi \int_0^b x^2 dy$$

$$= 2\pi \int_0^b \frac{a^2}{b^2}(b^2 - y^2)dy$$

$$= 2\pi \cdot \frac{a^2}{b^2}\int_0^b (b^2 - y^2)dy$$

$$= \frac{2\pi a^2}{b^2}\left[b^2 y - \tfrac{1}{3}y^3\right]_0^b$$

$$= \frac{2\pi a^2}{b^2}(b^3 - \tfrac{1}{3}b^3)$$

$$\therefore \quad V = \frac{4}{3}\pi a^2 b.$$

The solid formed by the rotation of the ellipse about

(1) The **major** axis is called a **prolate spheroid.**
(2) The **minor** ,, **an oblate spheroid.**

Note.—The solid, not of revolution, in which those sections which are perpendicular to the plane of XOY, as well as those which are parallel to it are **all ellipses,** is called an **ellipsoid.**

172. Paraboloid of revolution.

This is the solid generated by the rotation of a parabola about its axis. It is not a closed curve, consequently we can obtain only the solid generated by part of the curve.

There are two cases.

(1) **When the axis of the parabola coincides with OX.**

The general form of the equation in this case is

$$y^2 = 4ax.$$

OP in Fig. 69 represents part of the curve.

P is any point on the curve, its co-ordinates being (x, y). PA is the ordinate of P, and $OA = c$. OP rotates around OX, generating a solid, with a circular base PQR.

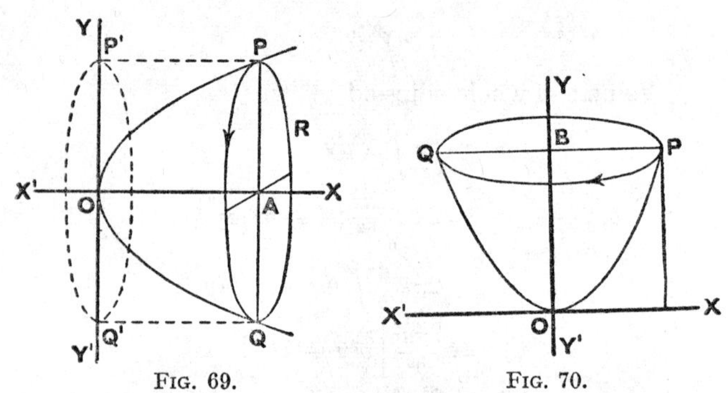

FIG. 69. FIG. 70.

As shown in § 167, the element of volume is πy^2dx, and the limits of x are 0 and c. Let V be the volume.

$$\therefore \quad V = \int_0^c \pi y^2 dx = \pi \int_0^c 4axdx$$
$$= \pi a \Big[2x^2\Big]_0^c$$
$$= 2\pi ac^2.$$

Note.—The cylinder indicated by the dotted lines in Fig. 69, having PRQ for one base, and a circle equal and parallel to it with O as centre, is the circumscribing cylinder of the paraboloid.

The volume of this cylinder $= \pi y^2 \times OA$
$$= \pi \times 4ac \times c$$
$$= 4\pi ac^2.$$

$\therefore$ Volume of the paraboloid equals half that of the circumscribing cylinder.

(2) When the axis of the parabola coincides with OY.

In Fig. 70, QOP represents part of a parabola, the equation of which is

$$y = ax^2.$$

Let P (x, y) be any point on the curve.
Let PB be its abscissa, so that

$$OB = b.$$

The element of volume as shown in § 168 (B) is

$$\pi x^2 dy.$$

The limits of y are 0 and b.
∴ using formula (2) of § 168

$$V = \int_0^b \pi x^2 dy$$
$$= \pi \int_0^b \frac{y}{a} dy$$
$$= \frac{\pi}{a}\left[\tfrac{1}{2}y^2\right]_0^b$$
$$= \tfrac{1}{2} \cdot \frac{\pi b^2}{a}.$$

Note.—Compare this with the volume of the circumscribing cylinder.

(3) **Parabola whose equation is $y = kx^2$ rotating about OX.**

The parabola does not rotate about its own axis, which coincides with OY, but with the other axis.

Let the curve OQP (Fig. 71) represent part of the curve of the function between the origin and $x = a$, where PM is the ordinate of P and $OM = a$.

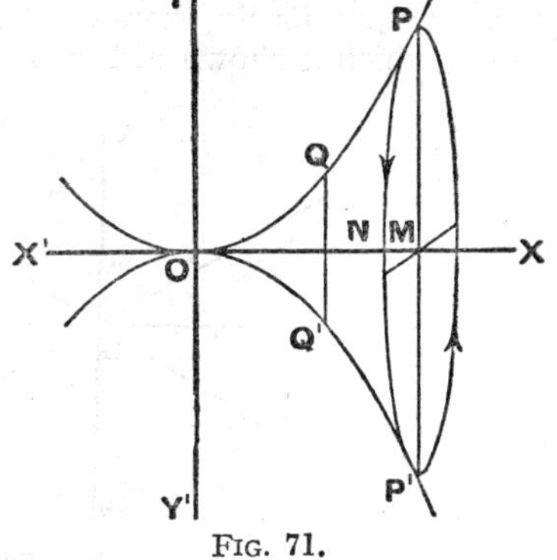

FIG. 71.

Let V be the volume generated by OP as the curve rotates around OX, occupying the position OQ^1P^1 after a half rotation.

Using formula (1) of § 168, the element of volume is $\pi y^2 dx$, and the limits of x are 0 and a.

$$\therefore \quad V = \int_0^a \pi y^2 dx$$
$$= \pi \int_0^a (kx^2)^2 dx$$
$$= \pi k^2 \left[\tfrac{1}{5}x^5\right]_0^a$$
$$= \tfrac{1}{5}\pi k^2 a^5.$$

If the part of the curve which is rotated is QP, where QN is the ordinate of Q and $ON = b$, then the volume generated is given by

$$V = \int_b^a \pi y^2 dx = \tfrac{1}{5}\pi k^2(a^5 - b^5).$$

173. Hyperboloid of Revolution.

This is the solid generated by the rotation of a hyperbola. It may take different forms.

(1) **Rotation about OX** of the curve whose equation is

$$\frac{x^2}{a^2} - \frac{y^2}{b^2} = 1.$$

Since there are two symmetrical branches of the curve, as shown previously, there will be two corresponding solids, one of which is shown in Fig. 72.

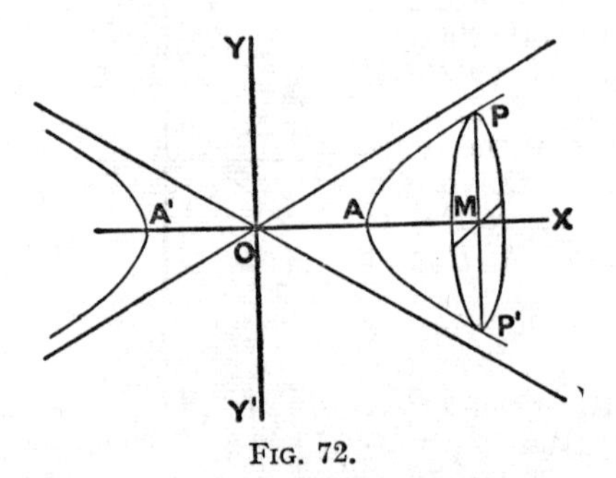

FIG. 72.

These two parts are called an **hyperboloid of two sheets.** Clearly, there will be no part of the solid between A

and A^1. Also there is no enclosed solid, but the volume can be found between sections corresponding to two values of x.

Let P be any point on the curve, and PM its ordinate.

Let $OM = c$.

Let V be the volume between the vertex A, where $x = a$ and $x = c$.

Then

$$V = \int_a^c \pi y^2 dx = \pi \int_a^c \frac{b^2(x^2 - a^2)}{a^2}$$

$$= \frac{\pi b^2}{a^2}\left[\tfrac{1}{3}x^3 - a^2x\right]_a^c = \frac{\pi b^2}{3a^2}(c^3 - 3a^2c - a^3 + 3a^3)$$

$$= \frac{\pi b^2}{3a^2}(c^3 - 3a^2c + 2a^3).$$

(2) **Rotation around OY.**

Let the equation be

$$\frac{x^2}{a^2} - \frac{y^2}{b^2} = 1.$$

The solid formed will be as indicated in Fig. 73.

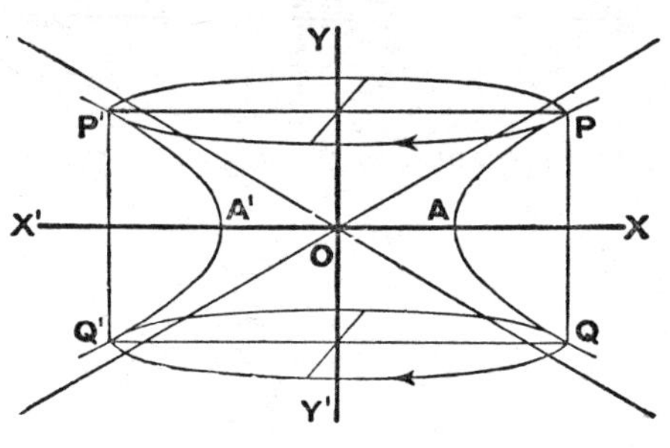

FIG. 73.

Since the two parts of the curve are symmetrical, any point P on the curve, after a half complete rotation, will coincide with the corresponding point P^1 on the other arm. This point, like every other point on the curve, will describe a circle.

The solid is therefore continuous, and is called a **hyperboloid of one sheet.** It stretches out infinitely around the y-axis, and any volume which has to be determined will be bounded by sections corresponding to two values of y, say y_1 and y_2.

This volume can be found as in former examples.

(3) **Rotation of rectangular hyperbola** about its asymptotes, which, as shown in § 151, Ex. 7, are the rectangular axes OX and OY. The equation of the curve is $xy = c^2$, and there are two parts of the solid, above and below OX.

The part of the volume contained between two sections parallel to one of the axes can be found in the usual way. Thus, if P and Q are two points on the curve (Fig. 74), and

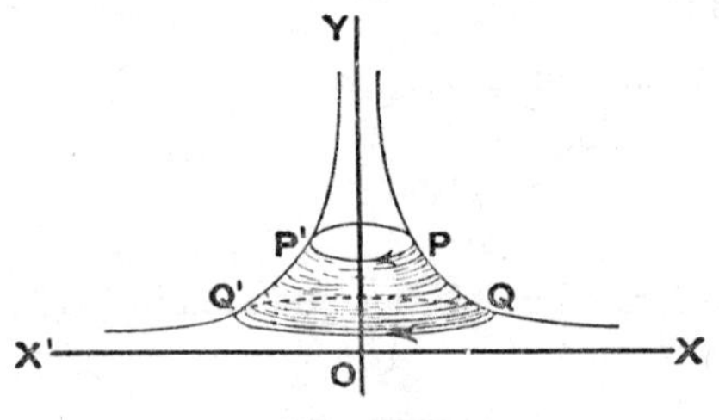

FIG. 74.

the corresponding values of y are y_1 and y_2, the volume would be given by

$$\int_{y_1}^{y_2} \pi x^2 dy.$$

Note.—Only that sheet of the hyperboloid which is above OX is shown. There is a second similar sheet below.

Exercise 36.

1. Find the volume generated by the arc of the curve $y = x^2$

(1) when it rotates round the x-axis between $x = 0$ and $x = 3$;

(2) when it rotates round the y-axis between $x = 0$ and $x = 2$.

2. Find the volume generated when an arc of the curve of $y = x^3$

(1) rotates round the x-axis between $x = 0$ and $x = 3$;

(2) rotates round the y-axis between $x = 0$ and $x = 2$.

3. Find the volume of the cone formed by the rotation round the x-axis of that part of the line $2x - y + 1 = 0$, intercepted between the axes.

4. The circle $x^2 + y^2 = 9$ rotates round a diameter which coincides with the x-axis. Find

(1) the volume of the segment between the planes perpendicular to OX whose distances from the centre, and on the same side of it, are 1 and 2;

(2) the volume of the spherical cap cut off by the plane whose distance from the centre is 2.

5. Find the volume generated by the rotation of the ellipse $x^2 + 4y^2 = 16$, about its major axis.

6. Find the volume generated by the rotation round the x-axis of the part of the curve $y^2 = 4x$ between the origin and $x = 4$.

7. Find the volume generated by rotating one branch of the hyperbola $x^2 - y^2 = a^2$ about OX, between the limits $x = 0$ and $x = 2a$.

8. Find the volume of the solid generated by the rotation round the y-axis of that part of the curve of $y^2 = x^3$ which is contained between the origin and $y = 8$.

9. Find the volume of the solid generated by the rotation about the x-axis of the part of the curve of $y = \sin x$, between $x = 0$ and $x = \pi$.

10. Find the volume generated by the rotation round the x-axis of the part of the curve of $y = x(x - 2)$ which lies below the x-axis.

11. If the curve of $xy = 1$ be rotated about the x-axis, find the volume generated by the part of the curve intercepted between $x = 1$, $x = 4$.

12. The parabolas $y^2 = 4x$ and $x^2 = 4y$ intersect and the area included between the curves is rotated round the x-axis. Find the volume of the solid thus generated.

174. Simpson's rule for volumes.

Simpson's rule for calculating the areas of irregular figures can be adapted to find the volume of an irregular solid. Thus, if the areas of the cross-sections of the solid at equal intervals are known, these can be plotted as ordinates of an irregular curve. For example, if in Fig. 58 of Exercise 34, each of the ordinates represents the area of a cross-section of the irregular solid and l represents the distance between the cross-sections, then the sum of their products, which are represented by areas in Fig. 58, will represent the volume of the solid. Just as by applying Simpson's rule in Example 34 we find the area of the irregular figure, so the products will now represent the volume of the irregular solid. In the particular example quoted the area was found to be 73·5 m², so, now, the **volume of the irregular solid is 73·5 m³.**

> *Note.*—When the values of the areas of sections are not known at equal intervals, those which are given should be drawn, the curve plotted and then the ordinates required should be drawn and measured.
>
> Examples can be found in books on practical mathematics, such as *National Certificate Mathematics*, Vol. II.

Areas of surfaces of solids of revolution.

175. Area of curved surface of right circular cone.

The curved surface of a right circular cone, if unrolled, is the sector of a circle. The problem is therefore that of determining the area of this sector, and this can be found by previous methods.

Let l = radius of the sector (*i.e.*, the slant side of the cone).

Let r = radius of circular base of cone.

Let A = area of curved surface of the cone.

Then it can readily be shown that

$$A = \pi r l.$$

Area of curved surface of a frustum of a cone.

Let the cone (Fig. 75) be cut by a plane, CD, parallel to the base.

Then $ABDC$ is a frustum of the cone.

The curved surface of the frustum can be considered as the limit of a very large number of small trapeziums, such as $PQRS$.

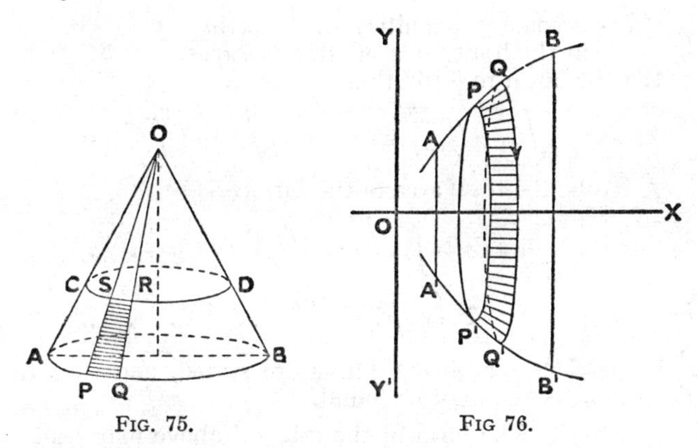

FIG. 75. FIG 76.

Using the formula for the area of a trapezium, in the limit, this sum—*i.e.*, the area of the curved surface of the frustum—is

$AC \times \frac{1}{2}$ (sum of circumferences of circles AB and CD).

$\therefore$ if $r =$ radius of base (AB)

and $r_1 =$ radius of section (CD).

$$\text{Area} = \tfrac{1}{2}AC \times 2\pi(r_1 + r)$$

176. General formula for area of a surface of revolution.

Let AB (Fig. 76) represent a portion of a curve which rotates round OX, generating a solid of revolution. We require to find an expression for the surface of this solid.

Let PQ be a small part of the curve, which on rotating generates a portion (shaded) of the surface of the whole.

Let $PQ = \delta s$.

P and Q, on rotation, describe circles, PP^1, QQ^1, with centres M and N on OX.

Lot PM y.

Then $QN = y + \delta y$.

If PQ be small the portion of the surface which it generates may be considered as surface of the frustum of a cone.

$\therefore$ as shown in § 175, its area is $2\pi \times \frac{y + (y + \delta y)}{2} \times \delta s$.

If PQ becomes indefinitely small so that $\delta y \longrightarrow 0$.
Then, in the limit, area of strip $= 2\pi y ds$.
It was shown in § 162 that

$$ds = \sqrt{1 + \left(\frac{dy}{dx}\right)^2} \cdot dx \quad \text{or} \quad \sqrt{1 + \left(\frac{dx}{dy}\right)^2} \cdot dy$$

$\therefore$ if s be the total area of the surface.

$$s = 2\pi \int y \sqrt{1 + \left(\frac{dy}{dx}\right)^2} \cdot dx \quad . \quad . \quad . \quad (1)$$

or

$$2\pi \int y \sqrt{1 + \left(\frac{dx}{dy}\right)^2} \cdot dy \quad . \quad . \quad . \quad (2)$$

In particular problems limits are stated, and thus the definite integrals may be found.

> *Note.*—The form of the integral above may lead to complicated integration, consequently only simple examples will be given.

Rotation round *OY*. The above formulae are used when rotation is round OX. If rotation be round OY, the following formulae may be used.

$$s = 2\pi \int x \sqrt{1 + \left(\frac{dy}{dx}\right)^2} \cdot dx \quad . \quad . \quad . \quad (3)$$

or

$$s = 2\pi \int x \sqrt{1 + \left(\frac{dx}{dy}\right)^2} \cdot dy \quad . \quad . \quad . \quad (4)$$

177. Area of the surface of a sphere.

Let $x^2 + y^2 = a^2$ be the equation of a circle which generates a sphere by rotation about OX, on which therefore lies a diameter.

Since
$$x^2 + y^2 = a^2$$
$$y = \sqrt{a^2 - x^2}.$$
$$\therefore \quad \frac{dy}{dx} = -\frac{x}{\sqrt{a^2 - x^2}}.$$

$$\therefore \quad ds = \sqrt{1 + \left(\frac{-x}{\sqrt{a^2 - x^2}}\right)^2} dx$$

$$= \sqrt{1 + \frac{x^2}{a^2 - x^2}} \cdot dx$$

$$= \sqrt{\frac{a^2}{a^2 - x^2}} \cdot dx$$

$$= \frac{a}{y} \cdot dx.$$

The limits of the integral when a quadrant rotates are o and a, giving rise to a hemisphere.

∴ using formula (1) above.

$$\text{Surface of hemi-sphere} = 2\pi \int_0^a y \times \frac{a}{y} dx$$

$$= 2\pi \int_0^a a dx$$

$$= 2\pi a \Big[x \Big]_0^a$$

$$= 2\pi a^2.$$

∴ Area of surface of sphere $= 4\pi a^2$.

Exercise 37.

1. Find the area of the surface of the solid generated by the rotation of the straight line $y = \frac{3}{4}x$ around the x-axis, between the values $x = 0$ and $x = 3$.

2. Find the area of the surface generated by the rotation about OX of the curve of $y = \sin x$, between $x = 0$ and $x = \pi$.

3. That part of the curve of $x^2 = 4y$ which is intercepted between the origin and the line $y = 8$ is rotated around OY. Find the area of the surface of the solid which is generated.

4. The curve of the function $x^{\frac{2}{3}} + y^{\frac{2}{3}} = a^{\frac{2}{3}}$ rotates around OX. Find the area of the surface of the solid which is formed between $x = 0$ and $x = a$.

5. Find the area of surface of the zone cut off from a sphere of radius r by two parallel planes, the distance between which is h.

6. Find the area of the surface of the solid generated by rotating round OX the part of the curve $y = x^3$, between $x = 0$ find $x = 1$.

CHAPTER XVII

USES OF INTEGRATION IN MECHANICS

I. Centre of Gravity.

178.

Integration, as a method of summation, can be applied to the solution of many problems in mechanics in which it is required to find the sum of an infinite number of infinitesimally small products. Some of these are included in this chapter, but in a volume of this size and purpose only a few of the simpler examples can be given.

179. The centre of gravity of a number of particles.

It is shown in treatises on mechanics that if a series of parallel forces acts upon a body, the point through which their resultant can be considered as acting is called the Centre of Force; also the resultant is the algebraical sum of these parallel forces.

This can be otherwise expressed as follows:

Let $m_1, m_2, m_3 \ldots$ be the masses of a number of particles.

Let $(x_1, y_1), (x_2, y_2), (x_3, y_3) \ldots$ be the co-ordinates of the positions of the particles with reference to two rectangular axes, OX, OY.

Each of the particles is acted upon by the **force of gravity**, this force being termed the **weight** of the particle and being proportional to its mass.

Since this force is always directed towards the centre of the earth, these forces, in a small system of particles, may be considered as a system of parallel forces, which can be denoted by

$$m_1g, m_2g, m_3g, \ldots$$

or

$$w_1, w_2, w_3, \ldots$$

where w represents the weight of a particle.

The centre of force of this system is the centre of gravity of the particles.

Let the co-ordinates of the centre of gravity be $(\bar{x}, \bar{y})$.

Let M be the sum of the masses of the particles.

i.e., $M = m_1 + m_2 + m_3 + \dots$

or $M = \Sigma(m)$.

The product of the mass and the distance of the particle from any point or axis, is called the moment of the force about that point or axis.

It is established in mechanics that **the moment about any axis of the resultant acting at the centre of force is equal to the sum of the moments of the particles about the same axis.**

$\therefore$ considering the system of particles above and taking moments about OY

$$Mg\bar{x} = m_1gx_1 + m_2gx_2 + m_3gx_3 = \dots$$

or, dividing throughout by g

$$M\bar{x} = m_1x_1 + m_2x_2 + m_3x_3 +$$

$\therefore$ with the usual algebraic notation

$$\bar{x} = \frac{\Sigma(mx)}{\Sigma(m)}.$$

Similarly, considering the moments about OX

$$\bar{y} = \frac{\Sigma(my)}{\Sigma(m)}.$$

The point $(\bar{x}, \bar{y})$, the moments of which we have found, is the **centre of mass** of the system, or considering the masses as acted upon by the force of gravity, the **centre of gravity** (c.g.) of the system.

180. The centre of gravity of a continuous body.

In the above section we have considered the c.g. of a system of particles irrespective of their distances from one another. But a continuous solid body can be regarded as made up of an infinite number of infinitely small particles, and the centre of gravity of these is the centre of gravity of the body.

As the moment of each of these particles about an axis is the product of its mass and its distance from the axis, the problem of finding the sum of these products at once suggests integration as the means of effecting it. The method of applying integration is most easily shown by examples, such as those which follow.

It should be noted that c.g. of a body must clearly lie

upon any axis of symmetry which the body possesses. For example, the c.g. of a solid of revolution must clearly lie on the axis about which the revolution takes place. This suggests that for the purpose of finding the c.g. it will generally be simpler to take the axis of revolution as a co-ordinate axis.

181. To find the centre of gravity of a uniform semi-circular lamina.

The c.g. evidently lies upon the radius which is perpendicular to the diameter of the semi-circle at its centre, *i.e.* on OA in Fig. 77. This line should therefore be taken as the x-axis and the diameter as the y-axis.

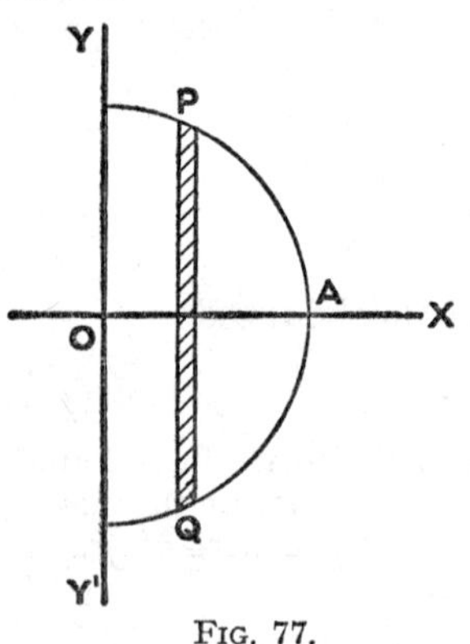

Fig. 77.

If the radius of the circle is a its equation is

$$x^2 + y^2 = a^2.$$

Since the lamina is uniform, its mass, or that of any part of it, can be represented by its area. If m be the mass of unit area, it will occur on both sides of the equations found in § 179, and so will cancel out.

Let $\bar{x}$ be the distance of the c.g. from O, along OX.

If a narrow strip of width δx be considered, at a distance x from OY, such as is indicated by PQ in Fig. 77, then

$$\text{area of the strip} = 2y \,.\, \delta x$$

and

$$\text{moment of the strip} = 2y\delta x \times x.$$

In the limit when the width of each strip becomes indefinitely small,

Sum of areas of strips, *i.e.*, area of semicircle

$$= \int_0^a 2y dx$$

also sum of moments of these strips

$$= \int_0^a 2y dx \times x = \int_0^a 2yx dx.$$

$\therefore$ by the principle of moments

$$\therefore\ \bar{x} \times \int_0^a 2y\,dx = \int_0^a 2xy\,dx.$$

$$\therefore\qquad \bar{x} = \int_0^a 2xy\,dx \div \int_0^a 2y\,dx.$$

But $$y = \sqrt{a^2 - x^2}.$$

$$\therefore\qquad \bar{x} = \int_0^a 2x\sqrt{a^2 - x^2}\,.\,dx \div \int_0^a 2\sqrt{a^2 - x^2}\,.\,dx$$

$$= \left[-\tfrac{2}{3}(a^2 - x^2)^{\frac{3}{2}}\right]_0^a \div \tfrac{1}{2}\pi a^2 \quad (\S\,151, \text{Ex. } 3)$$

$$= \tfrac{2}{3}a^3 \div \tfrac{1}{2}\pi a^2.$$

$$\therefore\qquad \bar{x} = \frac{4a}{3\pi}.$$

182. To find the centre of gravity of a solid hemisphere.

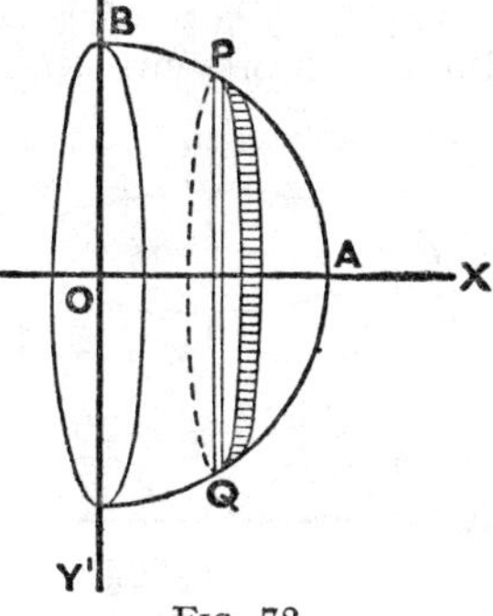

FIG. 78.

Let the semi-circle of the preceding example rotate about OX, thus generating a hemi-sphere. The c.g. will lie on the axis of rotation, OX.

Let $\bar{x}$ be its distance from O.

Equation of curve is

$$x^2 + y^2 = a^2.$$

$\therefore$ radius of circle $= a$.

The rectangle PQ of the preceding example on rotating will generate a slab, which, when the width of the rectangle is very small can be considered as cylindrical.

$$\therefore \text{ in the limit this volume} = \pi y^2 dx.$$

$$\therefore \text{ volume of hemi-sphere} = \int_0^a \pi y^2 dx$$

$$\text{moment of cylindrical slab} = \pi y^2 dx \times x.$$

$$\therefore \text{ sum of moments of all such slabs} = \int_0^a \pi y^2 x\,dx.$$

$$\text{Also moment of hemi-sphere} = \bar{x} \times \int_0^a \pi y^2 dx$$

But these are equal.

$$\therefore \bar{x} \times \int_0^a \pi y^2 dx = \int_0^a \pi y^2 x dx.$$

$$\therefore \quad \bar{x} = \pi \int_0^a x(a^2 - x^2)dx \div \int_0^a (a^2 - x^2)dx$$

$$= \pi\left[\tfrac{1}{2}a^2x^2 - \tfrac{1}{4}x^4\right]_0^a \div \tfrac{2}{3}\pi a^3 \qquad (\S\ 169)$$

$$= \pi(\tfrac{1}{2}a^4 - \tfrac{1}{4}a^4) \div \tfrac{2}{3}\pi a^3$$

$$= \tfrac{1}{4}\pi a^4 \div \tfrac{2}{3}\pi a^3$$

$$\therefore \quad \bar{x} = \tfrac{3}{8}a.$$

183. Centre of gravity of paraboloid generated by the rotation of the curve of $y = x^2$, about OY.

Let the limits of x be 0 and 2. When $x = 2$, $y = 4$. Fig. 79 represents the solid generated by the rotation about OY of that part of the parabola $y = x^2$ between the values $x = 0$ and $x = 2$ (see § 172).

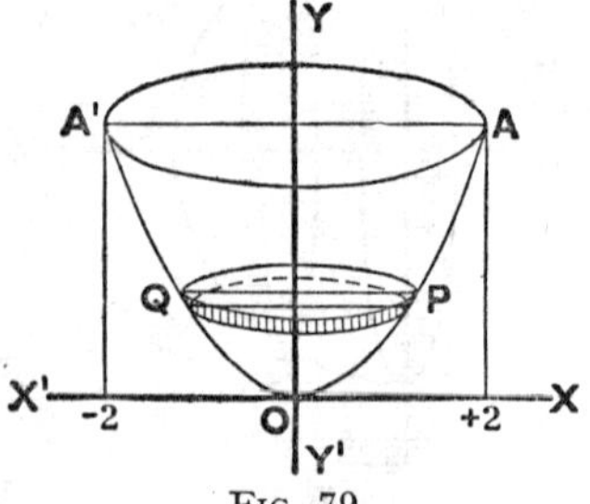

Fig. 79.

The c.g. lies on OY.

Let its distance from O be $\bar{y}$.

PQ represents a small cylindrical slab, formed, as in the preceding example, by the rotation of a rectangle of very small width.

Let the co-ordinates of P be (x, y).

In the limit when width of rectangle becomes infinitely small

$$\text{Volume of slab} = \pi x^2 dy.$$

$$\text{Moment of slab about } OX = \pi x^2 dy \times y$$

$$\therefore \text{sum of moments for all such slabs} = \int_{y=0}^{y=4} \pi x^2 dy \times y \quad (1)$$

$$\text{Volume of the whole solid} = \int_0^4 \pi x^2 dy.$$

$$\therefore \text{moment of whole solid} = \bar{y} \times \int_0^4 \pi x^2 dy \quad (2)$$

Equating (1) and (2)

$$\bar{y} \times \int_0^4 \pi x^2 dy = \int_0^4 \pi x^2 y dy.$$

$$\therefore \quad \bar{y} = \int_0^4 \pi y^2 dy \div \int_0^4 \pi y dy \quad (\text{since } x^2 = y)$$

$$= \pi\left[\tfrac{1}{3}y^3\right]_0^4 \div \pi\left[\tfrac{1}{2}y^2\right]_0^4$$

$$= (\tfrac{1}{3} \times 4^3) \div (\tfrac{1}{2} \times 16)$$

$$= \frac{8}{3}.$$

$\therefore$ the c.g. is $\frac{8}{3}$ units from 0 along OY.

Note.—This is $\frac{2}{3}$ the height of the solid.

184. Centre of gravity of a uniform circular arc.

Let BAC (Fig. 80) represent a circular arc.

Let r = radius of arc, centre O.

Let 2α = angle subtended at the centre.

Draw OA bisecting this angle.

Let OA be the x-axis.

The c.g. of the arc must lie on OA.

Let $\bar{x}$ = distance of c.g. from O.

Let P be the point (x, y), and PQ be a small arc subtending an angle $\delta\theta$ at O.

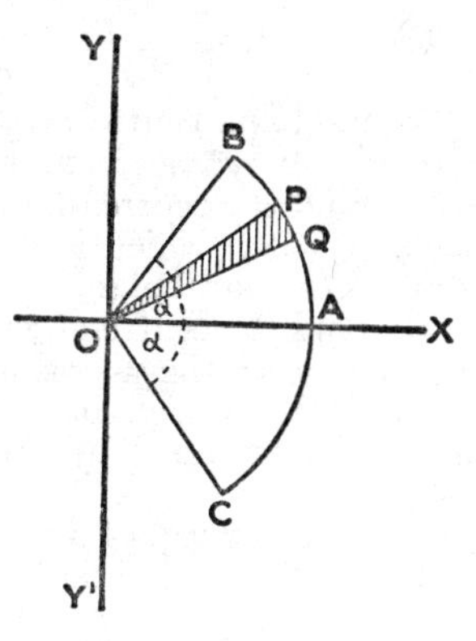

Fig. 80.

Then $PQ = r \, . \, \delta\theta$.

The c.g. of all such arcs as PQ must lie on OA.

$\therefore$ moment of PQ about $O = r\delta\theta \times x$ and $x = r \cos \theta$.

$\therefore$ „ PQ „ $O = r^2 \cos \theta \, . \, \delta\theta$.

In the limit when PQ is taken infinitely small

Moment of $PQ = r^2 \cos \theta d\theta$.

Mass of arc $BC = r \times 2\alpha$

(representing mass by length as arc is uniform)

$$\text{moment of arc} = \bar{x} \times r \times 2\alpha.$$

Equating moments

$$\bar{x} \times 2r\alpha = \int_{-\alpha}^{\alpha} r^2 \cos \theta \,.\, d\theta$$

$$= 2\int_{0}^{\alpha} r^2 \cos \theta \, d\theta$$

$$\therefore \quad \bar{x} \times 2r\alpha = 2r^2\Big[\sin \theta\Big]_0^\alpha$$

$$= 2r^2 \sin \alpha.$$

$$\therefore \quad \bar{x} = \frac{2r^2 \sin \alpha}{2r\alpha}.$$

$$\therefore \quad \bar{x} = \frac{r \sin \alpha}{\alpha}.$$

Exercise 38.

1. Find the centre of gravity of the parabolic segment bounded by $y^2 = 4ax$ and the line $x = b$.
2. Find the centre of gravity of the segment of the parabola $y^2 = 8x$, which is cut off by the line $x = 5$ and the axis of the parabola.
3. Find the centre of gravity of the area bounded by the curve $y = x^3$, the y-axis, and the line $y = 1$.
4. Find the c.g. of the parabolic segment of $y = x^2$, which is contained by the curve, the y-axis, and the line $y = 9$.
5. Find the c.g. of a quadrant of a circle, radius r.
6. Find the c.g. of the area between the curve of $y = \sin x$, and the x-axis from $x = 0$ to $x = \pi$.
7. Find the c.g. of a thin uniform wire in the shape of a semi-circle, radius r.
8. Find the c.g. of a thin uniform wire in the shape of a quadrant of a circle, radius r.
9. Find the c.g. of the circular sector shown in Fig. 80 as $OBAC$.
10. Find the c.g. of the right circular cone formed by the rotation of the line $y = mx$ about the origin to $x = h$.
11. Find the c.g. of a quadrant of an ellipse whose diameters are $2a$ and $2b$.

12. Find the c.g. of the area bounded by the hyperbola $xy = \kappa^2$, the x-axis, and the ordinates $x = a$, $x = b$.

13. Find the c.g. of the solid formed by the rotation of $y = x^2$ about the x-axis between the origin and $x = 3$.

14. If the portion of the curve of $ay^2 = x^3$ which is bounded by the curve, the x-axis and the ordinate $x = b$, rotates about the x-axis, find the c.g. of the solid thus generated.

MOMENTS OF INERTIA AND RADIUS OF GYRATION

185. Moments of Inertia.

Let $m_1, m_2, m_3, \ldots$ be the masses of a series of particles forming a system.

Let $r_1, r_2, r_3, \ldots$ be their distances from a given straight line or axis.

Then the sum of the products

$$m_1r_1^2, m_2r_2^2, m_3r_3^2 \ldots \text{ or } \Sigma(mr^2)$$

is called the **moment of inertia of the system,** and is usually denoted by M.I. or I.

It is also called the **second moment** of the system, while $\Sigma(mr)$, which was defined in § 179, is called the **first moment.**

As was pointed out when considering centre of gravity (§ 179), a continuous rigid body can be regarded as a system of infinitely small particles which, with the usual notation, can be expressed by dm.

The sum of the products or second moments then becomes Σr^2dm. This sum, taken throughout the body, becomes in the limit the integral $\int r^2dm$.

$$\therefore \quad \text{M.I.} = \int r^2dm.$$

The moment of inertia becomes of great importance when the body is rotating about an axis.

Suppose a body of mass M to be moving in a straight line with velocity v. Then its

$$\text{Kinetic Energy} = \tfrac{1}{2}Mv^2.$$

Thus the Kinetic Energy of any particle is $\frac{1}{2}v^2dm$.

Now suppose a body of mass M to be rotating with angular velocity ω about an axis.

Then a particle dm is moving at any given instant with linear velocity v where $v = r\omega$.

Its kinetic energy is $\frac{1}{2}dmv^2$

i.e., $$\frac{1}{2}dm(r\omega)^2$$

$\therefore$ the total kinetic energy of the body is

$$\text{K.E.} = \int \tfrac{1}{2}(r\omega)^2 dm = \tfrac{1}{2}\omega^2 \int r^2 dm$$
$$= \tfrac{1}{2}\omega^2 \times \text{M.I.}$$

$\therefore$ **Total kinetic energy $= \frac{1}{2}$(moment of inertia) $\times\ \omega^2$.**

186. Radius of gyration.

If the moment of inertia be written in the form

$$I = M\kappa^2$$

so that $$\kappa = \sqrt{I \div M}.$$

then κ **is called the radius of gyration of the body.**

From these statements it is clear that—

The kinetic energy of a body and the moment of inertia are the same as if the whole mass were supposed to be concentrated at a point whose distance from the axis of rotation is κ.

187. Worked examples.

Example 1. *Find the moment of inertia and the radius of gyration of a uniform straight rod about an axis perpendicular to the rod at its centre.*

Let M be the mass of the rod.

Let $2a$ be its length.

Since the rod is uniform, its mass per unit length is $\dfrac{M}{2a}$.

Let O (Fig. 81) be the centre of the rod, and OY the perpendicular through O.

It is required to find the M.I. of the rod about OY.

Let PQ be a small element δx, and $OP = x$. Then PQ has mass $\dfrac{M}{2a}\delta x$.

$\therefore$ M.I. of the element PQ about O is $\dfrac{M\delta x}{2a}x^2$.

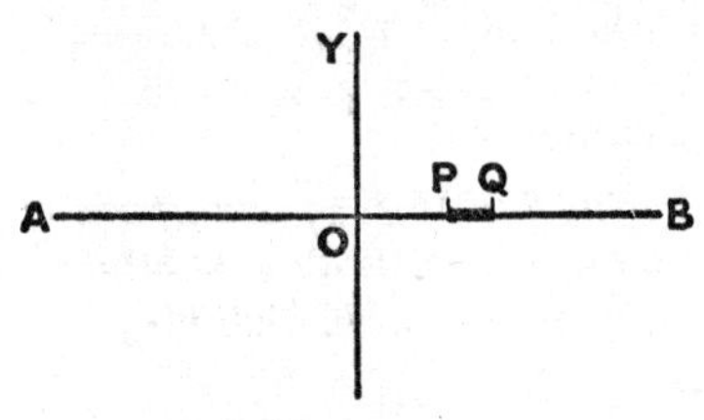

FIG. 81.

In the limit, when this element becomes indefinitely small,

$$\text{M.I. of the whole rod} = \int_{-a}^{a} \frac{Mx^2}{2a} dx$$

$$= \frac{M}{2a}\left[\frac{x^3}{3}\right]_{-a}^{a}$$

$$= \tfrac{1}{3}Ma^2$$

$$\therefore \quad I = \tfrac{1}{3}Ma^2.$$

Since $$Mk^2 = \tfrac{1}{3}Ma^2$$

$$\therefore \quad k = \frac{a}{\sqrt{3}}.$$

Example 2. *Find the M.I. of a uniform rectangular lamina of mass M, about an axis which bisects two opposite sides.*

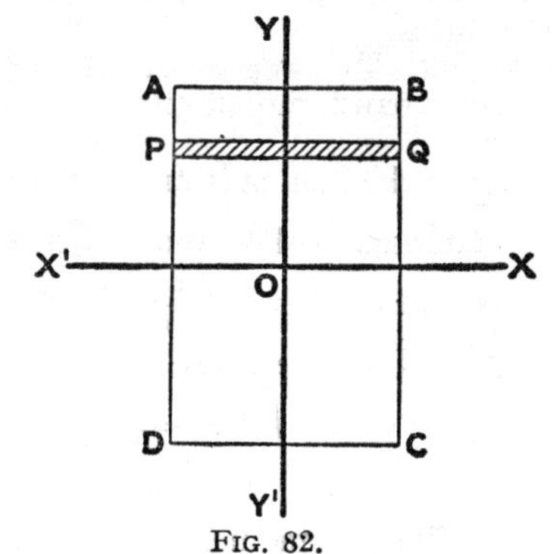

FIG. 82.

Let $ABCD$ (Fig. 82) represent the rectangle.

Let YOY^1 be the axis about which the M.I. is to be found.

Let $AB = 2a$.

Consider a thin strip PQ of mass M_1.

By Example 1 its M.I. $= \frac{1}{3}M_1a^2$.

The M.I. of the whole rectangle is equal to the sum of all such strips.

i.e., $\text{M.I.} = \frac{1}{3}(M_1a^2 + M_2a^2 + M_3a^2 + \ldots)$
$= \frac{1}{3}a^2(M_1 + M_2 + M_3 + \ldots$
$= \frac{1}{3}Ma^2.$

Example 3. *Find the M.I. of a uniform circular lamina of radius* r *and mass M, about an axis through its centre and perpendicular to the plane of the lamina.*

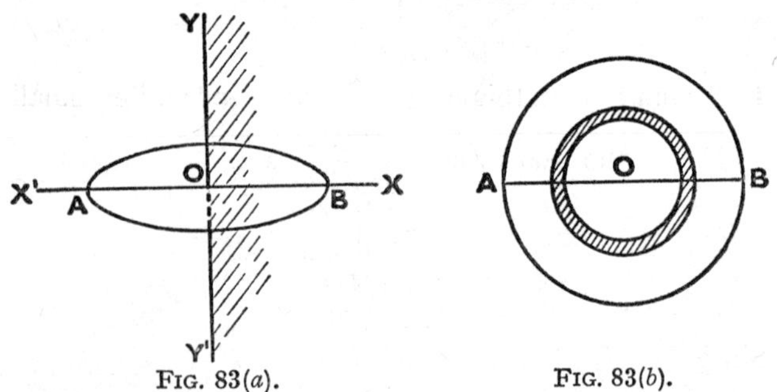

FIG. 83(*a*). FIG. 83(*b*).

Fig. 83(*a*) represents the circle, centre *O*, *OY* being the axis, perpendicular to the plane of the circle, about which it rotates.

Fig. 83(*b*) represents the plan of the circle. A small circular band, radii x and $x + \delta x$, represents the element of area.

Since the lamina is uniform, the mass per unit area is $\frac{M}{\pi r^2}$.

The mass of the band is $2\pi x\delta x \cdot \frac{M}{\pi r^2}$ (to the first order).

M.I. of the band is $\frac{2M}{r^2}x^3\delta x$.

$\therefore$ M.I. of the whole

$$= \frac{2M}{r^2}\int_0^r x^3dx = \frac{2M}{r^2}\left[\frac{x^4}{4}\right]_0^r = \frac{2M}{r^2} \cdot \frac{r^4}{4}$$

Exercise 39.

1. Find the moment of inertia and the radius of gyration of a uniform straight rod, length l, about an axis perpendicular to its length at one end of the rod.

2. Find the M.I. of a uniform rectangular lamina of sides $2a$ and $2b$ about the side of length $2b$.

3. Find the M.I. of a uniform circular lamina of radius r about a diameter.

4. Find the M.I., about OX, of the ellipse whose equation is

$$\frac{x^2}{a^2}+\frac{y^2}{b^2}=1.$$

5. Find the M.I. of an isosceles triangle, height h about

(1) its base;

(2) an axis through its vertex parallel to the base.

6. Find the M.I. of a right circular cone, radius of base r about its axis.

7. Find the M.I. of a uniform circular cylinder, radius of base r, about its axis.

8. Find the M.I. of a fine circular wire, radius a, about a diameter.

9. Find the M.I. about OY of the area of the segment of the parabola $y^2 = 4ax$ between the origin and the double ordinate corresponding to $x = b$.

10. Find the M.I. and radius of gyration of a uniform sphere, radius r, about a diameter.

188. Theorems on moments of inertia.

The following theorems are helpful in the calculation of moments of inertia in certain cases.

I. The moment of inertia of a lamina about an axis OZ perpendicular to its plane, is equal to the sum of the moments of inertia about any pair of rectangular axes OX and OY in the plane of the lamina.

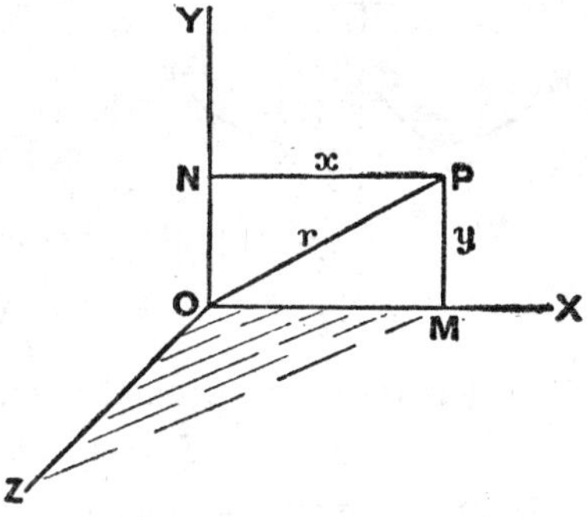

FIG. 84.

Let P be a particle of mass m in the plane of OX, OY (Fig. 84).

Let its co-ordinates with regard to these axes be (x, y).
Join OP. Let $OP = r$.
Let OZ be an axis perpendicular to the plane XOY.
Then POZ is a right angle.
$\therefore$ moment of inertia of particle at P about axis $OZ = mr^2$.
Let PM, PN be drawn perpendicular to OX, OY.

Then $$OP^2 = OM^2 + MP^2$$
$$= x^2 + y^2$$
or $$r^2 = x^2 + y^2.$$

But M.I. of mass m at P about OZ

$$= mr^2$$
$$= m(x^2 + y^2)$$
$$= mx^2 + my^2$$
$$= I_x + I_y$$
or $$I_z = I_x + I_y.$$

where I_x, I_y, I_z are the moments of inertia of m about the areas OX, OY and OZ, respectively.

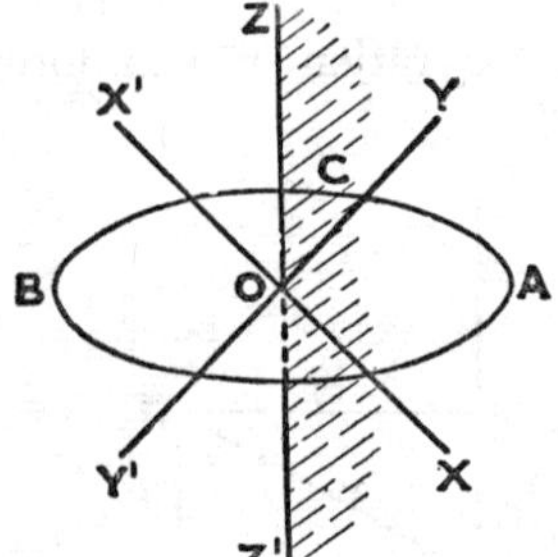

FIG. 85.

This is true for all particles of a lamina of which the particle at P is a part, and is therefore true for the whole lamina.

As an example let us consider the case of the circular lamina described in Exercise 39, question 3.

Let ABC (Fig. 85) represent a circular lamina.

Let XOX^1 be a diameter.

If I_x be the M.I. about this diameter, then it was found in question 3, that

$$I_x = \tfrac{1}{4}Mr^2.$$

If YOY^1 be another diameter at right angles to XOX^1,

then $$I_y = \tfrac{1}{4}Mr^2$$
$$\therefore \quad I_x + I_y = \tfrac{1}{4}Mr^2 + \tfrac{1}{4}Mr^2$$
$$= \tfrac{1}{2}Mr^2.$$

If OZ be an axis perpendicular to the lamina and therefore perpendicular to OX and OY, then, it was shown in Example 3, p. 310, that

$$I_z = \tfrac{1}{2}Mr^2.$$

Hence $$I_z = I_x + I_y.$$

II. Theorem of parallel axes.

Let I_c be the M.I. of a mass M about an axis through its centre of gravity; let a be the distance of a parallel axis from the centre of gravity. Then

$$\text{M.I.} = I_c + Ma^2.$$

This may be defined as follows:

The moment of inertia of a body about any axis is equal to the sum of—

(1) the moment of inertia about a parallel axis, and

(2) the product of the mass and the square of the distance of the axis from the centre of gravity.

It is evident that (2), *i.e.*, Ma^2, is the same as the M.I. of the whole mass, collected at the centre of gravity, about the selected axis.

189. Worked examples.

Example 1. *Find the M.I. of a uniform circular lamina, radius a, about a tangent.*

In Fig. 86 the tangent to the circular lamina centre C is taken as OY.

BC is an axis parallel to OY through C, which is, of course, the c.g.

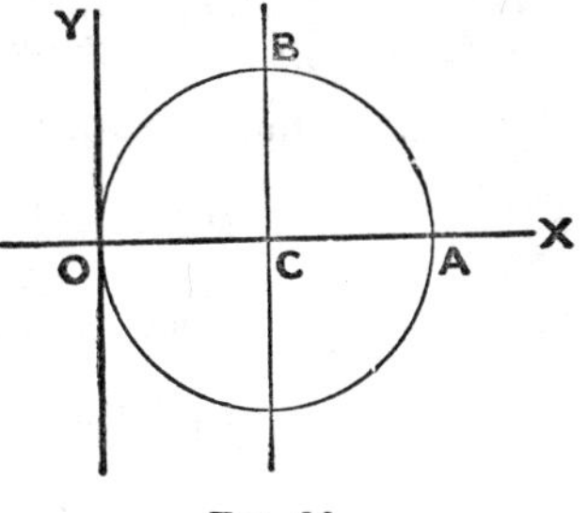

FIG. 86.

Then, by the above theorem

$$\text{M.I. about } OY = \text{M.I. about } BC + Ma^2.$$

But M.I. about $BC = \tfrac{1}{4}Ma^2$. (Ex. 39, (3), and example on Theor. I.)

$$\therefore \text{ M.I. about } OY = \tfrac{1}{4}Ma^2 + Ma^2$$
$$= \tfrac{5}{4}Ma^2.$$

Example 2. *Find the M.I. of a uniform lamina in the shape of an isosceles triangle, height* h *and vertical angle* 2α, *about—*

(1) *An axis through the vertex parallel to the base.*
(2) *A line through the c.g. parallel to the base.*
(3) *The base.*

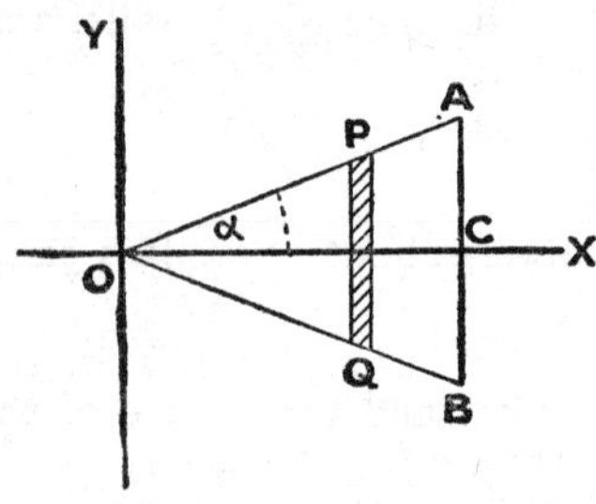

FIG. 87.

Arrange the triangle so that its axes of symmetry lies along OX (as in Fig. 87).

Then $OC = h$, $AC = h \tan \alpha$.

Let $P (x, y)$ be any point on OA.

The strip PQ represents an element of area

and $y = x \tan \alpha$.
δx = width of strip

(1) *To find M.I. about OY.*

Let m = mass of unit area.
M.I. of strip $= 2my\delta x \times x^2$

$\therefore$ in the limit, M.I. of triangle about

$$OY = m\int_0^h 2yx^2dx$$

but $y = x \tan \alpha$.

$$\therefore \quad \text{M.I.} = m\int_0^h 2x^3 \tan \alpha dx = 2m \tan \alpha \int_0^h x^3dx$$

$$= 2m \tan \alpha \left[\tfrac{1}{4}x^4\right]_0^h$$

$$= \tfrac{1}{2}mh^4 \tan \alpha$$

but mass of triangle, *i.e.*,

$$M = mh \times h \tan \alpha$$

$$= mh^2 \tan \alpha.$$

$$\therefore \quad \text{M.I.} = \tfrac{1}{2}Mh^2.$$

(2) *M.I. about an axis through the centre of gravity and parallel to AB.*

Let I_c = M.I. about an axis through c.g.
Let a = distance of c.g. from O.
In this case $a = \frac{2}{3}h$.
Using $I = I_c + Ma^2$, and substituting.

$$I_c = I - M(\tfrac{2}{3}h)^2$$
$$= \tfrac{1}{2}Mh^2 - \tfrac{4}{9}Mh^2$$
$$= \tfrac{1}{18}Mh^2.$$

(3) *M.I. about the base.*

Distance of c.g. from base $= \frac{1}{3}h$.
By the theorem of parallel axes
M.I. about base

$$= \text{(M.I. about axis through c.g.)} + \left(M \times \left(\frac{h}{3}\right)^2\right)$$
$$= \tfrac{1}{18}Mh^2 + \tfrac{1}{9}Mh^2$$
$$= \tfrac{1}{6}Mh^2.$$

Exercise 40.

1. Find the M.I. of a uniform rod, length $2a$, about an axis perpendicular to the rod through one extremity.
2. Find the M.I. of a uniform square lamina about an axis perpendicular to the plane of the square, at one corner.
3. Find the M.I. of a uniform lamina in the shape of an equilateral triangle of side a.

   (1) About a line parallel to the base through the centre of gravity.

   (2) About an axis through the centre of gravity and perpendicular to the plane of the triangle.

   (3) About a line perpendicular to the plane of the triangle and through a vertex.

4. Find the M.I. of a uniform circular lamina of radius a about an axis perpendicular to the plane of the disc through a point on the circumference.
5. Find the M.I. of a uniform right circular cylinder about a line through the centre of the axis of the cylinder and perpendicular to it. Length of cylinder is $2a$ and radius of base b.
6. Find the M.I. of a uniform thin spherical shell, radius

a, about a diameter. [Hint—see problem of finding surface of a sphere (§ 177)].

7. Find the M.I. of a solid sphere, radius a, about a diameter. [Hint.—divide the sphere into thin concentric shells and use the result of the previous question.]

8. Find the M.I. of a right circular cone, height h, about an axis drawn through the vertex parallel to the base, the radius of which is r.

9. Find the M.I. of an elliptic lamina, axes $2a$ and $2b$, about an axis drawn through the centre of the ellipse and perpendicular to its plane.

10. Find the M.I. of a uniform rectangular lamina, sides $2a$ and $2b$.

(1) About a side.
(2) About a diagonal.
(3) About an axis perpendicular to the plane of the rectangle and passing through a corner.

CHAPTER XVIII

PARTIAL DIFFERENTIATION

190. Functions of more than one variable.

THUS far we have been concerned only with functions of one independent variable. It was pointed out, however, in § 12, that a quantity may be a function of two or more independent variables. Examples were given in illustration.

We must now consider, very briefly, the problem of differentiation in such cases. An adequate treatment is not possible in an introductory book on the subject, but some simple aspects of the problem can be examined.

191. Partial differentiation.

We will begin with an example referred to in § 12, viz. that the volume of a gas is dependent upon both pressure and temperature.

Let V represent the volume of a gas.
,, p be the pressure on it.
,, t be the absolute temperature.

The law connecting these can be expressed by the formula

$$V = k \cdot \frac{t}{p}$$

where k is a constant.

(1) Suppose the **temperature to vary,** the pressure remaining constant.

Then $$\frac{dV}{dt} = k \cdot \frac{1}{p}.$$

(2) Suppose the **pressure to vary,** the temperature remaining constant.

Then $$\frac{dV}{dp} = k \cdot \frac{t}{-p^2} \quad \text{or} \quad -k\frac{t}{p^2}.$$

Thus the existence of two independent variables gives rise to **two differential coefficients.**

These are called **Partial derivatives** or **Partial Differential coefficients.** For the sake of simplicity the ordinary notation was employed above, but special symbols are employed to indicate partial coefficients. Instead of the letter "*d*," the symbol: ∂, read as "partial d," is employed. Thus the partial differential coefficients above would be written:

$$(1)\quad \frac{\partial V}{\partial t} = k \cdot \frac{1}{p}.$$

$$(2)\quad \frac{\partial V}{\partial p} = -k \cdot \frac{t}{p^2}.$$

Thus, (1) indicates that V is differentiated with respect to t (hence ∂t), while p is constant. Similarly, with (2).

In general—if z be a function of x and y, the partial differential coefficients are written:

(1j $\frac{\partial z}{\partial x}$, when x is variable and y constant.

(2) $\frac{\partial z}{\partial y}$, when y is variable and x constant.

Using the form referred to in § 33, of defining the differential coefficient, the partial differential coefficients can be expressed thus:

$$\frac{\partial z}{\partial x} = \underset{\delta x \to 0}{Lt} \frac{f(x + \delta x, y) - f(x, y)}{\delta x}$$

$$\frac{\partial z}{\partial y} = \underset{\delta y \to 0}{Lt} \frac{f(x, y + \delta y) - f(x, y)}{\delta y}.$$

Examples.

(1) $z = 2x^3 + 5x^2y + xy^2 + y^3.$

$\therefore \quad \frac{\partial z}{\partial x} = 6x^2 + 10xy + y^2$ (y constant)

$\frac{\partial z}{\partial y} = 5x^2 + 2xy + 3y^2$ (x constant)

(2) $z = \sin y + x^2 \cos y + e^{2x}$

$\frac{\partial z}{\partial x} = 2x \cos y + 2e^{2x}$ (y constant)

$\frac{\partial z}{\partial y} = \cos y - x^2 \sin y$ (x constant)

192. Graphical illustration of partial derivatives.

We have seen that a function with one independent variable can be represented by a plane curve. If, however, there are two independent variables, the dependent function can be represented by a surface, *i.e.*, co-ordinates in three dimensions are employed. This can be illustrated as follows.

In Fig. 88, let *XOY* represent a plane with *OX*, *OY* as co-ordinate axis at right angles to one another. Values of

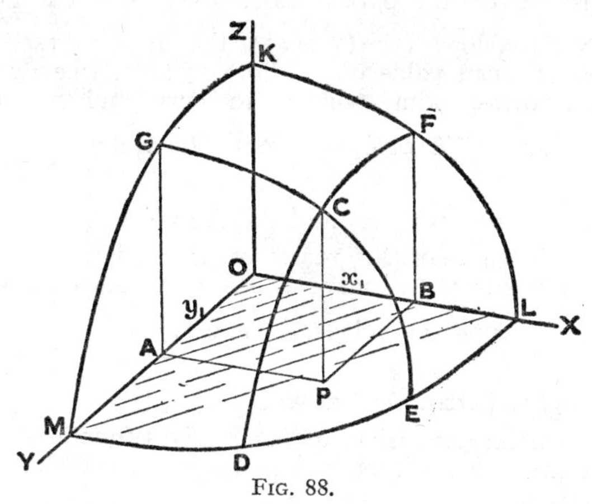

FIG. 88.

two variables x and y can be represented along *OX* and *OY* as heretofore. This we call the xy plane.

Draw *OZ* at right angles to the plane from *O*.

Thus the planes *XOZ*, *YOZ*, are perpendicular to the plane. *XOZ* is the (x, z) plane and *YOZ* is the (y, z) plane.

Values of z, corresponding to values of x and y, are marked on *OZ*.

Let *P* be a point in the plane of *XOY* with co-ordinates (x_1, y_1).

Along *OX* mark $OB = x_1$ and along *OY*, $OA = y_1$.

Then *P* is the position of the point in the plane *XOY*.

From *P* draw *PC* parallel to *OZ* and equal to z_1, where z_1 is the value of z corresponding to x_1 for x and y_1 for y.

Then C represents the position of the point in space when the co-ordinates are (x_1, y_1, z_1).

If other values of **x** and **y** are taken, with the corresponding values of **z**, we shall obtain an assemblage of points such as C, which will lie on a surface.

(1) Let **y be constant** and have the value y_1.

GCE will now represent the variations of z relative to x when y is constant.

Consequently the partial differential coefficient $\frac{\partial z}{\partial x}$ will represent the slope of the tangent to the curve, corresponding to any assigned value of x. For example, when $x = x_1$, C is the corresponding point on the curve and the tangent to the curve GCE at C represents the value of $\frac{\partial z}{\partial x}$ when $x = x_1$.

(2) Let **x be constant** and have the value x_1.

Then the curve of DCF represents the variations of z to y.

The tangent to the curve at any point on it represents $\frac{\partial z}{\partial y}$ for corresponding values of **y** and **z**.

193. Higher Partial Derivatives.

The partial derivatives are themselves functions of the variables concerned, and thus may have their partial derivatives.

(1) Thus if $\frac{\partial z}{\partial x}$ be differentiated with respect to x (y being constant), this is indicated by $\frac{\partial}{\partial x}\left(\frac{\partial z}{\partial x}\right)$ and denoted by $\frac{\partial^2 z}{\partial x^2}$.

(2) Since it is also a function of **y**, it can be differentiated with respect to y, x being constant. Thus we have:

$$\frac{\partial}{\partial y}\left(\frac{\partial z}{\partial x}\right) \text{ denoted by } \frac{\partial^2 z}{\partial y \partial x}.$$

(3) Similarly $\frac{\partial z}{\partial y}$ can be differentiated with respect

to x and y, so that when it is differentiated with respect to x, y being constant, we have:

$$\frac{\partial}{\partial x}\left(\frac{\partial z}{\partial y}\right) \text{ denoted by } \frac{\partial^2 z}{\partial x \partial y}.$$

(4) When differentiated with respect to y, x being constant, we have:

$$\frac{\partial}{\partial y}\left(\frac{\partial z}{\partial y}\right) \text{ denoted by } \frac{\partial^2 z}{\partial y^2}.$$

It will be seen that (2) and (3) are the same, except for the order of the differentials in the denominators. These indicate the order of differentiation.

In (2) we differentiate with respect to y first and then x.

In (3) we differentiate with respect to x first and then y.

It can be shown that these are commutative—*i.e.*, the order of differentiation is immaterial—*i.e.*, the result is the same

or $$\frac{\partial^2 z}{\partial y \partial x} = \frac{\partial^2 z}{\partial x \partial y}.$$

Similarly there may be third and higher derivatives.

194. Total differential.

When a function of a single variable such as $y = f(x)$ is differentiated, the result is expressed by

$$\frac{dy}{dx} = f'(x).$$

If this be written in the form

$$dy = f'(x)dx$$

the **differential** dy of the dependent variable y is thus expressed in terms of the **differential** dx of the independent variable x (see § 33).

We now proceed to find a similar expression, when z is a function of the independent variables x and y; *i.e.*, we require to obtain the relation between dz, dx, and dy.

Let $$z = f(x, y) \quad \ldots\ldots \quad (1)$$

Let x receive an increment δx.

Let y receive an increment δy.

And z receive a corresponding increment δz.

Then $$z + \delta z = f(x + \delta x, y + \delta y) \quad . \quad . \quad (2)$$

Subtracting (1) from (2),

$$\delta z = f(x + \delta x, y + \delta y) - f(x, y) \quad . \quad . \quad (A)$$

If y **only varies,** and is increased by δy, the result can be expressed by

$$f(x, y + \delta y) \quad . \quad . \quad . \quad . \quad . \quad (3)$$

If x **only varies** and is increased by δx, the result can be expressed by

$$f(x + \delta x, y) \quad . \quad . \quad . \quad . \quad . \quad (4)$$

If (3) be added to and subtracted from (A)

$$\delta z = \{f(x + \delta x, y + \delta y) - f(x, y + \delta y)\} + \{f(x, y + \delta y) - f(x, y)\}$$

$$\therefore \quad \delta z = \frac{\{f(x + \delta x, y + \delta y) - f(x, y + \delta y)\}\delta x}{\delta x}$$

$$+ \frac{\{f(x, y + \delta y) - f(x, y)\}\delta y}{\delta y} \quad (B)$$

(1) Considering the first part of B.

If δx and δy tend to become zero, then

$$\frac{f(x + \delta x, y + \delta y) - f(x, y + \delta y)}{\delta x}$$

in the limit becomes the partial differential coefficient of $f(x, y + \delta y)$, when x alone varies and y remains constant.

But in this expression δy ultimately vanishes, and thus it takes the form

$$\frac{f(x + \delta x, y) - f(x, y)}{\delta x}.$$

Thus it becomes the partial differential coefficient of $f(x, y)$, when x varies and y is constant,

i.e., $$\frac{\partial z}{\partial x}.$$

(2) Considering the second part of B.

In the limit this represents the partial differential coefficient of $f(x, y)$, when y alone varies,

i.e., $$\frac{\partial z}{\partial y}.$$

Also, in the limit, with the usual notation, δx, δy, δz, become the differentials dx, dy, dz.

$\therefore$ substituting for the corresponding parts of (B) they become

$$dz = \frac{\partial z}{\partial x}dx + \frac{\partial z}{\partial y}dy \quad . \quad . \quad . \quad (C)$$

This is called the **total differential of z**, where z is a function of the variables x and y.

A similar expression may be obtained when z is a function of three variables.

195. Total differential coefficient.

Let x and y, and consequently z, be functions of a variable t.

In equation (B) above, divide throughout by δt.

On proceeding to limits in the same way as was adopted above with (B), then in the limit we reach the result:

$$\frac{dz}{dt} = \frac{\partial z}{\partial x}\cdot\frac{dx}{dt} + \frac{\partial z}{\partial y}\cdot\frac{dy}{dt} \quad . \quad . \quad (D)$$

This is termed the total **differential coefficient** of z with regard to x and y, these being variables dependent on t.

If y is a function of x, and the total differential coefficient of dz is found by replacing t by x in the above, we get:

$$\frac{dz}{dx} = \frac{\partial z}{\partial x} + \frac{\partial z}{\partial y}\cdot\frac{dy}{dx}.$$

This may be obtained independently in the same way as the above.

196. A geometrical illustration.

The following geometrical illustration will probably be helpful to many students in realising the meaning and significance of the above results.

The area of a rectangle is a function of two variables, the lengths of its two unequal sides.

Fig. 89 represents a rectangle, with sides x and y.

Let A be its area.

Then $$A = xy.$$

Let x be variable and receive an increment δx, while y remains constant.

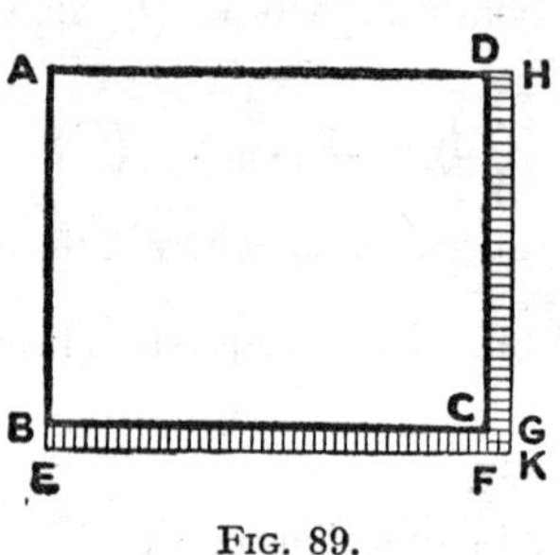

FIG. 89.

Then $A + \delta A = (x + \delta x)y$.

Subtracting $\delta A = y\delta x$, *i.e.*, rectangle $CGHD$.

The **rate of increase** of A with regard to x, y being constant, is the partial differential coefficient $\frac{\partial A}{\partial x}$, *i.e.*, $\frac{\partial}{\partial x}(xy) = y$.

Similarly, if y be variable, x being constant

$$\delta A = \text{the rectangle } BEFC = x\delta y$$

and rate of increase $= \frac{\partial A}{\partial y} = \frac{\partial}{\partial y}(xy) = x$.

If both x and y vary, then by formula C the **total differential** increase, in the limit, when δx and δy proceed to zero, is

$$dA = \frac{\partial A}{\partial x} \cdot dx + \frac{\partial A}{\partial y} dy.$$

Substituting the values of the partial differential coefficients, we get:

$$dA = ydx + xdy.$$

Comparing with Fig. 89, it is seen that the total increase in area, due to small increases of x and y, is rectangle $BEFC$ + rectangle $CGHD$ + the small rectangle $CFKG$, *i.e.*, in the limit

$$ydx + xdy + dxdy.$$

But $dxdy$ is **the product of two infinitesimals** and is called an infinitesimal of the second order. It can be disregarded in comparison with ydx and xdy, which are infinitesimals of the first order.

∴ total differential increase of area is $ydx + xdy$.

Total differential coefficient.

Now suppose $y = 8$ m, and at a given instant is increasing at the rate of 2 ms^{-1}.

Also let $x = 5$ *m, and be increasing at the same instant at* 3 *ms*$^{-1}$.

At what rate is A increasing at the given instant?

In this problem another variable, time (t), is introduced, so that x and y, and consequently A, vary with time.

The rate of increase of A is clearly given by the total differential coefficient as stated in formula (D).

This becomes

$$\frac{dA}{dt} = \frac{\partial A}{\partial x} \cdot \frac{dx}{dt} + \frac{\partial A}{\partial y} \cdot \frac{dy}{dt}.$$

We know that
$$\frac{\partial A}{\partial x} = y = 8$$
$$\frac{\partial A}{\partial y} = x = 5$$
$$\frac{dx}{dt} = 3$$
$$\frac{dy}{dt} = 2.$$

$\therefore$ substituting
$$\frac{dA}{dt} = (8 \times 3) + (5 \times 2)$$
$$= 34 \text{ m}^2\text{s}^{-1}.$$

197. Worked examples.

If $z = \tan^{-1}\frac{y}{x}$, *find the total differential dz.*

If
$$z = \tan^{-1}\frac{y}{x}$$
$$\frac{\partial z}{\partial x} = \frac{1}{1 + \left(\frac{y}{x}\right)^2} \times \left(-\frac{y}{x^2}\right)$$
$$= \frac{x^2}{x^2 + y^2} \times \left(-\frac{y}{x^2}\right) = \frac{-y}{x^2 + y^2}$$
$$\frac{\partial z}{\partial y} = \frac{1}{1 + \left(\frac{y}{x}\right)^2} \times \frac{1}{x}$$
$$= \frac{x^2}{x^2 + y^2} \times \frac{1}{x} = \frac{x}{x^2 + y^2}.$$

Substituting in formula D

$$dz = \frac{-y}{x^2 + y^2} \cdot dx + \frac{x}{x^2 + y^2}dy$$
$$= \frac{xdy - ydx}{x^2 + y^2}.$$

198. Implicit functions.

Partial derivatives will, no doubt, have reminded students of the method of differentiating implicit functions as described in § 48. The connection will be made clear by a modification of formula (C), § 194.

Let $z = f(x, y) =$ a constant, say c.
Then its differentials equal zero.
∴ Formula (C) becomes

$$dz = \frac{\partial z}{\partial x}dx + \frac{\partial z}{\partial y}dy = 0.$$

$$\therefore \quad \frac{\partial z}{\partial y} \cdot dy = -\frac{\partial z}{\partial x}dx$$

$$\therefore \quad \frac{dy}{dx} = -\frac{\frac{\partial z}{\partial x}}{\frac{\partial z}{\partial y}}.$$

It will be noted that though the total differential coefficient of z is zero, this was not the case with the partial differential coefficients.

Referring to § 48 it will be seen that the results are, in principle, identical.

Worked example. *If* $z = 4x^3 - xy^2 + y^3 = 0$, *find* $\frac{dy}{dx}$.

From above

$$\frac{dy}{dx} = -\left(\frac{\partial z}{\partial x} \div \frac{\partial z}{\partial y}\right)$$

but
$$\frac{\partial z}{\partial x} = 12x^2 - y^2$$
$$\frac{\partial z}{\partial y} = -2xy + 3y^2.$$

$\therefore$ substituting

$$\frac{dy}{dx} = -\frac{12x^2 - y^2}{-2xy + 3y^2}$$
$$= \frac{12x^2 - y^2}{2xy - 3y^2}.$$

Exercise 41.

A. Find the partial differential coefficients $\frac{\partial z}{\partial x}, \frac{\partial z}{\partial y}$, in questions 1 to 7.

1. $z = x^y$.
2. $z = \cos(x^2 + y^2)$.
3. $z = \frac{xy}{x^2 + y^2}$.
4. $z = x^3 + 3x^2y + 6xy^2 + 2y^3$.
5. $z = \sin^{-1}\frac{x}{y}$.
6. $z = \tan^{-1}\frac{x}{y}$.
7. $z = \frac{ax}{y^2}$.
8. If $z = \log(e^x + e^y)$, show that $\frac{\partial z}{\partial x} + \frac{\partial z}{\partial y} = 1$.

B. Find the total differentials in questions 9 to 14.

9. $z = \frac{x}{y}$.
10. $z = ax^2 + 2bxy + cy^2$.
11. $z = \log x^y$.
12. $z = x^2y + xy^3$.
13. $z = e^{xy}$.
14. $z = a^x e^y$.

---

15. If $u = 2x^2 + 3y^2$, find du, when $x = 1$, $y = 3$, $dx = 0{\cdot}01$, and $dy = 0{\cdot}02$.

16. If the law of an ideal gas be $V = \frac{kT}{p}$, where V represents the volume, p the pressure, and T the absolute temperature, find the relation between dV, dT, and dp.

17. If $u = x^5y - \sin y$, find $\frac{\partial^2 y}{\partial x dy}$, and show that it is equal to $\frac{\partial^2 y}{\partial y \partial x}$.

18. In the solid representing $z = a^2 - x^2 - 2y^2$ what is the slope at a point of the curve along a section for which y is constant? What is the slope at a point along a section for which x is constant?

19. The radius of the base of a right cylinder is increasing at a given instant at the rate of 1 ms^{-1}, while the height is increasing at 2 ms^{-1}. At the same instant the height is 10 m, and the radius of the base 5 m. At what rate is the volume increasing?

CHAPTER XIX

SERIES.

TAYLOR'S AND MACLAURIN'S THEOREMS

199. Infinite series.

WHEN studying algebra the student has become acquainted with certain "*series*," as, for example, geometric progression or series, arithmetical progression, and the binomial series.

In the first of these he will have considered the important problem of the *sum* of the series, when the number of terms is increased without limit, *i.e.*, becomes "**infinite.**"

Two cases arise:

(1) When the common ratio r is numerically greater than unity, as the number of terms increases the terms increase individually and so does their sum. If the number of terms becomes infinitely great, their sum also becomes infinite, *i.e.*, if S_n represent the sum of n terms, then, when $n \longrightarrow \infty$, $S_n \longrightarrow \infty$.

(2) If, however, the common ratio be less than unity, the terms continually decrease and the question of what happens to S_n when n becomes infinitely great is a matter for investigation.

In this case it is readily shown that when $n \longrightarrow \infty$, **S_n approaches a finite limit.**

200. Convergent and divergent series.

In general when considering any kind of series, it becomes a problem to be investigated as to whether

(1) S_n approaches a finite limit when $n \longrightarrow \infty$, or
(2) S_n approaches infinity when $n \longrightarrow \infty$.

If a series is of the first kind it is said to be **convergent,** if of the second, it is called **divergent.**

There is also a third type of series called *oscillating*, but we shall not consider it in this chapter.

For theoretical and practical purposes it is very important to know whether a given series is convergent or divergent.

There is no universal method of determining this, but there are various tests which can be applied for certain kinds of series. A consideration of such tests is, however, beyond the scope of this volume. Students who desire, or need to study, this important matter, should consult a book on Higher Algebra.

In this brief treatment of infinite series by the use of the Calculus, the series considered will be assumed, without proof, to be convergent.

201. Taylor's theorem.

In the binomial theorem it is stated that the function $(x + a)^n$ can be expanded in a series of descending powers of x and ascending powers of a. Many other functions can be similarly expanded, and various methods are employed for this purpose. In this chapter, however, it is proposed to investigate a general method of expanding functions in series.

Briefly, we shall see that $f(x + h)$ can, in general, be expanded in a series of ascending powers of h. Such an expansion is not possible for all functions, and there are limitations to the application of the theorem which defines the form of the expansion.

We will begin by stating the theorem which is known as **Taylor's Theorem**, and proceed afterwards to demonstrate the truth of it.

Taylor's Theorem.

$$f(x + h) = f(x) + hf'(x) + \frac{h^2}{\underline{|2}}f''(x)$$

$$+ \frac{h^3}{\underline{|3}}f'''x + \ldots + \frac{h^n}{\underline{|n}}f^n(x) + \ldots \text{ ad inf.}$$

The following assumptions will be made:

(1) *That any function which will be considered is capable of being expanded in this form.*

(2) *That subject to certain conditions in some cases, the series is convergent.*

(3) *That the successive differential coefficients,* $f^{\mathrm{I}}(x)$, $f^{\mathrm{II}}(x)$, $f^{\mathrm{III}}(x)$, . . . $f^n(x)$ *all exist.*

In accordance with (1) we will assume that $f(x + h)$ can be expanded in ascending powers of h as follows:

$$f(x + h) = A_0 + A_1h + A_2h^2 + A_3h^3 + \ldots \quad \text{(B)}$$

where the coefficients $A_0, A_1, A_2, \ldots$ are functions of x but do not contain h.

Since this is to be true for all values of h

let $$h = 0.$$

Then on substitution in (1), we have:

$$A_0 = f(x).$$

Since the series (*B*) is an identity, it may be assumed that if both sides be differentiated with respect to h, keeping x constant, the result in each case will be another identity.

Repeating the process, we get:

$$(1)\ f^{I}(a + h) = A_1 + (A_2 \times 2h) + (A_3 \times 3h^2) + (A_4 \times 4h^3) + \ldots$$

since $f(x) = 0$, where x is constant.

Similarly:

$$(2)\ f^{II}(x + h) = 2A_2 + (3.2A_3h) + (4.3A_4h^2) + \ldots$$
$$(3)\ f^{III}(x + h) = 3.2.1A_3 + 4.3.2A_4h + \ldots$$

and so for higher differential coefficients.

In all of these results put $h = 0$.

Then from

$$(1)\ f^{I}(x) = A_1.$$
$$(2)\ f^{II}(x) = 2.1A_2.$$
$$(3)\ f^{III}(x) = 3.2.1A_3.$$
$$f^{IV}(x) = 4.3.2.1A_4.$$

. . . . . .

i.e.,

$$A_1 = f^{I}(x)$$
$$A_2 = \frac{f^{II}(x)}{\lfloor 2}$$
$$A_3 = \frac{f^{III}(x)}{\lfloor 3}$$
$$A_n = \frac{f^{n}(x)}{\lfloor n} \quad \text{and so on.}$$

Substituting for these in (B) we obtain the theorem, viz.,

$$f(x+h) = f(x) + hf^{I}(x) + \frac{h^2}{\underline{|2}}f^{II}(x) + \frac{h^3}{\underline{|3}}f^{III}(x) + \ldots + \frac{h^n}{\underline{|n}}f^n(x) + \ldots$$

202. Application to the binomial theorem.

To expand $(x+h)^n$ (by Taylor's theorem).

$$f(x+h) = (x+h)^n$$
$$= f(x) + hf^{I}(x) + \frac{h^2}{\underline{|2}}f^{II}(x) + \frac{h^3}{\underline{|3}}f^{III}(x) + \ldots$$

When $h = 0$,

$$f(x) = x^n$$
$$f^{I}(x) = nx^{n-1}$$
$$f^{II}(x) = n(x-1)x^{n-2}$$
$$f^{III}(x) = n(n-1)(n-2)x^{n-3}$$
$$\cdot \quad \cdot \quad \cdot \quad \cdot \quad \cdot \quad \cdot$$

Substituting in Taylor's expansion

$$(x+h)^n = x^n + h \,.\, nx^{n-1} + \frac{h^2}{\underline{|2}}n(n-1)x^{n-2} + \frac{h^3}{\underline{|3}}n(n-1)(n-2)x^{n-3} + \ldots$$

or with the usual arrangement

$$(x+h)^n = x^n + nx^{n-1}h + \frac{n(n-1)}{\underline{|2}}x^{n-2}h^2 + \frac{n(n-1)(n-2)}{\underline{|3}}x^{n-3}h^3 + \ldots$$

203. Maclaurin's theorem (or Stirling's theorem).

This is another form of Taylor's theorem. It is obtained by putting $x = 0$, and for convenience replacing h by x. This is possible since Taylor's theorem is true for all values of x and h.

$\therefore$ let $\qquad x = 0, \quad$ and $\quad h = x.$

Then Taylor's theorem becomes

$$f(x) = f(0) + xf^{I}(0) + \frac{x^2}{\underline{|2}}f^{II}(0) + \ldots + \frac{x^n}{\underline{|n}}f^n(0) \ldots$$

In this form $f^n(0)$ means that in the nth differential coefficient of $f(x)$, x is replaced by 0.

204. Worked examples.

Example 1. *Expand* $\log(1 + x)$.

Since $f(x) = \log(1 + x)$; $\quad f(0) = \log(1) = 0$.

$$\therefore \quad f^{I}(x) = \frac{1}{1+x}; \qquad f^{I}(0) = \frac{1}{1} = 1$$

$$f^{II}(x) = -\frac{1}{(1+x)^2}; \qquad f^{II}(0) = -\frac{1}{1} = -1$$

$$f^{III}(x) = \frac{1.2}{(1+x)^3}; \qquad f^{III}(0) = 1.2$$

$$f^{IV}(x) = -\frac{1.2.3}{(1+x)^4}; \qquad f^{IV}(0) = -\underline{|3}$$

. . . . . .

$$f^{n}(x) = (-1)^{n-1} . \frac{\underline{|n-1}}{(1+x)^n}; \; f^{n}(0) = (-1)^{n-1}\underline{|n-1}.$$

Substituting these values in Maclaurin's series, viz.

$$f(x) = \log(1 + x) = f(0) + xf^{I}(0) + \frac{x^2}{\underline{|2}}f^{II}(0) + \ldots$$

we have:

$$\log(1 + x) = x - \frac{x^2}{2} + \frac{x^3}{3} - \frac{x^4}{4} + \ldots + (-1)^{n-1}x^n + \ldots$$

It should be remembered that the base employed throughout has been **e**. Consequently the above series may be used to calculate logarithms to that base. From these the logs to any other base, such as 10, can be obtained.

Example 2. *Expand* $\sin x$ *in a series involving powers of* x.

$$f(x) = \sin x. \qquad \therefore f(0) = 0$$

$$f^{I}(x) = \cos x = \sin\left(x + \frac{\pi}{2}\right). \qquad \therefore f^{I}(0) = 1$$

$$f^{II}(x) = -\sin x. \qquad \therefore f^{II}(0) = 0$$

$$f^{III}(x) = -\cos x = \sin\left(x + \frac{3\pi}{2}\right). \qquad \therefore f^{III}(0) = -1$$

. . . . . .

$$f^{n}(x) = \sin\left(x + \frac{n\pi}{2}\right). \qquad \therefore f^{n}(0) = \sin\frac{n\pi}{2}.$$

Substituting in Maclaurin's series we have:

$$\sin x = x - \frac{x^3}{\lfloor 3} + \frac{x^5}{\lfloor 5} - \frac{x^7}{\lfloor 7} + \ldots + \frac{x^n \sin \frac{n\pi}{2}}{\lfloor n}.$$

In this series x is measured in **radians.**

If now we put $x = 1$, we may readily calculate the value of a radian to as great a degree of accuracy as may be desired, by taking sufficient terms of the series. It will be noted that the terms decrease rather rapidly, or the series is said to converge rapidly.

It should be further noted that the series contains only odd powers of x, *i.e.*, it is an **odd function.** The series for $\cos x$ will be found to contain only even powers of x, *i.e.*, it is an **even function.**

Example 3. *Expand e^x in a series involving powers of x.*

We have
$$f(x) = e^x. \quad \therefore f(0) = 1.$$
$$\therefore f^{\mathrm{I}}(x) = e^x. \quad \therefore f^{\mathrm{I}}(0) = 1.$$
$$f^{\mathrm{II}}(x) = e^x. \quad \therefore f^{\mathrm{II}}(0) = 1.$$

Substituting in Maclaurin's series we get:

$$e^x = 1 + x + \frac{x^2}{\lfloor 2} + \frac{x^3}{\lfloor 3} + \frac{x^4}{\lfloor 4} + \ldots$$

Compare § 83.

205. Expansion by the differentiation and integration of known series.

The method may be illustrated by the following example:

By division.

$$\frac{1}{1 + x^2} = 1 - x^2 + x^4 - x^6 + \ldots$$

It may be proved that when a function is represented by a series, and the function and the series are integrated throughout, the results are equal.

$$\therefore \int \frac{dx}{1 + x^2} = \int dx - \int x^2 dx + \int x^4 dx - \ldots$$

$$\therefore \tan^{-1} x = x - \frac{x^3}{3} + \frac{x^5}{5} - \frac{x^7}{7} + \ldots$$

This is known as **Gregory's series.** It is convergent and **can be used to calculate the value of** π.

Thus, in the series let $x = 1$.

Then $\tan^{-1}(1) = \frac{\pi}{4}$.

Substituting in Gregory's series

$$\frac{\pi}{4} = 1 - \frac{1}{3} + \frac{1}{5} - \frac{1}{7} + \frac{1}{9} - \ldots$$

Hence, by taking sufficient terms, the value of π can be found to any required degree of accuracy. It converges slowly, however, and consequently other series which converge rapidly are employed for the calculation.

Exercise 42.

Expand the following functions in powers of x:

1. (*a*) $\sin(a + x)$; (*b*) $\cos(a + x)$.
2. e^{x+h}.
3. $\tan^{-1}(x + h)$.
4. $\log(1 + \sin x)$.
5. $\cos x$.
6. $\tan x$.
7. $\log(1 + e^x)$.
8. a^x.
9. e^{-kx}.
10. $e^{\sin x}$.
11. $\sec x$.
12. $\log \sec x$.
13. $\sin^{-1} x$.
14. $\log(1 - x)$.
15. $\sinh x$.
16. $e^x \sin x$.
17. $\tanh x$.

CHAPTER XX

ELEMENTARY DIFFERENTIAL EQUATIONS

206. Meaning of a differential equation

A differential equation is one which involves an independent variable, a dependent variable and one or more of their differential coefficients.

These equations are of great importance in Physics, Engineering of all kinds, and other applications of Mathematics. Although it is not possible in this volume to give more than a very brief introduction to what is a big subject, the elementary forms which are dealt with in this chapter may prove valuable to many students.

Examples of differential equations have already appeared in this book, as, for example, questions 49–54 in Exercise 16.

Again, as illustrated in § 100

If $$\frac{dy}{dx} = 2x \quad . \quad . \quad . \quad . \quad . \quad (1)$$

or $$dy = 2xdx \quad . \quad . \quad . \quad . \quad (2)$$

we obtain by integration the relation :

$$y = x^2 + c \quad . \quad . \quad . \quad . \quad (3)$$

(1) and (2) are differential equations, and (3) is their solution. Thus a differential equation is solved when, by integration, we find the relations between the two variables x and y.

This process involves the introduction of an undetermined constant. Thus the solution (3) is the general equation, or the relation between y and x for the whole family of curves represented in Fig. 28.

207. Formation of differential equations.

Differential equations arise or may be derived in a variety of ways.

For example, it is shown in mechanics that if s be the distance passed over in time t by a body moving with uniform acceleration, a, then

$$\frac{d^2s}{dt^2} = a \quad . \quad . \quad . \quad . \quad . \quad . \quad (1)$$

By integration

$$\frac{ds}{dt} = at + c_1 \quad . \quad . \quad . \quad . \quad . \quad (2)$$

Integrating again $\qquad s = \frac{1}{2}at^2 + c_1t + c_2 \quad . \quad . \quad (3)$

Of these (1) contains a **second** derivative, (2) the first derivative, while (3) is the general solution of (1) and (2).

Differential equations may also be formed by direct differentiation. Thus, let

$$y = x^3 + 7x^2 + 3x + 7 \quad . \quad . \quad . \quad (a)$$

then

$$\frac{dy}{dx} = 3x^2 + 14x + 3 \quad . \quad . \quad . \quad . \quad . \quad (b)$$

$$\frac{d^2y}{dx^2} = 6x + 14 \quad . \quad . \quad . \quad . \quad . \quad . \quad (c)$$

$$\frac{d^3y}{dx^3} = 6 \quad . \quad . \quad . \quad . \quad . \quad . \quad . \quad . \quad (d)$$

(*a*) is called the complete **primitive** of (*d*).

208. Kinds of differential equations.

(A) There are two main types of differential equations:

(1) **Ordinary differential equations**, involving only one independent variable.

(2) **Partial differential equations**, which involve more than one independent variable.

In this chapter we shall concern ourselves with (1) only.

(B) **Orders.** Differential equations of both types are classified according to the highest derivative which occurs in them. Thus of the differential equations (*b*), (*c*), (*d*) in § 207:

(*b*) is of the **first** order, having only the first derivative.

(*c*) is of the **second** order.

(*d*) is of the **third** order.

(C) **Degree.** The degree of a differential equation is that of the highest power of the highest differential which the equation contains after it has been simplified by clearing radicals and fractions.

Thus the equation $\left(\frac{d^2y}{dx^2}\right)^3 + 3\frac{dy}{dx} = 0$ is of the second order and third degree; $s = \sqrt{1 + \left(\frac{dy}{dx}\right)^2}\,dx$ (§ 162) is of the first order and second degree.

209. Solutions of a differential equation.

A solution which is **complete** or **general** must contain a number of arbitrary constants which is equal to the order of the equation. Thus in § 207 (3) contains two arbitrary constants and is the solution of (1), an equation of the second order.

Solutions which are obtained by assigning particular values to the constants, as in Exercise 16, question 54, are called **particular solutions.**

This chapter will be concerned only with equations of the first order and first degree.

Differential equations of the first order and first degree.

210. Since solutions of differential equations involve integration, it is not possible in consequence to formulate rules, as with differentiation, which will apply to any type of equation. Some indeed it is not possible to solve. But a large number of equations, including very many of practical importance, can be classified into various *types*, solutions for which can be found by established methods. Some of these types we will proceed to consider.

211. I. One variable absent.

There may be two forms:

(1) **When y is absent.**

The general form is $dy = f(x)dx$

and the solution is $y = \int f(x)dx.$

This requires ordinary integration for its solution.

Example. Solve the equation

$$dy = (x^4 + \sin x)dx.$$

Then
$$y = \int(x^4 + \sin x)dx.$$
$$\therefore \quad y = \tfrac{1}{5}x^5 - \cos x + c.$$

(2) **When x is absent.**

The general form is
$$\frac{dy}{dx} = f(y)$$
or
$$dy = f(y)dx.$$

This may be written in the form:
$$\frac{dx}{dy} = \frac{1}{f(y)}$$
or
$$dx = \frac{dy}{f(y)}$$
whence
$$\int dx = \int \frac{dy}{f(y)}.$$

The solution is then obtained by direct integration.

Example. *Solve the equation* $\frac{dy}{dx} = \tan y$.

Hence
$$\frac{dx}{dy} = \frac{1}{\tan y}$$
$$dx = \frac{dy}{\tan y}.$$
$$\therefore \quad \int dx = \int \frac{dy \cos y}{\sin y}.$$
$$\therefore \quad x = \log \sin y + c.$$

212. II. Variables separable.

If it is possible to re-arrange the terms of the equation in two groups, each containing only one variable, the variables are said to be separable. Then the equation takes the form
$$F(x)dx + f(y)dy = 0$$
in which $F(x)$ is a function of x only, and $f(y)$ a function of y only.

The general solution then is:
$$\int F(x)dx + \int f(y)dy = c.$$

213. Worked examples.

Example 1. *Solve the differential equation*

$$xdy + ydx = 0.$$

To separate the variables divide throughout by xy.

Then
$$\frac{dy}{y} + \frac{dx}{x} = 0.$$

$$\therefore \int \frac{dy}{y} + \int \frac{dx}{x} = 0.$$

$$\therefore \log y + \log x = c_1.$$

If the constant c_1 be written in the form $\log c$

Then $\log y + \log x = \log c$

whence $xy = c.$

The factor $\frac{1}{xy}$ used to multiply throughout to separate the variables is called **an integrating factor.**

Example 2. *Solve the equation*

$$(1 + x)ydx + (1 - y)xdy = 0.$$

Multiplying throughout by $\frac{1}{xy}$, we get:

$$\frac{1 + x}{x}dx + \frac{1 - y}{y}dy = 0.$$

or
$$\left(\frac{1}{x} + 1\right)dx + \left(\frac{1}{y} - 1\right)dy = 0.$$

$$\therefore \int\left(\frac{1}{x} + 1\right)dx + \int\left(\frac{1}{y} - 1\right)dy = c.$$

$$\therefore \log x + x + \log y - y = c.$$

or
$$\log xy + (x - y) = c.$$

Exercise 43.

Solve the following differential equations:

1. $\frac{dy}{dx} + \frac{k}{x^2} = 0.$ 2. $\frac{dy}{dx} = \frac{y}{a}.$ 3. $\frac{dy}{dx} = \frac{y}{x}.$
4. $(1 + y)dx - (1 - x)dy = 0.$
5. $(x + 1)dy - ydx = 0.$
6. $\sin x \cos ydx = \sin y \cos xdy.$

7. $(y^2 - x^2)dy + 2xydx = 0.$

8. $xy\frac{dy}{dx} = \frac{y^2 + 1}{x^2 + 1}.$

9. $2ydx = x(y - 1)dy.$

10. $y^2 + \sin 2x \cdot \frac{dy}{dx} = 1.$

11. $\frac{1 + y^2}{(1 + x^2)xy} = \frac{dy}{dx}.$ 12. $\frac{dy}{dx} = x^2y.$

13. $x\sqrt{y^2 - 1}dx - y\sqrt{x^2 - 1}dy = 0.$

14. $\frac{1 + x^2}{1 + y} = xy\frac{dy}{dx}.$ 15. $\frac{dy}{dx} = 2xy.$

16. The slope of a family of curves is $-\frac{y}{x}$. What is the equation of the set?

214. III. Linear equations.

An equation of the form

$$\frac{dy}{dx} + Py = Q,$$

where P and Q are constants, or functions of x only, is called a **linear differential equation.**

It is so called because y and its derivatives are of the first degree.

It has been discovered that if such an equation is multiplied throughout by the **integrating factor** $e^{\int Pdx}$, an equation is obtained which can be solved.

When multiplied by this factor, the equation becomes:

$$e^{\int Pdx}\left(\frac{dy}{dx} + Py\right) = Qe^{\int Pdx}.$$

It may now be seen that the integral of the left-hand side is $ye^{\int Pdx}$. This is evident on differentiating $ye^{\int Pdx}$. Therefore the solution is:

$$ye^{\int Pdx} = \int Qe^{\int Pdx}dx \quad . \quad . \quad . \quad (A)$$

The procedure in solving this type of differential equation is to begin by finding the integral $\int Pdx$, then substitute in (A).

Examples will illustrate the method more clearly.

215. Worked examples.

Example 1. *Solve the equation*

$$(1 - x^2)\frac{dy}{dx} - xy = 1.$$

Transforming this to the general form, viz.,

$$\frac{dy}{dx} + Py = Q$$

we get

$$\frac{dy}{dx} - \frac{x}{1 - x^2}y = \frac{1}{1 - x^2}.$$

Since the integrating factor is $e^{\int P dx}$, we proceed first to find $\int P dx$ in this case, noticing that $P = \frac{-x}{1 - x^2}$, $Q = \frac{1}{1 - x^2}$.

Comparing with the equation above we have,

$$\begin{aligned}\int P dx &= -\int \frac{x}{1 - x^2} dx \\ &= \tfrac{1}{2} \log (1 - x^2) \\ &= \log \sqrt{1 - x^2}.\end{aligned}$$

$\therefore$ integrating factor $= e^{\log \sqrt{1 - x^2}} = \sqrt{1 - x^2}$.

Using the form (A) in § 214, we have:

$$\begin{aligned}y\sqrt{1 - x^2} &= \int \frac{1}{1 - x^2} \times \sqrt{1 - x^2} dx \\ &= \int \frac{dx}{\sqrt{1 - x^2}} \\ &= \sin^{-1} x + c.\end{aligned}$$

$\therefore$ the solution is:

$$y\sqrt{1 - x^2} = \sin^{-1} x + c.$$

Example 2. *Solve the equation*

$$\cos x \frac{dy}{dx} + y \sin x = 1.$$

Dividing by $\cos x$,

$$\frac{dy}{dy} + y \tan x = \sec x.$$

Comparing with the type equation

$$P = \tan x.$$

$$\therefore \quad \int P dx = \int \tan x dx = \log \sec x.$$

$$\therefore \quad e^{\int P dx} = e^{\log \sec x} = \sec x.$$

Using formula (A) and substituting

$$y \sec x = \int \sec x \sec x dx$$

$$= \int \sec^2 x dx$$

$$= \tan x + c.$$

$\therefore$ the solution is:

$$y = \cos x \tan x + c \cos x$$

or

$$y = \sin x + c \cos x.$$

Example 3. *Solve the equation*

$$\frac{dy}{dx} + 2xy = 1 + 2x^2.$$

Comparing with the type equation

$$P = 2x; \; Q = 1 + 2x^2.$$

$$\therefore \quad \int P dx = \int 2x dx = x^2.$$

$\therefore$ integrating factor is e^{x^2}.
$\therefore$ using formula (A) and substituting

$$ye^{x^2} = \int (1 + 2x^2)e^{x^2} dx$$

$$= \int (e^{x^2} + 2x^2 e^{x^2}) dx$$

$$= xe^{x^2} + c.$$

$\therefore$ the solution is:

$$ye^{x^2} = xe^{x^2} + c$$

or

$$y = x + ce^{-x^2}.$$

EXAMPLE 44.

Solve the following differential equations:

1. $\dfrac{dy}{dx} - 2xy = 2x.$

2. $x\dfrac{dy}{dx} + x + y = 0.$

3. $\frac{dy}{dx} = y - x.$ 4. $\frac{dy}{dx} + xy = x.$

5. $\frac{dy}{dx} + ay = e^x.$ 6. $\frac{dy}{dx} + y \tan x = 1.$

7. $\frac{dy}{dx} - \frac{ay}{x} = \frac{x+1}{x}.$ 8. $\tan x \frac{dy}{dx} = 1 + y.$

9. $e^x dy = (1 - e^x y)dx.$ 10. $xdy - aydx = (x+1)dx.$

11. $\cos^2 x \cdot \frac{dy}{dx} + y = \tan x.$

12. $x^2 \cdot \frac{dy}{dx} + xy + 1 = 0.$

216. IV. Homogeneous equations.

These equations are of the form

$$P + Q\frac{dy}{dx} = 0$$

where P and Q are homogeneous functions of the same degree in x and y.

Then $\frac{P}{Q}$ is a function of $\frac{y}{x}$.

Such equations can be solved by using the substitution

$$\frac{y}{x} = v \quad \text{or} \quad y = vx.$$

Thus the two variables x and v are separable, and the solution can be found as before.

When the solution has been found, using these variables substitute $\frac{y}{x}$ for v and so reach the final solution.

217. Worked examples.

Example 1. *Solve the differential equation*

$$(x + y) + x \cdot \frac{dy}{dx} = 0.$$

In this example P and Q, *i.e.*, $x + y$ and x, are each functions of the first degree throughout in x and y.

Let $\qquad \frac{y}{x} = v \quad \text{or} \quad y = vx.$

Then $\qquad dy = vdx + xdv \qquad$ (d.c. of a product)

Substituting in equation above

$$(x + vx) + x \cdot \frac{vdx + xdv}{dx} = 0.$$

$$\therefore \quad (x + vx)dx + x(vdx + xdv) = 0$$

and $$xdx + 2vxdx + x^2dv = 0.$$

Separating the variables

$$(1 + 2v)dx + xdv = 0.$$

$$\therefore \quad \frac{dv}{1 + 2v} + \frac{dx}{x} = 0.$$

Integrating,

$$\tfrac{1}{2}\log(1 + 2v) + \log x = c_1$$

and $$\log(1 + 2v) + 2\log x = c_2.$$

$$\therefore \quad x^2(1 + 2v) = c.$$

Substituting $$x^2\left(1 + 2\frac{y}{x}\right) = c.$$

∴ solution is $$x^2 + 2xy = c.$$

Example 2. *Solve the equation*

$$(x^2 - y^2)dy = 2xydx.$$

Put $$y = vx$$

then $$dy = vdx + xdv.$$

Substituting

$$(x^2 - v^2x^2)(vdx + xdv) = 2vx^2dx.$$

Dividing by x^2

$$(1 - v^2)(vdx + xdv) = 2vdx$$

whence $$(1 - v^2)xdv = v(1 + v^2)dx.$$

Separating variables

$$\frac{1 - v^2}{v(1 + v^2)}dv = \frac{dx}{x}.$$

∴ by partial fractions

$$\left\{\frac{1}{v} - \frac{2v}{1 + v^2}\right\}dv = \frac{dx}{x}.$$

Integrating

$$\log v - \log (1 + v^2) = \log x + c_1.$$

$$\therefore \quad \log \frac{v}{1 + v^2} = \log x + \log c.$$

$$\therefore \quad \frac{v}{1 + v^2} = cx$$

replacing v by $\frac{y}{x}$
$$\frac{\frac{y}{x}}{1 + \frac{y^2}{x^2}} = cx$$

whence
$$\frac{xy}{x^2 + y^2} = cx$$

and the solution is
$$x^2 + y^2 = cy.$$

Exercise 45.

Solve the following equations:

1. $(x + y)dx + xdy = 0.$
2. $(x + y)dx - xdy = 0.$
3. $(x + y)dx + (y - x)dy = 0.$
4. $(x - 2y)dx + ydy = 0.$
5. $(x^2 + y^2) = 2xy\frac{dy}{dx}.$
6. $y^2 - x^2\frac{dy}{dx} = 0.$
7. $(y^2 - 2xy)dx = (x^2 - 2xy)dy.$
8. $x^2dy + y^2dx + xydy = 0.$
9. $y^2dx + (x^2 - xy)dy = 0.$
10. $y^2dx + (xy + x^2)dy = 0.$
11. $(x - 2y)dy + xdx = 0.$

218. V. Exact differential equations.

The equation $\quad Mdx + Ndy = 0$

is called an **exact differential equation**, when it is formed from its complete primitive by simple differentiation.

Thus, if the complete primitive be

$$x^3 + 3x^2y + y^3 = c \quad . \quad . \quad . \quad . \quad (A)$$

Then, on differentiation

$$(3x^2 + 6xy)dx + (3x^2 + 3y^2)dy = 0 \qquad (\S\ 198)$$

This is an exact differential equation. Consequently

(1) $(3x^2 + 6xy)$ is the partial differential coefficient $\frac{\partial u}{\partial x}$, and

(2) $(3x^2 - 3y^2)$ is the partial differential coefficient $\frac{\partial u}{\partial y}$.

The first is obtained by differentiating (A) with x variable and y constant, the second by differentiating with y variable and x constant.

In general the result is of the form

$$\frac{\partial u}{\partial x}dx + \frac{\partial u}{\partial y}dy = 0 \qquad (\S\ 198)$$

Comparing with the form

$$Mdx + Ndy = 0$$

it is evident that
$$M = \frac{\partial u}{\partial x}$$
$$N = \frac{\partial u}{\partial y}.$$

219. Test for an exact differential equation.

In § 193 it was shown that if $\frac{\partial u}{\partial x}, \frac{\partial u}{\partial y}$, be the first partial differential coefficients, that of the second derivatives are

$$\frac{\partial}{\partial y}\left(\frac{\partial u}{\partial x}\right) \quad \text{and} \quad \frac{\partial}{\partial x}\left(\frac{\partial u}{\partial y}\right).$$

These are denoted by

$$\frac{\partial^2 u}{\partial y \partial x} \quad \text{and} \quad \frac{\partial^2 u}{\partial x \partial y}.$$

It was further shown that these are equal.

Consequently, if the equation $Mdx + Ndy = 0$ is an exact differential equation

$$\frac{\partial}{\partial y}(M) = \frac{\partial}{\partial x}(N).$$

$\therefore$ if the function M be differentiated on the assumption that y is variable and x constant, and N be differentiated with x variable and y constant

the results are equal.

Thus, in the example above

$$(3x^2 + 6xy)dx + (3x^2 + 3y^2)dy = 0$$

$$\frac{\partial}{\partial y}(3x^2 + 6xy) = 6x$$

$$\frac{\partial}{\partial x}(3x^2 + 3y^2) = 6x.$$

Hence the equation is exact.

220. Solution of an exact differential equation.

The integral $\int M dx$, *i.e.*, M, integrated assuming x variable and y constant, will contain those terms in $N dy$ which contain x. Hence the following rule:

(1) *Integrate* $\int M dx$, *assuming* y *is constant.*

(2) ,, $\int N dy$,, x ,,

Add the results, but the terms common to both are written down once only.

Thus in the above example

$$\int (3x^2 + 6xy)dx = x^3 + 3x^2y$$

$$\int (3x^2 + 3y^2)dy = 3x^2y + y^3.$$

Since $3x^2y$ occurs in each, it is written down once only.

$\therefore$ the solution is

$$x^3 + 3x^2y + y^3 = c \qquad \text{(see § 218)}$$

221. Integrating factors.

Equations which are not exact may often be made so by multiplying throughout by a suitable function of x and y.

Such a factor is an integrating factor (see § 213).

It represents common factors which have been cancelled out during the process by which the equation was obtained from its primitive. This factor is not always easily obtained. In some cases it may be found by inspection; sometimes by the method of trial; in others there are rules for obtaining it. The work in this chapter will be confined to the simpler cases only.

222. Worked examples.

Example 1. *Solve the differential equation*

$$(x + y)dx + (x + 3y)dy = 0.$$

Applying the test of § 219, the second partial differential coefficient in each case is 1.

$\therefore$ the equation is exact.

Applying the rule of § 220

$$\int (x + y)dx = \tfrac{1}{2}x^2 + xy$$

$$\int (x + 3y)dy = xy + \tfrac{3}{2}y^2$$

$\therefore$ the solution is

$$\tfrac{1}{2}x^2 + xy + \tfrac{3}{2}y^2 = c_1$$

or

$$x^2 + 2xy + 3y^2 = c.$$

Example 2. *Solve the differential equation*

$$(6x^2 - 10xy + 3y^2)dx + (-5x^2 + 6xy - 3y^2)dy = 0.$$

Testing

$$\frac{\partial}{\partial y}(6x^2 - 10xy + 3y^2) = -10x + 6y$$

$$\frac{\partial}{\partial x}(-5x^2 + 6xy - 3y^2) = -10x + 6y.$$

Hence the equation is exact.

Solving by method of § 220

$$\int (6x^2 - 10xy + 3y^2)dx = 2x^3 - 5x^2y + 3xy^2.$$

$$\int (-5x^2 + 6xy - 3y^2)dy = -5x^2y + 3xy^2 - y^3.$$

Writing down the common terms $3xy^2$ and $-5x^2y$ once only the solution is

$$2x^3 - 5x^2y + 3xy^2 - y^3 = c.$$

Example 3. *Solve the differential equation*

$$2ydx + xdy = 0.$$

Applying the test, it is seen that this is not an exact equation.

Multiply throughout by the integrating factor $\frac{1}{xy}$.

Then

$$\frac{2}{x}dx + \frac{1}{y}dy = 0.$$

This is exact.

Solving $\int \frac{2}{x} dx = 2 \log x = \log x^2$

$\int \frac{1}{y} dy = \log y.$

$\therefore$ the solution is

$$\log x^2 + \log y = c^1$$

or $$x^2y = \log c^1$$

or $$x^2y = c.$$

Exercise 46.

Solve the differential equations.

1. $(x + y)dx + (x + 4y)dy = 0.$
2. $(2x + y + 1)dx + (x + 2y - 1)dy = 0.$
3. $2xdy + ydy = 3x^2dx.$
4. $(x^2 - y)dx + (x - y^2)dy = 0.$
5. $(2xy - y^2 + 2x)dx + (x^2 - 2xy + 2y)dy = 0.$
6. $ydx - (x + y^2)dy = 0$ $\left(\text{Integrating factor } \frac{1}{y^2}.\right)$
7. $xdy - ydx = x^2dx.$ $\left(\text{Integrating factor } \frac{1}{x^2}.\right)$
8. $x(1 - y^3)dy + ydx = 0.$
9. $(x^2 - y^2)dx + xydy = 0.$
10. $(y^2 - x^2)dy + 2xydx = 0.$
11. $xdy + ydx = xy^3dy.$

INTEGRALS OF STANDARD FORMS AND OTHER USEFUL INTEGRALS

I. Algebraic functions.

(1) $\int x^n dx = \frac{1}{n+1} x^{n+1}.$

(2) $\int \frac{dx}{x} = \log_e x.$

(3) $\int a^x dx = a^x \times \log_a e.$

(4) $\int e^x dx = e^x.$

II. Trigonometric functions.

(5) $\int \sin x dx = -\cos x.$

$$\int \sin ax dx = -\frac{1}{a}\cos ax.$$

(6) $\int \cos x dx = \sin x.$

$$\int \cos ax dx = \frac{1}{a}\sin ax.$$

(7) $\int \tan x dx = -\log \cos x = \log \sec x.$

$$\int \tan ax dx = \frac{1}{a}\log \sec ax.$$

(8) $\int \cot x dx = \log \sin x.$

$$\int \log \cot ax dx = \frac{1}{a}\log \sin ax.$$

III. Hyperbolic functions.

(9) $\int \sinh x dx = \cosh x.$

$$\int \sinh ax dx = \frac{1}{a}\cosh ax.$$

(10) $\int \cosh x dx = \sinh x.$

$$\int \cosh ax dx = \frac{1}{a}\sinh ax.$$

(11) $\int \tanh x dx = \log \cosh x.$

$$\int \tanh ax dx = \frac{1}{a}\log \cosh ax.$$

(12) $\int \coth x dx = \log \sinh x.$

$$\int \coth ax dx = \frac{1}{a}\log \sinh ax.$$

IV. Inverse trigonometrical functions.

(14) $\int \frac{dx}{\sqrt{a^2 - x^2}} = \sin^{-1}\frac{x}{a}$ or $-\cos^{-1}\frac{x}{a}.$

(15) $\int \frac{dx}{a^2 + x^2} = \frac{1}{a}\tan^{-1}\frac{x}{a}$ or $-\frac{1}{a}\cot^{-1}\frac{x}{a}.$

(16) $\int \frac{dx}{x\sqrt{x^2 - a^2}} = \frac{1}{a}\sec^{-1}\frac{x}{a}$ or $-\frac{1}{a}\text{cosec}^{-1}\frac{x}{a}.$

V. Inverse hyperbolic functions.

(17) $\displaystyle\int \frac{dx}{\sqrt{x^2+a^2}} = \sinh^{-1}\frac{x}{a}$ or $\log\{x+\sqrt{x^2+a^2}\}$.

(18) $\displaystyle\int \frac{dx}{\sqrt{x^2-a^2}} = \cosh^{-1}\frac{x}{a}$ or $\log\{x+\sqrt{x^2-a^2}\}$.

---

(19) $\displaystyle\int \frac{dx}{a^2-x^2} = \frac{1}{a}\tanh^{-1}\frac{x}{a}$ or $\displaystyle\frac{1}{2a}\log\frac{a+x}{a-x}$.

(20) $\displaystyle\int \frac{dx}{x^2-a^2} = -\frac{1}{a}\coth^{-1}\frac{x}{a}$ or $\displaystyle\frac{1}{2a}\log\frac{x-a}{x+a}$.

---

(21) $\displaystyle\int \frac{dx}{x\sqrt{a^2-x^2}} = -\frac{1}{a}\operatorname{sech}^{-1}\frac{x}{a}$ or

$$-\frac{1}{a}\log\left\{\frac{a+\sqrt{a^2-x^2}}{x}\right\}.$$

(22) $\displaystyle\int \frac{dx}{x\sqrt{a^2+x^2}} = -\frac{1}{a}\operatorname{cosech}^{-1}\frac{x}{a}$ or

$$-\frac{1}{a}\log\left\{\frac{a+\sqrt{a^2+x^2}}{x}\right\}.$$

---

(17) (*a*) $$\int \frac{dx}{\sqrt{b^2x^2+a^2}} = \frac{1}{b}\sinh^{-1}\frac{bx}{a}$$

$$= \frac{1}{b}\log\{bx+\sqrt{b^2x^2+a^2}\}.$$

(18) (*a*) $$\int \frac{dx}{\sqrt{b^2x^2-a^2}} = \frac{1}{b}\cosh^{-1}\frac{bx}{a}$$

$$= \frac{1}{b}\log\{bx+\sqrt{b^2x^2-a^2}\}.$$

---

(19) (*a*) $$\int \frac{dx}{a^2-b^2x^2} = \frac{1}{ba}\tanh^{-1}\frac{bx}{a}$$

$$= \frac{1}{2ba}\log\frac{a+bx}{a-bx}.$$

(20) (*a*) $$\int \frac{dx}{b^2x^2-a^2} = -\frac{1}{ba}\coth^{-1}\frac{bx}{a}$$

$$= \frac{1}{2ba}\log\frac{bx-a}{bx+a}.$$

---

(21) (a) $\int \frac{dx}{x\sqrt{a^2 - b^2x^2}} = -\frac{1}{a}\operatorname{sech}^{-1}\frac{bx}{a}$

$= -\frac{1}{a}\log\left\{\frac{a + \sqrt{a^2 - b^2x^2}}{bx}\right\}.$

(22) (a) $\int \frac{dx}{x\sqrt{a^2 + b^2x^2}} = -\frac{1}{a}\operatorname{cosech}^{-1}\frac{bx}{a}$

$= -\frac{1}{a}\log\left\{\frac{a + \sqrt{a^2 + b^2x^2}}{bx}\right\}.$

Squares of the circular functions.

(23) $\int \sin^2 x dx = \frac{1}{2}(x - \frac{1}{2}\sin 2x).$

(24) $\int \cos^2 x dx = \frac{1}{2}(x + \frac{1}{2}\sin 2x).$

(25) $\int \tan^2 x dx = \tan x - x.$

(26) $\int \cot^2 x dx = -(\cot x + x).$

(27) $\int \sec^2 x dx = \tan x$

(28) $\int \operatorname{cosec}^2 x dx = -\cot x.$

Other useful integrals.

(29) $\int \sqrt{a^2 - x^2} dx = \frac{a^2}{2}\sin^{-1}\frac{x}{a} + \frac{1}{2}x\sqrt{a^2 - x^2}.$

(30) $\int \sqrt{x^2 - a^2} = \frac{1}{2}x\sqrt{x^2 - a^2} - \frac{a^2}{2}\cosh^{-1}\frac{x}{a}$

or $\frac{1}{2}x\sqrt{x^2 - a^2} - \frac{a^2}{2}\log\frac{x + \sqrt{x^2 - a^2}}{a}.$

(31) $\int \sqrt{x^2 + a^2} dx = \frac{1}{2}x\sqrt{x^2 + a^2} + \frac{a^2}{2}\sinh^{-1}\frac{x}{a}$

or $\frac{1}{2}x\sqrt{x^2 + a^2} + \frac{a^2}{2}\log\frac{x + \sqrt{x^2 + a^2}}{a}.$

(32) $\int \sec x dx = \log\tan\left(\frac{\pi}{4} + \frac{x}{2}\right)$

or $\log(\sec x + \tan x).$

(33) $\int \operatorname{cosec} x dx = \log\tan\frac{x}{2}.$

(34) $\int \log x dx = x(\log x - 1).$

ANSWERS

p. 17. Exercise 1.

1. $-1, 1, 1, 17, 2a^2 - 4a + 1, 2(x + \delta x)^2 - 4(x + \delta x) + 1$.
2. $7, 0, -5, a(a + 6), \dfrac{(1 - a)(1 + 5a)}{a^2}, 0$.
3. $0, 1, \frac{1}{2}, \dfrac{\sqrt{3}}{2}, -1$.
4. 9, 9·61, 9·0601, 9·006001, 6·001.
5. 1, 2, 8, 1·414 . . .
6. $7, 0, -11, -x^3 - 5x^2 + 3x + 7$.
7. $3(t + \delta t)^2 + 5\ t + \delta t) - 1$. 8. $2x \,.\, \delta x + 2 \,.\, \delta x + (\delta x)^2$.
9. (1) $x^3 + 3x^2 \,.\, \delta x + 3x(\delta x)^2 + (\delta x)^3$.
   (2) $3x^2 \,.\, \delta x + 3x(\delta x)^2 + (\delta x)^3$.
   (3) $3x^2 + 3x(\delta x) + (\delta x)^2$.
10. (1) $2x^2 + 4hx + 2h^2$. (2) $4hx + 2h^2$.
    (3) $4x + 2h$.

p. 31. Exercise 2.

1. (*a*) 0; (*b*) values less than 1;
   (*c*) $1, 1·25, 2, 5, 10, -2, -1, -\frac{1}{2}, -\frac{1}{3}$;
   (*d*) infinity
   (*e*) the graph is a hyperbola similar to that of Fig. 3, but the y axis is at $x = 1$.
2. (*a*) 3·1, 3·01, 3·001, 3·000001; (*b*) 3;
3. (*a*) 5; (*b*) infinity.
4. (*a*) 11, 5, 3, 2·5, 2·1, 2·01; (*b*) 2.
5. 1. 6. 2. 7. $3x^2$. 8. $\frac{1}{2}$. 9. $\frac{4}{3}$.

p. 43. Exercise 3.

1. 1·5; 1·2.
2. (*a*) 2·5; (*b*) − 0·8; (*c*) $-\dfrac{b}{a}$.
3. $y = 1·2x + 4$.
4. $\delta s = 9·8t \times (\delta t) + 4·9(\delta t)^2$; $\dfrac{s}{t} = 9·8t + 4·9(\delta t)$;
   (1) 20·58; (2) 20·09; (3) 19·649;
   (4) 19·6049; Velocity is 19·60 ms^{-1}.
5. 6.
6. $\delta y = 3x^2(\delta x) + 3x(\delta x)^2 + (\delta x)^3$;
   $\dfrac{\delta y}{\delta x} = 3x^2 + 3x(\delta x) + (\delta x)^2$; 12.
7. $\delta y = \dfrac{-\delta x}{x^2 + x\delta x}$; $\dfrac{\delta y}{\delta x} = -\dfrac{1}{x^2 + x\delta x}$; gradient = − 1;
   slope = 135°.
8. (1) 2; (2) 2. 9. (1) 12; (2) 8.

p. 56. **Exercise 4.**

1. $7x^6$; 5; $\frac{1}{3}$; $0{\cdot}06$; $\frac{5}{4}x^4$; $60x^3$; $4x^5$; $4{\cdot}5x^2$; $32x$.
2. $4bx^3$; $\dfrac{6ax^5}{b}$; apx^{p-1}; $2ax^{2a-1}$; $2(2b+1)x^{2b}$; $8\pi x$.
3. 6; $0{\cdot}54$; -3; p.
4. $\frac{1}{2}x^2$; $\frac{3}{2}x^2$; $\frac{1}{3}x^3$; $\frac{1}{12}x^3$; $\frac{1}{6}x^6$; $\dfrac{x^{n+1}}{n+1}$; $\dfrac{x^{2a+1}}{2a+1}$; $\dfrac{x^4}{6}$; $\frac{4}{3}ax^3$.
5. a. 6. $20t$. 7. $2\pi r$. 8. $4\pi r^2$.
9. $\dfrac{5}{2\sqrt{x}}$; $-\dfrac{5}{x^2}$; $-\dfrac{5}{2x^{\frac{3}{2}}}$; $\dfrac{2}{3\sqrt[3]{x}}$; $\frac{3}{4}\sqrt[4]{\dfrac{2}{x}}$.
10. $\dfrac{0{\cdot}4}{x^{0{\cdot}6}}$; $\dfrac{1{\cdot}6}{x^{0{\cdot}8}}$; $-\dfrac{1{\cdot}6}{x^{1{\cdot}2}}$; $\dfrac{-24}{x^5}$; $-\dfrac{p}{x^{p+1}}$.
11. $19{\cdot}2x^{2{\cdot}2}$; $\dfrac{-3}{x^{2{\cdot}5}}$; $\dfrac{20{\cdot}3}{x^{0{\cdot}3}}$; $-\dfrac{18}{5x^{\frac{8}{5}}}$. 12. $-\dfrac{40}{v^3}$. 13. $1{\cdot}5$; 0.
14. 24. 15. $-0{\cdot}02$, $-0{\cdot}5$, -2, -8. 16. $-\dfrac{2}{x^3}$.
17. $x = 1$. 18. $x = \dfrac{1}{\sqrt{3}}$. 19. $x = \frac{1}{16}$. 20. $x = \frac{1}{2}$.

p. 60. **Exercise 5.**

1. $12x + 5$. 2. $9x^2 + 1$. 3. $16x^3 + 6x - 1$.
4. $x + \frac{1}{7}$. 5. $-\dfrac{5}{x^2} + 4$. 6. $-\dfrac{4}{x^2} + \dfrac{4}{x^3}$.
7. $5 - 2x + 9x^2$. 8. $\dfrac{4}{\sqrt{x}}$. 9. $u + at$.
10. $5 + 32t$. 11. $6t - 4$. 12. $3ax^2 + 2bx + c$.
13. $2x - \dfrac{2}{x^3}$. 14. $\dfrac{1}{2\sqrt{x}}\left(1 - \dfrac{1}{x}\right)$. 15. $3(1 + x)^2$.
16. $2nx^{2n-1} - 2nx$. 17. $\dfrac{1}{2\sqrt{x}} + \dfrac{1}{3\sqrt[3]{x^2}} - \dfrac{2}{x^2}$.
18. 3; $x = \frac{3}{4}$. 19. $x = +2$ or -2.
20. $2, -1, 2$. 21. $x = +1$ or $x = -1$.

p. 62. **Exercise 6.**

1. $12x + 5$. 2. $\frac{3}{2}x^2 + 2x + \frac{1}{2}$. 3. $9x^2 + 2x - 10$.
4. $8x^3 + 10x$. 5. $12x^3 + 33x^2 - 8x$. 6. $3x^2$.
7. $3x^2$. 8. $4x^3 + 12x^2 + 6x - 8$. 9. $4x^3$.
10. $4x^3 - 2x + 2$. 11. $3x^2$. 12. $24x^3 + 6x^2 - 22x - 3$.
13. $4x^3$. 14. $18x^2 + 26x + 9$.
15. $(2ax + b)(px + q) + p(ax^2 + bx + c)$.
16. $\dfrac{1}{2\sqrt{x}}(2x - 1)(x^2 + x + 1) + 2\sqrt{x}(x^2 + x + 1)$
$+ \sqrt{x}(2x - 1)(2x + 1)$.
17. $3\sqrt{x}(\sqrt{x} + 2)(\sqrt{x} - 1) + x(2\sqrt{x} + 1)$.

p. 65. Exercise 7.

1. $\dfrac{-6}{(2x-1)^2}$.
2. $\dfrac{6x}{(1-3x^2)^2}$.
3. $\dfrac{2}{(x+2)^2}$.
4. $\dfrac{1}{(x+2)^2}$.
5. $\dfrac{11}{(2x+3)^2}$.
6. $\dfrac{-2b}{(x-b)^2}$.
7. $\dfrac{2b}{(x+b)^2}$.
8. $\dfrac{x^2-8x}{(x-4)^2}$.
9. $\dfrac{-8x}{(x^2-4)^2}$.
10. $\dfrac{1-x}{2\sqrt{x}(x+1)^2}$.
11. $\dfrac{x-1}{2x^{\frac{3}{2}}}$.
12. $\dfrac{-1}{\sqrt{x}(\sqrt{x}-1)^2}$.
13. $\dfrac{6x^2}{(x^3+1)^2}$.
14. $\dfrac{-2x^2+2}{(x^2-x+1)^2}$.
15. $\dfrac{5x^2-10x}{(3x^2+x-1)^2}$.
16. $\dfrac{x^2-1}{x^2}$.
17. $\dfrac{4x^3(2a^2-x^2)}{(a^2-x^2)^2}$.
18. $\dfrac{-5}{(2-3x)^2}$.
19. $\dfrac{x^2-4x+2}{x^2-4x+4}$.
20. $\dfrac{-(x+3x^{\frac{1}{2}})}{x^3}$.

p. 71. Exercise 8.

1. $4(2x+5)$; $-20(1-5x)^3$; $(3x+7)^{-\frac{2}{3}}$.
2. $\dfrac{2}{(1-2x)^2}$; $-4(1-2x)$; $\dfrac{-1}{\sqrt{1-2x}}$.
3. $10x(x^2-4)^4$; $-3x\sqrt{1-x^2}$; $\dfrac{3x}{\sqrt{3x^2-7}}$.
4. $\dfrac{4x}{(1-2x^2)^2}$; $\dfrac{-2x}{\sqrt{1-2x^2}}$; $\dfrac{1-2x^2}{\sqrt{1-x^2}}$.
5. $\dfrac{1}{(4-x)^2}$; $\dfrac{1}{2(4-x)^{\frac{3}{2}}}$; $\dfrac{2}{(4-x)^3}$.
6. $\dfrac{-2x}{(x^2-1)^2}$; $\dfrac{-x}{(x^2-1)^{\frac{3}{2}}}$; $\dfrac{1}{(1+x^2)^{\frac{3}{2}}}$.
7. $\dfrac{x}{(1-x^2)^{\frac{3}{2}}}$; $\dfrac{1}{2\sqrt{\{x(1-x)^3\}}}$; $\dfrac{-1}{(1+x)^{\frac{3}{2}}(1-x)^{\frac{1}{2}}}$.
8. $\dfrac{1-x-x^2}{(1+x)\sqrt{1-x^2}}$; $\dfrac{2x}{3(x^2+1)^{\frac{2}{3}}}$.
9. $\dfrac{x}{\sqrt{a^2+x^2}}$; $\dfrac{-x}{(a^2+x^2)^{\frac{3}{2}}}$.
10. $\dfrac{2x-1}{2\sqrt{1-x+x^2}}$; $-4nx(1-2x^2)^{n-1}$.

11. $\dfrac{x(2a^2 - x^2)}{(a^2 - x^2)^{\frac{3}{2}}}$; $2\left(x + \dfrac{1}{x}\right)\left(1 - \dfrac{1}{x^2}\right)$.

12. $\dfrac{-3x^2}{2(1 + x^3)^{\frac{3}{2}}}$; $\dfrac{-(1 + x)}{x^2\sqrt{1 + 2x}}$.

13. $\dfrac{1}{(1 + x^2)^{\frac{3}{2}}}$; $\dfrac{-1}{x^2\sqrt{1 + x^2}}$.

14. $\dfrac{-4x + 3}{2(2x^2 - 3x + 4)^{\frac{3}{2}}}$; $\dfrac{4x - 5x^2}{2\sqrt{1 - x}}$.

15. $\dfrac{1}{(1 - x)\sqrt{1 - x^2}}$; $\dfrac{3(x + 1)}{\sqrt{2x + 3}}$.

16. $\dfrac{-6x + 7y}{7x + 18y}$.

17. $-\dfrac{2x(x^2 + y^2) - x}{2y(x^2 + y^2) + y}$.

18. $-\dfrac{x^2 - y}{y^2 - x}$.

19. $-\dfrac{x^{n-1}}{y^{n-1}}$.

20. $\dfrac{dy}{dx} = \dfrac{-2x + 3}{2y + 4} = \frac{1}{6}$ at (1, 1).

p. 76. Exercise 9.

1. $x(3x - 2)$; $2(3x - 1)$; 6.
2. $2bx^{2b-1}$; $2b(2b - 1)x^{2b-2}$; $2b(2b - 1)(2b - 2)x^{2b-3}$.
3. $20x^3 - 9x^2 + 4x - 1$; $60x^2 - 18x + 4$; $120x - 18$.
4. $50x^4 - 12x^2 + 5$; $200x^3 - 24x$; $600x^2 - 24$.
5. $-\dfrac{1}{x^2}$; $\dfrac{2}{x^3}$; $-\dfrac{6}{x^4}$.
6. $\dfrac{1}{2x^{\frac{1}{2}}}$; $-\dfrac{1}{4x^{\frac{3}{2}}}$; $\dfrac{3}{8x^{\frac{5}{2}}}$.
7. $\dfrac{1}{\sqrt{2x + 1}}$; $\dfrac{-1}{\sqrt{(2x + 1)^3}}$; $\dfrac{3}{\sqrt{(2x + 1)^5}}$.
8. $-\dfrac{2}{x^3}$; $\dfrac{6}{x^4}$; $-\dfrac{24}{x^5}$.
9. $\dfrac{\lfloor n}{2a}\left\{\dfrac{1}{(a - x)^{n+1}} + \dfrac{(-1)^n}{(a + x)^{n+1}}\right\}$.
10. -5; $\frac{5}{12}$; the lowest point on the curve.
11. -7; 2; 0 and $\frac{10}{3}$.
12. $x = 3$, $x = 2$; 2·5 = 0·25 (the lowest point on the curve).

p. 100 Exercise 10.

1. $\dfrac{dy}{dx} = 2x - 2$; -4, -2, 2, 4; $x = 1$; $\dfrac{d^2y}{dx^2}$ is positive; point is a minimum.
2. $\dfrac{dy}{dx} = 3 - 2x$; 3, 1, -1, -3; 1·5; negative; maximum.
3. (1) $x = \frac{1}{4}$; minimum. (2) $x = \frac{1}{3}$; maximum.
   (3) $x = -2$; minimum. (4) $x = -\frac{1}{4}$; minimum.
4. (1) Min. value -16, $x = 2$; max. value $+16$, $x = -2$.
   (2) Max. value 5, $x = 1$; min. value 4, $x = 2$.

(3) Max. value 12, $x = 0$; min. value -20, $x = 4$.
(4) Max. value 41, $x = -2$; min. value $9\frac{3}{4}$, $x = \frac{1}{2}$.
(5) Max. value 2, $x = 3$; min. value -2, $x = 1$.

5. Max. value 4, $x = 0$; min. value 0, $x = 2$.
6. Max. value -4, $x = -\frac{1}{2}$; min. value 4, $x = \frac{1}{2}$.
7. 5, 5.
8. $\dfrac{u^2 \sin^2 \theta}{2g}$; $\dfrac{u^2 \sin^2\theta}{g}$.
9. height = diameter.
10. 2·52 m; depth 1·26 m.
11. $s = 3 + 4{\cdot}8t - 1{\cdot}6t^2$; 6·6.
12. 4·5.
13. 4·42 cm (approx.).
14. 1·5 m deep, 0·5 m broad.
15. (1) $x = -3$. (2) $x = 2$. (3) $x = -\frac{1}{2}$.
16. Max. $= +0{\cdot}385$; min. $= -0{\cdot}385$; gradient $= -1$.
17. $x = 0$.
18. Height: 734·69; Time: 11·24.
19. Centre of beam.

p. 108. Exercise 11.

1. $3 \cos x$.
2. $3 \cos 3x$.
3. $-\frac{1}{2} \sin \dfrac{x}{2}$.
4. $\frac{1}{3} \sec.^2 \dfrac{x}{3}$.
5. $0{\cdot}6 \sec 0{\cdot}6x \tan 0{\cdot}6x$.
6. $-\frac{1}{6} \operatorname{cosec} \dfrac{x}{6} \cot \dfrac{x}{6}$.
7. $2(\cos 2x - \sin 2x)$.
8. $3(\cos 3x + \sin 3x)$.
9. $\sec x(\tan x + \sec x)$.
10. $4 \cos 4x - 5 \sin 5x$.
11. $-\frac{1}{2} \sin \frac{1}{2}\theta + \frac{1}{4} \cos \frac{1}{4}\theta$.
12. $2 \cos \left(2x + \dfrac{\pi}{2}\right)$.
13. $\sin (3\pi - x)$.
14. $\frac{1}{2} \operatorname{cosec} (a - \frac{1}{2}x) \cot (a - \frac{1}{2}x)$.
15. $3 \sin^2 x \cos x$.
16. $3x^2 \cos x^3$.
17. $-6 \cos^2 (2x) \sin (2x)$.
18. $2x \sec (x^2) \tan (x^2)$.
19. $\dfrac{-\sec^2 (\sqrt{1 - x})}{2\sqrt{1 - x}}$.
20. $n(a \cos nx - b \sin nx)$.
21. $a \sin x$.
22. $\sec^2 \dfrac{x}{2}$.
23. $-2 \sin \left(2x + \dfrac{\pi}{2}\right)$.
24. $2 \sec^2 2x - 2 \tan x \sec^2 x$.
25. $2x + \frac{3}{2} \cos \frac{1}{2}x$.
26. $\dfrac{a}{x^2} \sin \dfrac{a}{x}$.
27. $\sin x + x \cos x$.
28. $\dfrac{\sin x - x \cos x}{\sin^2 x}$.
29. $x \sec^2 x + \tan x$.
30. $\dfrac{\tan x - x \sec^2 x}{\tan^2 x}$.
31. $\dfrac{x \sec^2 x - \tan x}{x^2}$.
32. $2 \cos 2x + 8x \cos (2x)^2$.
33. $-6x \cos^2 (x^2) \sin (x^2)$.
34. $2x \tan x + x^2 \sec^2 x$.
35. $-5 \operatorname{cosec}^2 (5x + 1)$.
36. $-6 \cot 3x \operatorname{cosec}^2 3x$.
37. $-\dfrac{\sin x}{2\sqrt{\cos x}}$.
38. $2(\cos^2 2x - \sin^2 2x)$
39. 0.
40. $4 \sin x \cos x$.

41. $\dfrac{\sin x}{(1 + \cos x)^2}$.
42. $\dfrac{2 \sin x}{(1 + \cos x)^2}$.
43. $\dfrac{\sin x - 2x \cos x}{2\sqrt{x} \sin^2 x}$.
44. $2x(\cos 2x - x \sin 2x)$.
45. $\dfrac{2x(\cos 2x + x \sin 2x)}{\cos^2 2x}$.
46. $\sin x + \cos x$.
47. $\dfrac{2 \sin x + x \cos x}{2\sqrt{\sin x}}$.
48. $\dfrac{\sin x \cos x(2 + \sin x)}{(1 + \sin x)^2}$.
49. $\dfrac{\sec^2 x}{(1 - \tan x)^2}$.
50. $\sec x(2 \sec^2 x - \operatorname{cosec}^2 x)$.

p. 116. Exercise 12.

1. Max., $x = \dfrac{\pi}{6}$; min., $x = -\dfrac{\pi}{6}$.
2. Max., $x = \dfrac{\pi}{4}$.
3. Max., $x = \dfrac{\pi}{3}$.
4. Max., $x = \dfrac{\pi}{4}$.
5. Max., $x = \tan^{-1} 2$.
6. Max., $x = \dfrac{\pi}{2}$ or $\sin^{-1} \frac{1}{4}$.
7. Max. $1{\cdot}5\sqrt{3}$ when $x = \frac{2}{3}\pi$; min. $-1{\cdot}5\sqrt{3}$ when $x = \frac{4}{3}\pi$.
8. Max. when $\sin x = +\sqrt{\frac{2}{3}}$; min. when $x = -\sqrt{\frac{2}{3}}$.
9. $33° 42'$ (approx.).
10. Min. -1, when $x = \dfrac{\pi}{4}$.

p. 123. Exercise 13.

1. (a) $\dfrac{4}{\sqrt{1 - 16x^2}}$; (b) $\dfrac{1}{\sqrt{4 - x^2}}$.
2. (a) $\dfrac{-b}{\sqrt{a^2 - x^2}}$; (b) $\dfrac{-1}{\sqrt{9 - x^2}}$.
3. (a) $\dfrac{a}{a^2 + x^2}$; (b) $\dfrac{-1}{1 + (a - x)^2}$.
4. (a) $\dfrac{-4x}{\sqrt{1 - 4x^4}}$; (b) $\dfrac{1}{2\sqrt{x - x^2}}$.
5. (a) $f'(x) = \sin^{-1} x + \dfrac{x}{\sqrt{1 - x^2}}$; (b) $-\dfrac{1}{x\sqrt{x^2 - 1}}$.
6. (a) $\dfrac{3}{\sqrt{6x - 9x^2}}$; (b) $\dfrac{-2}{x\sqrt{x^2 - 4}}$.
7. (a) $\dfrac{1}{x^2 + 2x + 2}$; (b) $2x \tan^{-1} x + 1$.
8. (a) $\dfrac{-1}{2(2 - x)\sqrt{1 - x}}$; (b) $\dfrac{-1}{\sqrt{1 - x^2}}$.
9. (a) $\dfrac{1}{x\sqrt{25x^2 - 1}}$; (b) $\dfrac{2}{x\sqrt{x^4 - 1}}$.

10. (a) 1; (b) $\frac{1}{2}\sqrt{1 + \operatorname{cosec} x}$.

11. (a) $\frac{2}{x\sqrt{a^2x^2-1}}$; (b) $\frac{1}{2\sqrt{x}(1+x)}$.

12. (a) $\frac{2}{1+x^2}$; (b) $\frac{-1}{\sqrt{1-x^2}}$.

13. (a) $\frac{1}{\sqrt{a^2-x^2}}$; (b) $\frac{-2}{x^2+1}$.

14. (a) $\frac{1}{1+x^2}$; (b) $\frac{2}{\sqrt{1-x^2}}$.

15. (a) $f'(x) = \tan^{-1}x + \frac{x}{1+x^2}$; (b) $\sec^2 x \sin^{-1} x + \frac{\tan x}{\sqrt{1-x^2}}$.

p. 136. **Exercise 14.**

1. (a) $5e^{5x}$. (b) $\frac{1}{2}e^{\frac{x}{2}}$. (c) $\frac{1}{2\sqrt{x}}e^{\sqrt{x}}$.
2. (a) $-2e^{-2x}$. (b) $-\frac{5}{2}e^{-\frac{5x}{2}}$. (c) $-2e^{5-2x}$.
3. (a) $-pe^{-px}$. (b) $\frac{1}{a}e^{\frac{x}{a}}$. (c) ae^{ax+b}.
4. (a) $\frac{e^x - e^{-x}}{2}$. (b) $\frac{e^x + e^{-x}}{2}$. (c) $2xe^{x^2}$.
5. (a) $(x+1)e^x$. (b) $(1-x)e^{-x}$. (c) $xe^{-x}(2-x)$.
6. (a) $e^x(x+5)$. (b) $e^x(\sin x + \cos x)$. (c) $10e^x$.
7. (a) $2^x \log_e 2$. (b) $\frac{2 \times 10^{2x}}{0{\cdot}4343}$. (c) $\cos x \times e^{\sin x}$.
8. (a) $x^{n-1}a^x(n + x\log a)$. (b) $2a^{2x+1}\log a$.
   (c) $-\sin x \,.\, e^{\cos x}$.
9. (a) $2bxa^{bx^2}\log a$. (b) $(a+b)^x \log(a+b)$.
   (c) $e^{\tan x}\sec^2 x$.
10. (a) $\frac{1}{x}$. (b) $\frac{2ax+b}{ax^2+bx+c}$.
11. (a) $\frac{2}{x}$. (b) $\frac{3x^2}{x^3+3}$.
12. (a) $1 + \log x$. (b) $\frac{p}{px+q}$.
13. (a) $\cot x$. (b) $-\tan x$.
14. (a) $\frac{2a}{a^2-x^2}$. (b) $\frac{e^x-e^{-x}}{e^x+e^{-x}}$.
15. (a) $\frac{1}{\sqrt{x^2+1}}$. (b) $\frac{1}{2(1+\sqrt{x})}$.
16. (a) $\frac{1}{\sin x}$. (b) $\frac{x}{x^2+1}$. (c) $\frac{e^x(2x-1)}{2x^{\frac{3}{2}}}$.

17. (a) $2xe^{4x}(1 + 2x)$. (b) $-ake^{-kx}(\sin kx - \cos kx)$.
18. (a) $\dfrac{e^{ax}(2ax - 1)}{2x^{\frac{3}{2}}}$. (b) $\frac{1}{2}\cot x$.
19. (a) $x^x(1 + \log x)$. (b) $\dfrac{1}{2\sqrt{x^2 - 1}}$.
20. (a) $\dfrac{1}{1 + e^x}$. (b) $\cos x\,(1 + \log \sin x)$.
21. (a) $\dfrac{\sqrt{a}}{\sqrt{x}(a - x)}$. (b) $e^{ax}\sin x(2\cos x + a\sin x)$.
22. (a) $6x \,.\, a^{3x^2}\log a$. (b) $\dfrac{4}{(e^x + e^{-x})^2}$.
23. (a) $\dfrac{1}{x\sqrt{1 - (\log x)^2}}$. (b) $\dfrac{e^{-x}}{\sqrt{1 - e^{-2x}}}$.
24. (a) $e^{ax}\{a\cos(bx + c) - b\sin(bx + c)\}$.
(b) $-e^{-ax}\{a\cos 3x + 3\sin 3x\}$.
(c) $-e^{-\frac{1}{2}x}\left\{\frac{1}{2}\sin\left(\pi x + \frac{\pi}{2}\right) - \pi\cos\left(\pi x + \frac{\pi}{2}\right)\right\}$.
25. (a) $\dfrac{-a}{x\sqrt{a^2 - x^2}}$. (b) $\dfrac{2}{e^x + e^{-x}}$.
26. (a) a^2e^{ax}, a^3e^{ax}, a^4e^{ax}, a^ne^{ax}.
(b) a^2e^{-ax}, $-a^3e^{-ax}$, a^4e^{-ax}, $(-1)^na^ne^{-ax}$.
(c) $-\dfrac{1}{x^2}$, $\dfrac{1 \times 2}{x^3}$, $-\dfrac{1 \times 2 \times 3}{x^4}$, $\dfrac{(-1)^{n-1}\underline{|n-1}}{x^n}$.

p. 149. **Exercise 15.**

1. (a) $\frac{1}{2}\cosh\dfrac{x}{2}$. (b) $2\cosh 2x$. (c) $\frac{1}{3}\sinh\dfrac{x}{3}$.
2. (a) $a\,\text{sech}^2\,ax$. (b) $\frac{1}{4}\text{sech}^2\dfrac{x}{4}$.
(c) $a(\cosh ax + \sinh ax)$.
3. (a) $-\dfrac{1}{x^2}\cosh\dfrac{1}{x}$. (b) $\sinh 2x$. (c) $3\cosh^2 x\sinh x$.
4. (a) $a\cosh(ax + b)$. (b) $4x\sinh 2x^2$.
(c) $na\sinh^{n-1}ax\cosh ax$.
5. (a) $\cosh 2x$. (b) $2\sinh 2x$. (c) $2\tanh x\,\text{sech}^2\,x$.
6. (a) $\dfrac{2}{\sinh 2x}$. (b) $x\cosh x$. (c) $\tanh x$.
7. (a) $3x^2\sinh 3x + 3x^3\cosh 3x$. (b) 1.
(c) $\cosh x e^{\sinh x}$.
8. (a) $\dfrac{\cosh x}{2\sqrt{\sinh x}}$. (b) 2. (c) $\text{sech}^2\,x e^{\tanh x}$.
9. (a) $\dfrac{1}{\sqrt{x^2 + 4}}$. (b) $\dfrac{1}{\sqrt{x^2 - 25}}$. (c) $\dfrac{-2}{(1 + x)\sqrt{2(1 + x^2)}}$.
10. (a) $\sec x$. (b) $\text{sech}\,x$. (c) $\sec x$.

11. (a) sech x. (b) sec x. (c) $\dfrac{2}{1-x^2}$.

12. (a) $\dfrac{2}{\sqrt{2x(2x+1)}}$. (b) $\dfrac{1}{\sqrt{1+x^2}}$. (c) $\dfrac{-1}{x(x+2)}$.

13. (a) $\dfrac{2}{1-x^4}$. (b) $\frac{1}{2}$ sec x. (c) $\frac{1}{2}$ sech x.

14. (a) $\log\left\{\dfrac{x+\sqrt{x^2+4}}{2}\right\}$. (b) $\log\left\{\dfrac{x+\sqrt{x^2-9}}{3}\right\}$.

(c) $\log\left\{\dfrac{2x+\sqrt{4x^2+9}}{3}\right\}$. (d) $\log\left\{\dfrac{3x+\sqrt{9x^2-4}}{2}\right\}$.

(e) $\frac{1}{2}\log\dfrac{4+x}{4-x}$.

15. (a) $\dfrac{1}{\sqrt{x^2+a^2}}$. (b) $\dfrac{1}{\sqrt{x^2-a^2}}$. (c) $\dfrac{a}{a^2-x^2}$.

p. 160. Exercise 16.

The constant of integration is not shown in the following answers after the first twelve.

1. $\frac{3}{2}x^2 + C$. 2. $\frac{5}{3}x^3 + C$. 3. $\frac{1}{8}x^4 + C$.
4. $0{\cdot}08\, x^5 + C$. 5. $\frac{4}{3}x^9 + C$. 6. $5t^3 + C$.
7. $\frac{1}{2}x + C$. 8. $\theta + C$. 9. $\frac{4}{3}x^3 - \frac{5}{2}x^2 + x + C$.
10. $\frac{3}{5}x^5 - \dfrac{5x^4}{4} + C$. 11. $\frac{8}{3}x^3 - \frac{1}{4}x^2 + C$.
12. $\frac{6}{5}x^5 + \frac{3}{2}x^4 + C$. 13. $\frac{1}{3}x^3 - 9x$. 14. $\frac{2}{3}x^3 + \frac{5}{2}x^2 - 12x$.
15. $-\dfrac{1}{x}$. 16. $-\dfrac{1}{x^3}$. 17. $-\dfrac{1}{0{\cdot}4x^{0{\cdot}4}}$.
18. $\frac{3}{4}x^{\frac{4}{3}}$. 19. $\sqrt{x}$. 20. $\frac{3}{2}x^{\frac{2}{3}}$.
21. $\frac{2}{3}x^{\frac{3}{2}} + 2x^{\frac{1}{2}}$. 22. $\frac{3}{5}x^{\frac{5}{3}} + x + 3x^{\frac{1}{3}}$. 23. $-\dfrac{1}{\sqrt{2x}}$.
24. $\dfrac{\pi}{2}x - 10x^{0{\cdot}5}$. 25. gt. 26. $-\dfrac{1}{2x^2} + \dfrac{1}{x} + \log x - x$.
27. $\frac{2}{3}t^{\frac{3}{2}}$. 28. $x - \frac{1}{9}x^3 - x^{\frac{1}{2}}$. 29. $1{\cdot}4 \log x$.
30. $\log(x+3)$. 31. $\dfrac{1}{a}\log(ax+b)$. 32. $\log\dfrac{(x-1)^3}{(x-2)^4}$.
33. $\log(x^2+4)$. 34. $-\frac{1}{2}\log(3-2x)$. 35. $x + 3\log x$.
36. $\frac{1}{3}x^3 - 7\log x$. 37. $\log x + \dfrac{1}{x} - \dfrac{1}{2x^2}$. 38. $\dfrac{2}{3a}(ax+b)^{\frac{3}{2}}$.
39. $\frac{1}{3}(2x+3)^{\frac{3}{2}}$. 40. $\frac{4}{3}\left(1+\dfrac{x}{2}\right)^{\frac{3}{2}}$. 41. $\dfrac{2}{a}\sqrt{ax+b}$.
42. $-2\sqrt{1-x}$. 43. $\dfrac{1}{3a}(ax+b)^3$. 44. $\dfrac{x^2}{2} + \dfrac{x^3}{3} + \dfrac{x^4}{4} + \dfrac{x^5}{5}$.
45. $\frac{1}{2}\log(x^2-1)$. 46. $-\dfrac{1}{a}\log(1+\cos ax)$.

47. $\frac{1}{3}\log(e^{32} + 6)$.
48. $\frac{1}{2}\log(2x + \sin 2x)$.
49. $y = \frac{1}{4}x^4 + C_1x$.
50. $y = 2x^3 + 3$.
51. $y = \frac{5}{6}x^3 + 2x - \frac{11}{6}$.
52. $y = 2x^2 - 5x + 6$.
53. $y = 3x^3 - 5x^2 + 4x + 4$.
54. $s = \frac{4}{3}t^3 + 8t + 10$.

p. 164. Exercise 17.

1. $\frac{3}{2}e^{2x}$.
2. $\frac{1}{3}e^{3x-1}$.
3. $\frac{1}{2}(e^{2x} - e^{-2x}) + 2x$.
4. $ae^{\frac{x}{a}}$.
5. $2\left(e^{\frac{x}{2}} - e^{-\frac{x}{2}}\right)$.
6. $\frac{1}{a}(e^{ax} + e^{-ax})$.
7. $\frac{1}{3}(e^{3x} + a^{3x}\log_a e)$.
8. $2^x \log_2 e$.
9. $\frac{1}{3}10^{3x}\log_{10} e$.
10. $(a^x - a^{-x})\log_a e$.
11. $\frac{1}{2}e^{x^2}$.
12. $- e^{\cos x}$.
13. $-\frac{1}{3}\cos 3x$.
14. $\frac{3}{5}\sin 5x$.
15. $-2\cos\frac{1}{2}\left(x + \frac{\pi}{3}\right)$.
16. $\frac{1}{2}\sin(2x + \alpha)$.
17. $-3\cos\frac{1}{3}x$.
18. $\frac{1}{3}\cos(\alpha - 3x)$.
19. $\frac{1}{a}\sin ax - \frac{1}{b}\cos bx$.
20. $-\frac{1}{2a}\cos 2ax$.
21. $\frac{1}{3}\sin 3x + 3\cos\frac{x}{3}$.
22. $\log(x + \sin x)$.
23. $\frac{1}{4}\sin^4 x$.
24. $e^{\tan x}$.
25. $\frac{1}{a}\log\sec ax + \frac{1}{b}\log\sin bx$.
26. $\log(1 + \sin^2 x)$.
27. $\frac{1}{2}\sinh 2x$.
28. $\frac{2}{a}\cosh\frac{a}{2}x$.
29. $\frac{1}{3}\log\cosh 3x$.
30. $\frac{1}{b}\{\sin(a - bx) - \cos(a + bx)\}$.
31. $\frac{2}{3}e^{\frac{3x}{2}} + 4e^{\frac{x}{2}} - 2e^{-\frac{x}{2}}$.
32. $\frac{2}{3}\log\sec\frac{3x}{2}$.
33. $3\tan\frac{x}{3}$.
34. $\log(1 + e^x)$.
35. $\log(1 + \tan x)$.
36. $\frac{2}{3}(\sin x)^{\frac{3}{2}}$.

p. 168. Exercise 18.

1. (*a*) $\sin^{-1}\frac{x}{3}$; (*b*) $\cosh^{-1}\frac{x}{3}$ or $\log\{x + \sqrt{x^2 - 9}\}$.
   (*c*) $\sinh^{-1}\frac{x}{3}$ or $\log\{x + \sqrt{x^2 + 9}\}$.
2. (*a*) $\frac{1}{3}\tan^{-1}\frac{x}{3}$. (*b*) $\frac{1}{3}\tanh^{-1}\frac{x}{3}$ or $\frac{1}{6}\log\frac{3 + x}{3 - x}$.
   (*c*) $-\frac{1}{3}\coth^{-1}\frac{x}{3}$ or $\frac{1}{6}\log\frac{x - 3}{x + 3}$.
3. (*a*) $\sin^{-1}\frac{x}{4}$. (*b*) $\frac{1}{4}\tanh^{-1}\frac{x}{4}$ or $\frac{1}{8}\log\frac{4 + x}{4 - x}$.

4. (*a*) $\cosh^{-1}\frac{x}{4}$ or $\log\{x + \sqrt{x^2 - 16}\}$.
   (*b*) $-\frac{1}{4}\coth^{-1}\frac{x}{4}$ or $\frac{1}{8}\log\frac{x-4}{x+4}$.

5. (*a*) $\sinh^{-1}\frac{x}{4}$ or $\log\{x + \sqrt{x^2 + 16}\}$. (*b*) $\frac{1}{4}\tan^{-1}\frac{x}{4}$.

6. (*a*) $\frac{1}{3}\sin^{-1}\frac{3x}{5}$.
   (*b*) $\frac{1}{3}\cosh^{-1}\frac{3x}{5}$ or $\frac{1}{3}\log\{3x + \sqrt{9x^2 - 25}\}$.
   (*c*) $\frac{1}{3}\sinh^{-1}\frac{3x}{5}$ or $\frac{1}{3}\log\{3x + \sqrt{9x^2 + 25}\}$.

7. (*a*) $\frac{1}{6}\tan^{-1}\frac{2x}{3}$. (*b*) $\frac{1}{6}\tanh^{-1}\frac{2x}{3}$ or $\frac{1}{12}\log\frac{3+2x}{3-2x}$.
   (*c*) $-\frac{1}{6}\coth^{-1}\frac{2x}{3}$ or $\frac{1}{12}\log\frac{2x-3}{2x+3}$.

8. (*a*) $\frac{1}{6}\tan^{-1}\frac{3x}{2}$. (*b*) $\frac{1}{3}\sinh^{-1}\frac{3x}{2}$ or $\frac{1}{3}\log\{3x + \sqrt{9x^2 + 4}\}$.
   (*c*) $\frac{1}{3}\cosh^{-1}\frac{3x}{2}$ or $\frac{1}{3}\log\{3x + \sqrt{9x^2 - 4}\}$.

9. (*a*) $\frac{1}{7}\sinh^{-1}\frac{7x}{5}$ or $\frac{1}{7}\log\{7x + \sqrt{49x^2 + 25}\}$.
   (*b*) $\frac{1}{\sqrt{2}}\sinh^{-1}\sqrt{\frac{2}{5}}\,.\,x$ or $\frac{1}{\sqrt{2}}\log\{x\sqrt{2} + \sqrt{2x^2 + 5}\}$.

10. (*a*) $\sin^{-1}\frac{x}{\sqrt{5}}$. (*b*) $\frac{1}{10}\tanh^{-1}\frac{2x}{5}$.

11. (*a*) $\frac{1}{2}\sinh^{-1}\frac{2x}{\sqrt{5}}$ or $\frac{1}{2}\log\{2x + \sqrt{5 + 4x^2}\}$.
    (*b*) $\frac{1}{2}\tanh^{-1}\frac{2x}{5}$.

12. (*a*) $\frac{1}{\sqrt{7}}\sinh^{-1}\frac{\sqrt{7}x}{6}$ or $\frac{1}{\sqrt{7}}\log\{\sqrt{7}x + \sqrt{7x^2 + 36}\}$.
    (*b*) $-\operatorname{cosech}^{-1}x$ or $-\log\left\{\frac{1+\sqrt{1+x^2}}{x}\right\}$.

13. (*a*) $\frac{1}{2}\sec^{-1}\frac{x}{2}$.
    (*b*) $-\frac{1}{2}\operatorname{cosech}^{-1}\frac{x}{2}$ or $-\frac{1}{2}\log\left\{\frac{2+\sqrt{x^2+4}}{x}\right\}$.

14. (*a*) $-\frac{1}{2}\operatorname{sech}^{-1}\frac{x}{2}$ or $-\frac{1}{2}\log\left\{\frac{2+\sqrt{4-x^2}}{x}\right\}$.
    (*b*) $\cosh^{-1}\frac{x}{3} + \frac{1}{3}\sec^{-1}\frac{x}{3}$.

p. 172. **Exercise 19.**

1. $\frac{1}{2}(x - \sin x)$. 2. $\frac{1}{2}(x + \sin x)$.
3. $2\tan\frac{x}{2} - x$. 4. $\frac{1}{4}\left\{\frac{3x}{2} + \sin 2x + \frac{1}{8}\sin 4x\right\}$.

5. $\frac{1}{4}\left\{\frac{3x}{2} - \sin 2x + \frac{1}{8}\sin 4x\right\}$. 6. $-\frac{1}{2}(\cot 2x + x)$.

7. $\frac{1}{2}x - \frac{1}{8}\sin 4x$. 8. $\frac{1}{2}x + \frac{1}{12}\sin 6x$.

9. $\frac{1}{2}x + \frac{1}{4a}\sin 2(ax + b)$. 10. $-\frac{3}{4}\cos x + \frac{1}{12}\cos 3x$.

11. $\frac{1}{12}\sin 3x + \frac{3}{4}\sin x$. 12. $\frac{1}{2}(\sin x - \frac{1}{5}\sin 5x)$.

13. $\frac{1}{4}(\sin 2x + \frac{1}{2}\sin 4x)$. 14. $-\frac{1}{4}(\cos 2x + \frac{1}{3}\cos 6x)$.

15. $-\left(\frac{1}{11}\cos\frac{11x}{2} + \frac{1}{5}\cos\frac{5x}{2}\right)$.

16. $-\frac{1}{2}\left\{\frac{\cos(a+b)x}{a+b} + \frac{\cos(a-b)x}{a-b}\right\}$.

17. $-\frac{1}{4}\cos 2\theta$. 18. $\frac{1}{8}(x - \frac{1}{4}\sin 4x)$.

19. $\tan x - \cot x$. 20. $2\tan x - x$.

21. $\frac{1}{2}\tan^2 x - \log\sec x$. 22. $-\frac{1}{16}(\frac{1}{3}\sin^3 2x - x + \frac{1}{4}\sin 4x)$.

23. $2\sqrt{2}\sin\frac{x}{2}$. 24. $\tan x + \frac{1}{3}\tan^3 x$.

p. 183. Exercise 20.

1. $\frac{9}{2}\sin^{-1}\frac{x}{3} + \frac{x\sqrt{9 - x^2}}{2}$. 2. $\frac{25}{2}\sin^{-1}\frac{x}{5} + \frac{x\sqrt{25 - x^2}}{2}$.

3. $\frac{1}{4}\sin^{-1}2x + \frac{x}{2}\sqrt{1 - 4x^2}$. 4. $\frac{9}{4}\sin^{-1}\frac{2x}{3} + \frac{x}{2}\sqrt{9 - 4x^2}$.

5. $\frac{x\sqrt{x^2 - 4}}{2} - 2\log\frac{x + \sqrt{x^2 - 4}}{2}$.

6. $\frac{1}{2}x\sqrt{x^2 - 25} - \frac{25}{2}\log\frac{x + \sqrt{x^2 - 25}}{5}$.

7. $\frac{1}{2}x\sqrt{x^2 + 49} + \frac{49}{2}\sinh^{-1}\frac{x}{7}$. 8. $\frac{1}{2}x\sqrt{x^2 + 5} + \frac{5}{2}\sinh^{-1}\frac{x}{\sqrt{5}}$.

9. $\frac{8}{5}\sinh^{-1}\frac{5x}{4} + \frac{1}{2}x\sqrt{25x^2 + 16}$.

10. $\frac{1}{2}x\sqrt{x^2 - 3} - \frac{3}{2}\cosh^{-1}\frac{x}{\sqrt{3}}$. 11. $-\frac{\sqrt{1 - x^2}}{x}$.

12. $\frac{1}{2}x\sqrt{1 + x^2} - \frac{1}{2}\sinh^{-1}x$. 13. $\sinh^{-1}x - \frac{\sqrt{1 + x^2}}{x}$.

14. $-\frac{\sqrt{a^2 + x^2}}{a^2 x}$. 15. $\frac{x}{\sqrt{1 - x^2}} - \sin^{-1}x$.

16. $\sqrt{\frac{1 + x}{1 - x}}$ (see formulæ, *Trigonometry*, § 83).

17. $2\log\tan\frac{x}{4}$. 18. $2\log\tan\left(\frac{\pi}{4} + \frac{x}{4}\right)$.

19. $\frac{1}{3}\log\tan\frac{3x}{2}$. 20. $\log\tan x$.

21. $\tan\frac{x}{2}$. 22. $\tan x - \sec x$.

23. $\tan x + \sec x$. 24. $\log\frac{1}{1 - \sin x}$.

25. $\frac{1}{2}\tan^{-1}\left(\frac{1}{2}\tan\frac{x}{2}\right)$.

26. $\frac{1}{2}\tan^{-1}\left(2\tan\frac{x}{2}\right)$.

27. $\frac{2}{3}\tanh^{-1}\left(\frac{1}{3}\tan\frac{x}{2}\right)$.

28. $\frac{1}{3}\log\left(\dfrac{\tan\frac{x}{2}-2}{2\tan\frac{x}{2}-1}\right)$.

p. 187. Exercise 21.

1. $\frac{1}{3}\sin x^3$.
2. $\frac{1}{6}\log\frac{1}{1-2x^3}$. (*Algebra*, p. 211.)
3. $\sqrt{1+x^2}$.
4. $-\frac{2}{5}\sqrt{2-5x}$.
5. $-2\cos\sqrt{x}$.
6. $\frac{2}{3}\sqrt{1+x^3}$.
7. $\frac{1}{2}\log\frac{1}{1+2\cos x}$.
8. $\frac{1}{2}(\log x)^2$.
9. $\frac{1}{3}(5+x^2)^{\frac{3}{2}}$.
10. $\tan^{-1}x^2$.
11. $\frac{1}{30}(x-2)^5(5x+2)$.
12. $\log(x+1)+\frac{4x+3}{2(x+1)^2}$.
13. $\frac{2}{3}(x+2)\sqrt{x-1}$.
14. $\frac{2}{15}(3x+2)(x-1)^{\frac{3}{2}}$.
15. $-\sqrt{5-x^2}$.
16. $\frac{1}{3}(x^2+2)\sqrt{x^2-1}$.
17. $\frac{3}{14}(x-2)^{\frac{4}{3}}(2x+3)$.
18. $\frac{1}{6}(2x^3+3x^2+6x-11)+\log(x-1)$.
19. $\frac{1}{15}(3x^2+4)(x^2-2)^{\frac{3}{2}}$.
20. $2\{\sqrt{x}+3\log(\sqrt{x}-3)\}$.
21. $2\{\frac{1}{2}x-\sqrt{x}+\log(\sqrt{x}+1)\}$.
22. $2\left\{\frac{x\sqrt{x}}{3}-\frac{x}{2}+\sqrt{x}-\log(\sqrt{x}+1)\right\}$.
23. $\frac{1}{5}\cos^5 x-\frac{1}{3}\cos^3 x$.
24. $\frac{1}{3}\sin^3 x-\frac{2}{5}\cos^5 x+\frac{1}{7}\sin^7 x$.
25. $\frac{3}{8}(x^2-3)(x^2+1)^{\frac{1}{3}}$.
26. $\frac{1}{2e^x}-\frac{x}{4}+\frac{1}{4}\log(e^x-2)$.
27. $\frac{1}{45}(1+2x^3)^{\frac{3}{2}}(3x^3-1)$.
28. $-\frac{\sqrt{1+x^2}}{x}$.
29. $\frac{-(1-x^2)^{\frac{3}{2}}}{3x^3}$.
30. $\frac{2}{3}(1+\log x)^{\frac{3}{2}}$.

p. 193. Exercise 22.

1. $\sin x - x\cos x$.
2. $\frac{1}{9}\sin 3x-\frac{1}{3}x\cos 3x$.
3. $(x^2-2)\sin x+2x\cos x$.
4. $x(x^2-6)\sin x+3(x^2-2)\cos x$.
5. $\frac{x^2}{2}(\log x-\frac{1}{2})$.
6. $\frac{x^3}{3}(\log x-\frac{1}{3})$.
7. $\frac{x^4}{4}(\log x-\frac{1}{4})$.
8. $\frac{2}{3}x^{\frac{3}{2}}(\log x-\frac{2}{3})$.
9. $e^x(x-1)$.
10. $e^x(x^2-2x+2)$.
11. $-e^{-ax}\left(\frac{ax+1}{a^2}\right)$.

12. $\frac{1}{5}e^x(\cos 2x + 2 \sin 2x)$.
13. $x \cos^{-1} x - \sqrt{1 - x^2}$.
14. $x \tan^{-1} x - \frac{1}{2} \log (1 + x^2)$.
15. $\frac{1}{2}(x^2 + 1) \tan^{-1} x - \frac{1}{2}x$.
16. $\frac{1}{2}e^x (\sin x - \cos x)$.
17. $\frac{1}{4}x^2 - \frac{1}{4}x \sin 2x - \frac{1}{8} \cos 2x$.
18. $-\frac{1}{4}x \cos 2x + \frac{1}{8} \sin 2x$.
19. $x \tan x - \log \sec x$.
20. $x \cosh x - \sinh x$.
21. $\frac{x^3}{3} \sin^{-1} x + \frac{x^2 + 2}{9} \sqrt{1 - x^2}$.
22. $\frac{x^4}{4} \{(\log x)^2 - \frac{1}{2} \log x + \frac{1}{8})\}$.

p. 195. Exercise 23.

1. $x - 2 \log (x + 2)$.
2. $-\{x + \log (1 - x)\}$.
3. $\frac{1}{b^2}\{a + bx - a \log (a + bx)\}$.
4. $x + 2 \log (x - 1)$.
5. $-x + 2 \log (x + 1)$.
6. $x - 2 \log (2x + 3)$.
7. $\frac{1}{2}x^2 - 2x + 4 \log (x + 2)$.
8. $-x - \frac{1}{2}x^2 + \log (1 - x)$.
9. $\frac{1}{9}\left\{\frac{3x^2}{2} + x + \frac{1}{3} \log (3x - 1)\right\}$.
10. $\frac{1}{b^3}\{\frac{1}{2}(a + bx)^2 - 2a(a + bx) + a^2 \log (a + bx)\}$.
11. $3\{\frac{1}{3}x^3 - x^2 + 4x - 8 \log (x + 2)\}$.
12. $\frac{1}{3}x^3 + \frac{1}{2}x^2 + x + \log (x - 1)$.

p. 200. Exercise 24.

1. $\frac{1}{2} \log \frac{x - 1}{x + 1}$.
2. $\frac{1}{2} \log \frac{1 + x}{1 - x}$.
3. $x + \log \frac{x - 2}{x + 2}$.
4. $\frac{1}{12} \log \frac{2x - 3}{2x + 3}$.
5. $3 \log (x + 2) - 2 \log (x + 4)$.
6. $2 \log (x + 3) + \log (x - 2)$.
7. $\log (2x + 5) + 3 \log (x - 7)$.
8. $\frac{2}{5} \log (x - 1) - \frac{1}{15} \log (3x + 2)$.
9. $3 \log (x + 1) - \frac{5}{4} \log (4x - 1)$.
10. $\log (1 - x) + \frac{2}{1 - x}$.
11. $2 \log (x + 2) + \frac{5}{x + 2}$.
12. $\frac{1}{2} \log (2x + 3) + \frac{1}{2x + 3}$.
13. $x + 2 \log (x - 4) - \log (x + 3)$.
14. $x + \frac{5}{3} \log (x - 2) - \frac{2}{3} \log (x + 1)$.
15. $x^2 + 2 \log (x + 2) - \log (x - 3)$.
16. $\frac{1}{2}x^2 - 2x + 2 \log (x + 1) - \log (x - 1)$.

p. 203. Exercise 25.

1. $-\log x + \frac{1}{2} \log (x + 1) + \frac{1}{2} \log (x - 1)$.
2. $\frac{1}{4}\{\log (x + 2) - \log x\} - \frac{1}{2x}$.

3. $-\frac{3}{2}\log x + \frac{5}{3}\log(x-1) - \frac{1}{6}\log(x+2)$.
4. $\frac{1}{2}\log(x-1) + \frac{1}{5}\log(x-2) + \frac{3}{10}\log(x+3)$.
5. $-\frac{3}{25}\log(x+2) + \frac{3}{25}\log(x-3) - \frac{7}{5(x-3)}$.
6. $-\frac{1}{2(x+1)} + \frac{1}{4}\{\log(x-1) - \log(x+1)\}$.
7. $-\log x + 2\log(x-1) - \frac{1}{x-1} - \frac{1}{(x-1)^2}$.
8. $\log x - \frac{1}{2}\log(x^2+1)$.
9. $\frac{1}{5}\{\log(x-2) - \frac{1}{2}\log(x^2+1)\} - \frac{2}{5}\tan^{-1}x$.
10. $\frac{1}{10}\{\log(x^2+4) - 2\log(x+1)\} + \frac{2}{5}\tan^{-1}\frac{x}{2}$.
11. $\frac{3}{10}\{\log(x^2+4) - 2\log(1-x)\} - \frac{1}{5}\tan^{-1}\frac{x}{2}$.
12. $\frac{1}{4}\{\log(x+1) + \log(x-1) - \log(x^2+1)\} + {}^{-1}$.
13. $\frac{1}{4}\{\log(x-1) - \log(x+1)\} + \frac{1}{2}\tan^{-1}x$.
14. $\log x + 2\tan^{-1}x$.

p. 209. Exercise 26.

1. $\frac{1}{\sqrt{8}}\tan^{-1}\frac{x+3}{\sqrt{8}}$.
2. $\frac{1}{2\sqrt{13}}\log\frac{(x+3)-\sqrt{13}}{(x+3)+\sqrt{13}}$.
3. $\frac{1}{\sqrt{2}}\tan^{-1}\frac{x+2}{\sqrt{2}}$.
4. $\frac{1}{\sqrt{13}}\tan^{-1}\frac{2x+1}{\sqrt{13}}$.
5. $-\frac{1}{2}\log(3x^2+4x+2) + \frac{3}{\sqrt{2}}\tan^{-1}\frac{3x+2}{\sqrt{2}}$.
6. $2\log(x^2-2x-1) - \frac{\sqrt{2}}{4}\log\frac{(x-1)-\sqrt{2}}{(x-1)+\sqrt{2}}$.
7. $\log(x^2+4x+5) + \tan^{-1}(x+2)$.
8. $\frac{1}{3}\log(x+1) - \frac{1}{6}\log(x^2-x+1) + \frac{1}{\sqrt{3}}\tan^{-1}\frac{2x-1}{\sqrt{3}}$.
9. $x - 2\log(x^2+2x+2) + 3\tan^{-1}(x+1)$.
10. $-\frac{3}{2}\log(1-2x-x^2) + \frac{1}{\sqrt{2}}\log\frac{\sqrt{2}+(1+x)}{\sqrt{2}-(1+x)}$.
11. $\frac{2}{3}\log(3x^2+x+3) + \frac{26}{3\sqrt{35}}\tan^{-1}\frac{6x+1}{\sqrt{35}}$.
12. $\frac{2}{3}\log(x+1) + \frac{1}{6}\log(x^2-x+1) + \frac{1}{\sqrt{3}}\tan^{-1}\frac{2x-1}{\sqrt{3}}$.

p. 213. Exercise 27.

1. $\sinh^{-1}(x+3)$ *or* $\log\{(x+3)+\sqrt{x^2+6x+10}\}$.
2. $\sinh^{-1}\frac{x+1}{\sqrt{3}}$ *or* $\log\{(x+1)+\sqrt{x^2+2x+4}\}$.

3. $\cosh^{-1}\frac{x-2}{\sqrt{2}}$ *or* $\log\{(x-2)+\sqrt{x^2-4x+2}\}$.

4. $\sin^{-1}\frac{2x+1}{\sqrt{5}}$.

5. $\frac{1}{\sqrt{5}}\cosh^{-1}\frac{5x-6}{4}$.

6. $\sin^{-1}\frac{x-2}{2}$.

7. $\sqrt{x^2+1}$.

8. $\sqrt{x^2+1}+\sinh^{-1}x$.

9. $\sqrt{x^2-1}+\cosh^{-1}x$.

10. $\sqrt{x^2-x+1}+\frac{1}{2}\sinh^{-1}\frac{2x-1}{\sqrt{3}}$.

11. $2\sqrt{x^2-2x+5}-\sinh^{-1}\frac{x-1}{2}$.

12. $-3\sin^{-1}\frac{x+2}{\sqrt{7}}-2\sqrt{3-4x-x^2}$.

13. $2\sqrt{x^2+x+1}+2\sinh^{-1}\frac{2x+1}{\sqrt{3}}$.

14. $\sqrt{x^2+2x-1}+\cosh^{-1}\frac{x+1}{\sqrt{2}}$.

p. 215.

Exercise 28.

1. $\sqrt{x^2-4}+2\cosh^{-1}\frac{x}{2}$.

2. $\frac{1}{2}\sqrt{4x^2-9}+\frac{3}{2}\cosh^{-1}\frac{2x}{3}$.

3. $\sqrt{x(x+3)}-\frac{3}{2}\cosh^{-1}\frac{2x+3}{3}$.

4. $\frac{1}{2}\sqrt{2x^2-x-3}+\frac{5}{4\sqrt{2}}\cosh^{-1}\frac{4x-1}{5}$.

5. $\frac{1}{2}x^2-\frac{1}{2}x\sqrt{x^2-1}+\frac{1}{2}\cosh^{-1}x$.

6. $-\frac{1}{\sqrt{10}}\sinh^{-1}\frac{3x+10}{x}$.

7. $-\sinh^{-1}\left(\frac{x+2}{x\sqrt{3}}\right)$.

8. $-\cosh^{-1}\left(\frac{2x+1}{x}\right)$.

9. $-\sin^{-1}\left\{\frac{-x}{(x+1)\sqrt{2}}\right\}$.

10. $-\operatorname{cosech}^{-1}x$ *or* $\log\left\{\frac{\sqrt{x^2+1}-1}{x}\right\}$.

11. $\sqrt{1+x^2}+\log\frac{\sqrt{1+x^2}-1}{x}$.

12. $-\frac{\sqrt{1+x^2}}{x}$.

13. $\log(x+\sqrt{1+x^2})-\frac{\sqrt{1+x^2}}{x}$.

14. $\frac{1}{2}x\sqrt{1+x^2}-\frac{1}{2}\log(x+\sqrt{1+x^2})$.

15. $-\frac{2}{3}(x+2)\sqrt{1-x}$.

16. $\log\frac{\sqrt{x+2}-1}{\sqrt{x+2}+1}$.

17. $\log\frac{\sqrt{x+1}-1}{\sqrt{x+1}+1}$.

p. 224. Exercise 29.

1. $\dfrac{3^{n+1}-1}{n+1}$.
2. $4\frac{1}{3}$.
3. $1\frac{5}{6}$.
4. 9.
5. 2·925.
6. $\frac{14}{3}$.
7. $\frac{28}{3}$.
8. $\frac{1}{3}$.
9. 0.
10. 0.
11. 2·16 (approx.).
12. $2(e-1) = 3{\cdot}436$ (approx.).
13. $\frac{1}{4}\pi r^2$.
14. $\dfrac{8\pi\rho a^5}{15}$.
15. $\dfrac{1}{k}(e^{kb} - e^{ka})$.
16. $\dfrac{\pi}{4}$.
17. $\log \sqrt{2}$.
18. 1.
19. $-\frac{1}{4}$.
20. $-\frac{1}{9}$.
21. $\dfrac{\pi}{2} - 1$.
22. $\dfrac{\pi}{4} - \frac{1}{2}\log 2$.
23. $\dfrac{\pi}{2}$.
24. $\frac{2}{9}(7\sqrt{7} - 8)$.
25. $\frac{10}{3}$.
26. $4 - 2\log 3$.
27. $\dfrac{\pi}{2}$.
28. $\dfrac{\pi}{4a}$.
29. $\dfrac{\pi}{2}$.
30. 1.
31. $\log(2+\sqrt{3})$.
32. ·9379.
33. $\dfrac{\pi}{2} - 1$.
34. $1 - \dfrac{\pi}{4}$.
35. $\sin^{-1}\frac{3}{4} - \sin^{-1}\frac{1}{2}$.
36. π.
37. $-\frac{7}{288}$.

p. 232. Exercise 30.

Note.—The omission of an answer indicates that no finite value of the integral exists.

2. $\frac{1}{8}$.
3. $\dfrac{\pi}{2}$.
4. $\frac{1}{2}\log 3$.
6. 1.
7. $\frac{1}{2}$.
9. $1 - \log 2$.
10. $\log 2 - \frac{1}{2}$.
12. $\dfrac{\pi}{2}$.
13. π.
14. $-\frac{1}{4}$.
16. 2.
17. -1.
18. 2.
20. 0.

p. 258. Exercise 31.

1. $152\frac{1}{4}$.
2. $36\frac{3}{4}$.
3. 4·047 (approx.).
4. $\frac{7}{3}$.
5. $25\frac{1}{3}$.
6. 4π.
7. 6·199 and 3·628 (both approx.).
8. 12π.
9. 4·982.
10. $4\log 2$.
11. $\frac{4}{27}$.
12. $e^3 - 1$.
13. $4\frac{1}{2}$.
14. $25\frac{3}{5}$.
15. $\frac{3}{2} - \log 2$.
16. Between -2 and 0, area $= 5\frac{1}{3}$; between 0 and 3 area $= 15\frac{3}{4}$.
17. $341\frac{1}{3}$.
18. $\frac{1}{3}$.
19. 2·3504.
20. 40.
21. $\frac{1}{32}$.

p. 265. Exercise 32.

1. $\frac{3\pi a^2}{2}$. 2. $\frac{1}{8}\pi a^2$; 4. 3. $\frac{1}{2}a^2$; 2.
4. $\frac{4a^2\pi^3}{3}$. 5. $\frac{4a^2}{3}$ (for the integral see Example 19).
6. $\frac{59\pi}{2}$.

p. 267. Exercise 33.

1. 0·637 (approx.). 2. 0·5. 3. 0·256.
4. $\frac{8}{3}$. 5. $\frac{b}{a}$. 6. $\frac{4}{\pi\sqrt{2}}$.
7. $\frac{2a}{\pi}$. 8. $\frac{2v_0^2}{\pi g}$.

p. 272. Exercise 34.

1. 260 m^2. 2. 6·24 m^2. 3. 60·7 mm^2.
4. 1437 m^2. 5. 73·5 m^2.

p. 280. Exercise 35.

1. $\sqrt{5} + \frac{1}{2}\log(2 + \sqrt{5})$. 2. $2\sqrt{5} + \log(2 + \sqrt{5})$.
3. $\frac{335}{27}$. 4. $\frac{1}{2}\left(e - \frac{1}{e}\right)$.
5. $(\sqrt{5} - \sqrt{2}) + \log\frac{\sqrt{5} - 1}{\sqrt{2} - 1}$. 6. 1·732.
7. $2\pi a$. 8. $6{\cdot}1a$.
9. $a\left\{(\sqrt{5} - \sqrt{2}) + \log(1 + \sqrt{2}) - \log\frac{1 + \sqrt{5}}{2}\right\}$.
10. $\frac{3\pi a}{2}$.

p. 294. Exercise 36.

1. (a) $\frac{243\pi}{5}$; (b) 8π. 2. (a) $\frac{2187\pi}{7}$; (b) $\frac{96\pi}{5}$.
3. $\frac{\pi}{6}$. 4. (1) $\frac{20\pi}{3}$; (2) $\frac{8\pi}{3}$.
5. $\frac{64\pi}{3}$. 6. 32π. 7. $\frac{4}{3}\pi a^3$. 8. $\frac{384\pi}{7}$.
9. $\frac{\pi^2}{2}$. 10. $\frac{16\pi}{15}$. 11. $\frac{3\pi}{4}$. 12. $\frac{96\pi}{5}$.

p. 299. **Exercise 37.**

1. $\dfrac{135\pi}{16}$.
2. $2\pi\{\sqrt{2} + \log(\sqrt{2} + 1)\}$.
3. $\dfrac{208\pi}{3}$.
4. $\dfrac{12\pi a^2}{5}$.
5. $2\pi rh$.
6. $\dfrac{\pi}{16}(\sqrt{1000} - 1)$.

p. 306. **Exercise 38.**

1. $\bar{x} = \frac{3}{5}b$; $\bar{y} = 0$.
2. $\bar{x} = 3$; $\bar{y} = \frac{3}{4}\sqrt{10}$.
3. $\bar{x} = \frac{2}{5}$; $\bar{y} = \frac{4}{7}$.
4. $\bar{x} = \frac{9}{8}$; $\bar{y} = \frac{27}{5}$.
5. $\dfrac{4r\sqrt{2}}{3\pi}$ from the centre along the middle radius.
6. $\bar{x} = \dfrac{\pi}{2}$; $\bar{y} = \dfrac{\pi}{8}$.
7. $\dfrac{2r}{\pi}$ from centre along radius at right angles to diameter.
8. $\bar{x} = \dfrac{2r}{\pi}$; $\bar{y} = \dfrac{2r}{\pi}$.
9. $\frac{2}{3} \cdot \dfrac{r \sin \alpha}{\alpha}$.
10. $\frac{3}{4}h$.
11. $\bar{x} = \dfrac{4a}{3\pi}$; $y = \dfrac{4b}{3\pi}$.
12. $\bar{x} = \dfrac{b - a}{\log b - \log a}$; $\bar{y} = \dfrac{k^2(b - a)}{2ab(\log b - \log a)}$.
13. $\bar{x} = 2 \cdot 5$; $\bar{y} = 0$.
14. $\frac{4}{5}b$ from 0.

p. 310. **Exercise 39.**

1. $\frac{1}{3}Ml^2$; $\dfrac{l}{\sqrt{3}}$.
2. $\frac{4}{3}Ma^2$.
3. $\frac{1}{4}Mr^2$.
4. $\frac{1}{4}Mb^2$.
5. (1) $\frac{1}{6}Mh^2$; (2) $\frac{1}{2}Mh^2$.
6. $\frac{3}{10}Mr^2$.
7. $\frac{1}{2}Mr^2$.
8. $\frac{1}{2}Ma^2$.
9. $\frac{3}{7}Mb^2$.
10. $\frac{2}{5}Mr^2$; $r\sqrt{\frac{2}{5}}$.

p. 315. **Exercise 40.**

1. $\frac{4}{3}Ma^2$.
2. $\frac{2}{3}Ma^2$.
3. (1) $\frac{1}{24}Ma^2$; (2) $\frac{1}{12}Ma^2$; (3) $\frac{5}{12}Ma^2$.
4. $\frac{3}{2}Ma^2$.
5. $M\left(\dfrac{a^2}{3} + \dfrac{b^2}{4}\right)$.
6. $\frac{2}{3}Ma^2$.
7. $\frac{2}{5}Ma^2$.
8. $\frac{3}{20}M(r^2 + 4h^2)$.
9. $\frac{1}{4}M(a^2 + b^2)$.
10. (1) $\frac{16}{3}ab^3$; (2) $\frac{8}{3}\dfrac{a^3b^3}{a^2 + b^2}$; (3) $\frac{16}{3}ab(a^2 + b^2)$.

p. 327.

Exercise 41.

1. yx^{y-1}; $x^y \log x$.
2. $-2x \sin(x^2 + y^2)$; $-2y \sin(x^2 + y^2)$.
3. $\dfrac{y}{x^2 + y^2}$; $-\dfrac{x}{x^2 + y^2}$.
4. $3x^2 + 6xy + 6y^2$; $3x^2 + 12xy + 6y^2$.
5. $\dfrac{1}{\sqrt{y^2 - x^2}}$; $\dfrac{-x}{y\sqrt{y^2 - x^2}}$.
6. $\dfrac{y}{x^2 + y^2}$; $\dfrac{-x}{x^2 + y^2}$.
7. $\dfrac{a}{y^2}$; $-\dfrac{2ax}{y^3}$.
9. $\dfrac{ydx - xdy}{y^2}$.
10. $2(ax + by)dx + 2(bx + cy)dy$.
11. $\dfrac{y}{x}dx + \log x dy$.
12. $(2xy + y^3)dx + (x^2 + 3xy^2)dy$.
13. $e^{xy}(ydx + xdy)$.
14. $a^x e^y(\log a dx + dy)$.
15. 0·40 (approx.).
16. $dV = \dfrac{k}{p}dT - \dfrac{kT}{p^2}dp$.
17. Each equals $5x^4$.
18. $-2x$; $-4y$.
19. $150\pi\delta$ m^{3}s^{-1}.

p. 334.

Exercise 42.

1. (a) $\sin a + x \cos a - \dfrac{x^2}{|\underline{2}} \sin a - \dfrac{x^3}{|\underline{3}} \cos a + \ldots$

   (b) $\cos a - x \sin a - \dfrac{x^2}{|\underline{2}} \cos a + \dfrac{x^3}{|\underline{3}} \sin a + \ldots$
2. $e^x\left(1 + h + \dfrac{h^2}{|\underline{2}} + \dfrac{h^3}{|\underline{3}} + \ldots\right)$.
3. $\tan^{-1} x + \dfrac{h}{1 + x^2} - \dfrac{xh^2}{(1 + x^2)^2} - \dfrac{1 - 3x^2}{(1 + x^2)^3} \cdot \dfrac{h^3}{3} + \ldots$
4. $x - \frac{1}{2}x^2 + \frac{1}{6}x^3 - \ldots$
5. $1 - \dfrac{x^2}{|\underline{2}} + \dfrac{x^4}{|\underline{4}} - \dfrac{x^6}{|\underline{6}} + \ldots + \dfrac{x^n}{|\underline{n}} \cos \dfrac{n\pi}{2}$.
6. $x + \frac{1}{3}x^3 + \frac{2}{15}x^5 + \ldots$
7. $\log 2 + \frac{1}{2}x + \frac{1}{8}x^2 - \dfrac{x^4}{192} + \ldots$
8. $1 + x \log a + \dfrac{x^2 (\log a)^2}{|\underline{2}} + \dfrac{x^3 (\log a)^3}{|\underline{3}} + \ldots$
9. $1 - kx + \dfrac{k^2x^2}{|\underline{2}} - \dfrac{k^3x^3}{|\underline{3}} + \ldots$
10. $1 + x + \dfrac{x^2}{2} - \dfrac{x^4}{8} + \ldots$
11. $1 + \dfrac{x^2}{|\underline{2}} + \dfrac{5x^4}{|\underline{4}} + \dfrac{61x^6}{|\underline{6}} + \ldots$

12. $\frac{x^2}{2} + \frac{x^4}{12} + \frac{x^6}{45} + \dots$ 13. $x + \frac{x^3}{6} + \frac{3x^5}{40} + \dots$

14. $-\left(x + \frac{x^2}{2} + \frac{x^3}{3} + \frac{x^4}{4} + \dots\right)$.

15. $x + \frac{x^3}{\underline{|3}} + \frac{x^5}{\underline{|5}} + \frac{x^7}{\underline{|7}} + \dots$

16. $x + x^2 + \frac{2x^3}{\underline{|3}} - \frac{2^2x^5}{\underline{|5}} - \frac{2^3x^6}{\underline{|6}} \dots$

17. $x - \frac{1}{3}x^3 + \frac{2}{15}x^5 - \dots$

p. 339. Exercise 43.

1. $y = \frac{k}{x} + c$. 2. $y = ce^{\frac{x}{a}}$. 3. $y = cx$.

4. $(1 + y)(1 - x) = c$. 5. $\frac{y}{x + 1} = c$.

6. $\sec x = c \sec y$. 7. $x^2 + y^2 - cy = 0$.

8. $(1 + y^2)(1 + x^2) = cx^2$. 9. $\log x^2y - y = c$.

10. $\frac{1 + y}{1 - y} = c \tan x$. 11. $(1 + y^2)(1 + x^2) = cx^2$.

12. $y = e^{\frac{1}{3}x^3 + c}$. 13. $\sqrt{x^2 - 1} - \sqrt{y^2 - 1} = c$.

14. $\frac{x^2}{2} + \log x - \frac{y^3}{3} - \frac{y^2}{2} = c$. 15. $y = ce^{x^2}$.

16. $xy = c$.

p. 342. Exercise 44.

1. $y + 1 = cx^2$. 2. $x^2 + 2xy = c$.

3. $y = x + 1 + ce^x$. 4. $y = ce^{-\frac{x^2}{2}} + 1$.

5. $y = \frac{e^x}{a + 1} + ce^{-ax}$.

6. $y \sec x = \log(\sec x + \tan x) + c$.

7. $y = cx^a + \frac{x}{1 - a} - \frac{1}{a}$. 8. $y + 1 = c \sin x$.

9. $ye^x = x + c$. 10. $y = \frac{x}{1 - a} - \frac{1}{a} + cx^a$.

11. $y = \tan x - 1 + ce^{-\tan x}$. 12. $xy + \log x = c$.

p. 345. Exercise 45.

1. $x^2 + 2xy = c$. 2. $y = x(\log x + c)$

3. $\log \sqrt{x^2 + y^2} - \tan^{-1}\frac{y}{x} = c$. 4. $\frac{x}{x - y} = \log \frac{y - x}{c}$.

5. $x^2 - y^2 = cx$.

6. $\frac{1}{x} - \frac{1}{y} = c$.

7. $xy(x - y) = c$.

8. $xy^2 = c(x + 2y)$.

9. $y = ce^{\frac{y}{x}}$.

10. $xy^2 = c(x + 2y)$.

11. $x^3 + 3x^2 y - 4y^3 = c$.

p. 349. Exercise 46.

1. $x^2 + 2xy + 4y^2 = c$.

2. $x^2 + xy + y^2 + x - y = c$.

3. $2xy - x^3 = c$.

4. $x^3 + y^3 - 3xy = c$.

5. $x^2y - xy^2 + x^2 + y^2 = c$.

6. $\frac{x}{y} - y = c$.

7. $y = x(x + c)$.

8. $\log xy - \frac{1}{3}y^3 = c$.

9. $\log x + \frac{1}{2}\frac{y^2}{x^2} = c$.

10. $x^2 + y^2 - cy = 0$.

11. $\log xy - \frac{y^3}{3} = c$.

CIRCULAR MEASURE OF ANGLES

Degree.	Radians	6′	12′	18′	24′	30′	36′	42′	48′	54
0°	00000	·00175	·00349	·00524	·00698	·00873	01047	·01222	·01396	·01571
1°	·01745	·01920	·02094	·02269	02443	·02618	·02793	·02967	·03142	·03316
2°	·03491	·03665	03840	·04014	04189	·04363	·04538	·04712	·04887	·05061
3°	·05236	·05411	·05585	·05760	·05934	·06109	·06283	·06458	·06632	·06807
4°	·06981	·07156	·07330	·07505	·07679	·07854	·08029	·08203	·08378	·08552
5°	·08727	·08901	·09076	09250	09425	·09599	·09774	·09948	·10123	10297
6°	·10472	10647	10821	10996	11170	·11345	·11519	·11694	·11868	·12043
7°	·12217	12392	·12566	12741	12915	13090	13265	·13439	13614	·13788
8°	·13963	14137	14312	14486	14661	14835	·15010	·15184	15359	·15533
9°	15708	·15882	·16057	16232	16406	·16581	·16755	16930	·17104	·17279
10°	17453	17628	17802	·17977	·18151	18326	·18500	·18675	·18850	·19024
11°	19199	19373	·19548	·19722	19897	·20071	·20246	·20420	·20595	·20769
12°	·20944	·21118	·21293	·21468	·21642	·21817	·21991	·22166	·22340	·22515
13°	·22689	·22864	·23038	·23213	·23387	·23562	·23736	·23911	·24086	·24260
14°	.24435	·24609	·24784	·24958	·25133	25307	·25482	·25656	·25831	·26005
15°	26180	·26354	·26529	·26704	·26878	27053	·27227	·27402	·27576	·27751
16°	·27925	·28100	·28274	·28449	·28623	28798	·28972	·29147	·29322	·29496
17°	29671	·29845	30020	·30194	30369	·30543	·30718	·30892	·31067	·31241
18°	·31416	·31590	·31765	·31940	·32114	·32289	·32463	·32638	·32812	·32987
19°	33161	33336	·33510	·33685	33859	·34034	·34208	·34383	·34558	·34732
20°	34907	·35081	35256	·35430	·35605	35779	35954	·36128	·36303	·36477
21°	·36652	·36826	·37001	·37176	·37350	37525	37699	·37874	·38048	·38223
22°	·38397	·38572	·38746	·38921	39095	39270	·39444	·39619	·39794	39968
23°	·40143	·40317	40492	·40666	40841	41015	41190	·41364	41539	·41713
24°	41888	·42062	·42237	42412	·42586	42761	42935	·43110	·43284	·43459
25°	43633	·43808	·43982	·44157	44331	44506	44680	44855	·45029	·45204
26°	·45379	45553	45728	·45902	46077	46251	46426	·46600	·46775	·46949
27°	·47124	47298	47473	47647	·47822	·47997	·48171	·48346	·48520	·48695
28°	48869	49044	·49218	49393	49567	49742	·49916	·50091	·50265	·50440
29°	50615	·50789	50964	·51138	51313	51487	51662	·51836	·52011	·52185
30°	52360	52534	52709	52883	53058	53233	53407	·53582	·53756	·53931
31°	·54105	54280	54454	54629	54803	54978	·55152	·55327	·55501	·55676
32°	55851	·56025	56200	·56374	56549	56723	·56898	·57072	·57247	·57421
33°	·57596	·57770	57945	58119	·58294	·58469	·58643	·58818	58992	·59167
34°	59341	59516	59690	59865	60039	·60214	·60388	·60563	·60737	·60912
35°	·61087	·61261	·61436	61610	·61785	·61959	·62134	·62308	·62483	·62657
36°	·62832	·63006	·63181	·63355	63530	·63705	·63879	·64054	·64228	·64403
37°	·64577	·64752	·64926	·65101	65275	·65450	·65624	·65799	·65973	·66148
38°	·66323	66497	·66672	·66846	67021	·67195	·67370	·67544	·67719	·67893
39°	·68068	·68242	68417	·68591	·68766	·68941	69115	·69290	·69464	·69639
40°	·69813	69988	70162	70337	70511	·70686	·70860	·71035	·71209	·71384
41°	71558	71733	71908	72082	72257	72431	·72606	·72780	·72955	·73129
42°	73304	73478	73653	73827	·74002	74176	·74351	·74526	·74700	·74875
43°	·75049	75224	75398	75573	75747	·75922	·76096	·76271	·76445	·76620
44°	76794	76969	77144	77318	77493	·77667	77842	·78016	·78191	·78365

Differences—

1	2′	3′	4′	5′
29	58	87	116	145

CIRCULAR MEASURE OF ANGLES

Degree.	Radians	6′	12′	18′	24′	30′	36′	42′	48′	54′
45°	·78540	·78714	·78889	·79063	·79238	·79412	·79587	·79762	·79936	·80111
46°	·80285	·80460	·80634	·80809	·80983	·81158	·81332	·81507	·81681	·81856
47°	·82030	·82205	·82380	·82554	·82729	·82903	·83078	·83252	·83427	·83601
48°	·83776	·83950	·84125	·84299	·84474	·84648	·84823	·84998	·85172	·85347
49°	·85521	·85696	·85870	·86045	·86219	·86394	·86568	·86743	·86917	·87092
50°	·87266	·87441	·87616	·87790	·87965	·88139	·88314	·88488	·88663	·88837
51°	·89012	·89186	·89361	·89535	·89710	·89884	·90059	·90234	·90408	·90583
52°	·90757	·90932	·91106	·91281	·91455	·91630	·91804	·91979	·92153	·92328
53°	·92502	·92677	·92852	·93026	·93201	·93375	·93550	·93724	·93899	·94073
54°	·94248	·94422	·94597	·94771	·94946	·95120	·95295	·95470	·95644	·95819
55°	·95993	·96168	·96342	·96517	·96691	·96866	·97040	·97215	·97389	·97564
56°	·97738	·97913	·98088	·98262	·98437	·98611	·98786	·98960	·99135	·99309
57°	·99484	·99658	·99833	1·00007	1·00182	1·00356	1·00531	1·00706	1·00880	1·01055
58°	1·01229	1·01404	1·01578	1·01753	1·01927	1·02102	1·02276	1·02451	1·02625	1·02800
59°	1·02974	1·03149	1·03323	1·03498	1·03673	1·03847	1·04022	1·04196	1·04371	1·04545
60°	1·04720	1·04894	1·05069	1·05243	1·05418	1·05592	1·05767	1·05941	1·06116	1·06291
61°	1·06465	1·06640	1·06814	1·06989	1·07163	1·07338	1·07512	1·07687	1·07861	1·08036
62°	1·08210	1·08385	1·08559	1·08734	1·08909	1·09083	1·09258	1·09432	1·09607	1·09781
63°	1·09956	1·10130	1·10305	1·10479	1·10654	1·10828	1·11003	1·11177	1·11352	1·11527
64°	1·11701	1·11876	1·12050	1·12225	1·12399	1·12574	1·12748	1·12923	1·13097	1·13272
65°	1·13446	1·13621	1·13795	1·13970	1·14145	1·14319	1·14494	1·14668	1·14843	1·15017
66°	1·15192	1·15366	1·15541	1·15715	1·15890	1·16064	1·16239	1·16413	1·16588	1·16763
67°	1·16937	1·17112	1·17286	1·17461	1·17635	1·17810	1·17984	1·18159	1·18333	1·18508
68°	1·18682	1·18857	1·19031	1·19206	1·19381	1·19555	1·19730	1·19904	1·20079	1·20253
69°	1·20428	1·20602	1·20777	1·20951	1·21126	1·21300	1·21475	1·21649	1·21824	1·21999
70°	1·22173	1·22348	1·22522	1·22697	1·22871	1·23046	1·23220	1·23395	1·23569	1·23744
71°	1·23918	1·24093	1·24267	1·24442	1·24617	1·24791	1·24966	1·25140	1·25315	1·25489
72°	1·25664	1·25838	1·26013	1·26187	1·26362	1·26536	1·26711	1·26885	1·27060	1·27235
73°	1·27409	1·27584	1·27758	1·27933	1·28107	1·28282	1·28456	1·28631	1·28805	1·28980
74°	1·29154	1·29329	1·29503	1·29678	1·29852	1·30027	1·30202	1·30376	1·30551	1·30725
75°	1·30900	1·31074	1·31249	1·31423	1·31598	1·31772	1·31947	1·32121	1·32296	1·32470
76°	1·32645	1·32820	1·32994	1·33169	1·33343	1·33518	1·33692	1·33867	1·34041	1·34216
77°	1·34390	1·34565	1·34739	1·34914	1·35088	1·35263	1·35438	1·35612	1·35787	1·35961
78°	1·36136	1·36310	1·36485	1·36659	1·36834	1·37008	1·37183	1·37357	1·37532	1·37706
79°	1·37881	1·38056	1·38230	1·38405	1·38579	1·38754	1·38928	1·39103	1·39277	1·39452
80°	1·39626	1·39801	1·39975	1·40150	1·40324	1·40499	1·40674	1·40848	1·41023	1·41197
81°	1·41372	1·41546	1·41721	1·41895	1·42070	1·42244	1·42419	1·42593	1·42768	1·42942
82°	1·43117	1·43292	1·43466	1·43641	1·43815	1·43990	1·44164	1·44339	1·44513	1·44688
83°	1·44862	1·45037	1·45211	1·45386	1·45560	1·45735	1·45910	1·46084	1·46259	1·46433
84°	1·46608	1·46782	1·46957	1·47131	1·47306	1·47480	1·47655	1·47829	1·48004	1·48178
85°	1·48353	1·48528	1·48702	1·48877	1·49051	1·49226	1·49400	1·49575	1·49749	1·49924
86°	1·50098	1·50273	1·50447	1·50622	1·50796	1·50971	1·51146	1·51320	1·51495	1·51669
87°	1·51844	1·52018	1·52193	1·52367	1·52542	1·52716	1·52891	1·53065	1·53240	1·53414
88°	1·53589	1·53764	1·53938	1·54113	1·54287	1·54462	1·54636	1·54811	1·54985	1·55160
89°	1·55334	1·55509	1·55683	1·55858	1·56032	1·56207	1·56382	1·56556	1·56731	1·56905

Differences—

1′	2′	3′	4′	5′
29	58	87	116	145

HYPERBOLIC LOGARITHMS

No.		Third significant figure.									Difference for 4th significant figure.								
	0	1	2	3	4	5	6	7	8	9	1	2	3	4	5	6	7	8	9
1·0	0·0000	0100	0198	0296	0392	0488	0583	0677	0770	0862	10	19	29	38	48	57	67	76	86
1·1	0·0953	1044	1133	1222	1310	1398	1484	1570	1655	1740	9	17	26	35	44	52	61	70	78
1·2	0·1823	1906	1989	2070	2151	2231	2311	2390	2469	2546	8	16	24	32	40	48	56	64	72
1·3	0·2624	2700	2776	2852	2927	3001	3075	3148	3221	3293	7	15	22	30	37	45	52	59	67
1·4	0·3365	3436	3507	3577	3646	3716	3784	3853	3920	3988	7	14	21	28	35	41	48	55	62
1·5	0·4055	4121	4187	4253	4318	4383	4447	4511	4574	4637	6	13	19	26	32	39	45	52	58
1·6	0·4700	4762	4824	4886	4947	5008	5068	5128	5188	5247	6	12	18	24	30	36	42	48	55
1·7	0·5306	5365	5423	5481	5539	5596	5653	5710	5766	5822	6	11	17	24	29	34	40	46	51
1·8	0·5878	5933	5988	6043	6098	6152	6206	6259	6313	6366	5	11	16	22	27	32	38	43	49
1·9	0·6419	6471	6523	6575	6627	6678	6729	6780	6831	6881	5	10	15	20	26	31	36	41	46
2·0	0·6931	6981	7031	7080	7129	7178	7227	7275	7324	7372	5	10	15	20	24	29	34	39	44
2·1	0·7419	7467	7514	7561	7608	7655	7701	7747	7793	7839	5	10	14	19	23	28	33	37	42
2·2	0·7885	7930	7975	8020	8065	8109	8154	8198	8242	8286	4	9	13	18	22	27	31	36	40
2·3	0·8329	8372	8416	8459	8502	8544	8587	8629	8671	8713	4	9	13	17	21	26	30	34	38
2·4	0·8755	8796	8838	8879	8920	8961	9002	9042	9083	9123	4	8	12	16	20	24	29	33	37
2·5	0·9163	9203	9243	9282	9322	9361	9400	9439	9478	9517	4	8	12	16	20	24	27	31	35
2·6	0·9555	9594	9632	9670	9708	9746	9783	9821	9858	9895	4	8	11	15	19	23	26	30	34
2·7	0·9933	9969	**0006**	**0043**	**0080**	**0116**	**0152**	**0188**	**0225**	**0260**	4	7	11	15	18	22	25	29	33
2·8	1·0296	0332	0367	0403	0438	0473	0508	0543	0578	0613	4	7	11	14	18	21	25	28	32
2·9	1·0647	0682	0716	0750	0784	0818	0852	0886	0919	0953	3	7	10	14	17	20	24	27	31
3·0	1·0986	1019	1053	1086	1119	1151	1184	1217	1249	1282	3	7	10	13	16	20	23	26	30
3·1	1·1314	1346	1378	1410	1442	1474	1506	1537	1569	1600	3	6	10	13	16	19	22	25	29
3·2	1·1632	1663	1694	1725	1756	1787	1817	1848	1878	1909	3	6	9	12	15	18	22	25	28
3·3	1·1939	1969	2000	2030	2060	2090	2119	2149	2179	2208	3	6	9	12	15	18	21	24	27
3·4	1·2238	2267	2296	2326	2355	2384	2413	2442	2470	2499	3	6	9	12	15	17	20	23	26
3·5	1·2528	2556	2585	2613	2641	2669	2698	2726	2754	2782	3	6	8	11	14	17	20	23	25
3·6	1·2809	2837	2865	2892	2920	2947	2975	3002	3029	3056	3	5	8	11	14	16	19	22	25
3·7	1·3083	3110	3137	3164	3191	3218	3244	3271	3297	3324	3	5	8	11	13	16	19	21	24
3·8	1·3350	3376	3403	3429	3455	3481	3507	3533	3558	3584	3	5	8	10	13	16	18	21	23
3·9	1·3610	3635	3661	3686	3712	3737	3762	3788	3813	3838	3	5	8	10	13	15	18	20	23
4·0	1·3863	3888	3913	3938	3962	3987	4012	4036	4061	4085	2	5	7	10	12	15	17	20	22
4·1	1·4110	4134	4159	4183	4207	4231	4255	4279	4303	4327	2	5	7	10	12	14	17	19	22
4·2	1·4351	4375	4398	4422	4446	4469	4493	4516	4540	4563	2	5	7	9	12	14	17	19	21
4·3	1·4586	4609	4633	4656	4679	4702	4725	4748	4770	4793	2	5	7	9	12	14	16	18	21
4·4	1·4816	4839	4861	4884	4907	4929	4951	4974	4996	5019	2	5	7	9	11	13	16	18	20
4·5	1·5041	5063	5085	5107	5129	5151	5173	5195	5217	5239	2	4	7	9	11	13	15	18	20
4·6	1·5261	5282	5304	5326	5347	5369	5390	5412	5433	5454	2	4	6	9	11	13	15	17	19
4·7	1·5476	5497	5518	5539	5560	5581	5602	5623	5644	5665	2	4	6	8	11	13	15	17	19
4·8	1·5686	5707	5728	5748	5769	5790	5810	5831	5851	5872	2	4	6	8	10	12	14	16	19
4·9	1·5892	5913	5933	5953	5974	5994	6014	6034	6054	6074	2	4	6	8	10	12	14	16	18
5·0	1·6094	6114	6134	6154	6174	6194	6214	6233	6253	6273	2	4	6	8	10	12	14	16	18
5·1	1·6292	6312	6332	6351	6371	6390	6409	6429	6448	6467	2	4	6	8	10	12	14	16	18
5·2	1·6487	6506	6525	6544	6563	6582	6601	6620	6639	6658	2	4	6	8	10	11	13	15	17
5·3	1·6677	6696	6715	6734	6752	6771	6790	6808	6827	6845	2	4	6	7	9	11	13	15	17
5·4	1·6864	6882	6901	6919	6938	6956	6974	6993	7011	7029	2	4	6	7	9	11	13	15	16

Note.—Heavy type indicates change of characteristic.

HYPERBOLIC LOGARITHMS

No.		Third significant figure.									Difference for 4th significant figure.								
	0	1	2	3	4	5	6	7	8	9	1	2	3	4	5	6	7	8	9
5·5	1·7047	7066	7084	7102	7120	7138	7156	7174	7192	7210	2	4	5	7	9	11	13	14	16
5·6	1·7228	7246	7263	7281	7299	7317	7334	7352	7370	7387	2	4	5	7	9	11	12	14	16
5·7	1·7405	7422	7440	7457	7475	7492	7509	7527	7544	7561	2	3	5	7	9	10	12	14	16
5·8	1·7579	7596	7613	7630	7647	7664	7681	7699	7716	7733	2	3	5	7	9	10	12	14	15
5·9	1·7750	7766	7783	7800	7817	7834	7851	7867	7884	7901	2	3	5	7	8	10	12	13	15
6·0	1·7918	7934	7951	7967	7984	8001	8017	8034	8050	8066	2	3	5	7	8	10	12	13	15
6·1	1·8083	8099	8116	8132	8148	8165	8181	8197	8213	8229	2	3	5	6	8	10	11	13	15
6·2	1·8245	8262	8278	8294	8310	8326	8342	8358	8374	8390	2	3	5	6	8	10	11	13	14
6·3	1·8405	8421	8437	8453	8469	8485	8500	8516	8532	8547	2	3	5	6	8	9	11	13	14
6·4	1·8563	8579	8594	8610	8625	8641	8656	8672	8687	8743	2	3	5	6	8	9	11	12	14
6·5	1·8718	8733	8749	8764	8779	8795	8810	8825	8840	8856	2	3	5	6	8	9	11	12	14
6·6	1·8871	8886	8901	8916	8931	8946	8961	8976	8991	9006	2	3	5	6	8	9	11	12	13
6·7	1·9021	9036	9051	9066	9081	9095	9110	9125	9140	9155	1	3	4	6	7	9	10	12	13
6·8	1·9169	9184	9199	9213	9228	9242	9257	9272	9286	9301	1	3	4	6	7	9	10	12	13
6·9	1·9315	9330	9344	9359	9373	9387	9402	9416	9430	9445	1	3	4	6	7	9	10	12	13
7·0	1·9459	9473	9488	9502	9516	9530	9544	9559	9573	9587	1	3	4	6	7	9	10	11	13
7·1	1·9601	9615	9629	9643	9657	9671	9685	9699	9713	9727	1	3	4	6	7	8	10	11	12
7·2	1·9741	9755	9769	9782	9796	9810	9824	9838	9851	9865	1	3	4	6	7	8	10	11	12
7·3	1·9879	9892	9906	9920	9933	9947	9961	9974	9988	0001	1	3	4	5	7	8	9	11	12
7·4	2·0015	0028	0042	0055	0069	0082	0096	0109	0122	0136	1	3	4	5	7	8	9	11	12
7·5	2·0149	0162	0176	0189	0202	0215	0229	0242	0255	0268	1	3	4	5	7	8	9	11	12
7·6	2·0281	0295	0308	0321	0334	0347	0360	0373	0386	0399	1	3	4	5	7	8	9	10	12
7·7	2·0412	0425	0438	0451	0464	0477	0490	0503	0516	0528	1	3	4	5	6	8	9	10	12
7·8	2·0541	0554	0567	0580	0592	0605	0618	0631	0643	0656	1	3	4	5	6	8	9	10	11
7·9	2·0669	0681	0694	0707	0719	0732	0744	0757	0769	0782	1	3	4	5	6	8	9	10	11
8·0	2·0794	0807	0819	0832	0844	0857	0869	0882	0894	0906	1	2	4	5	6	7	9	10	11
8·1	2·0919	0931	0943	0956	0968	0980	0992	1005	1017	1029	1	2	4	5	6	7	9	10	11
8·2	2·1041	1054	1066	1078	1090	1102	1114	1126	1138	1150	1	2	4	5	6	7	9	10	11
8·3	2·1163	1175	1187	1199	1211	1223	1235	1247	1258	1270	1	2	4	5	6	7	8	10	11
8·4	2·1282	1294	1306	1318	1330	1342	1353	1365	1377	1389	1	2	4	5	6	7	8	9	11
8·5	2·1401	1412	1424	1436	1448	1459	1471	1483	1494	1506	1	2	4	5	6	7	8	9	11
8·6	2·1518	1529	1541	1552	1564	1576	1587	1599	1610	1622	1	2	3	5	6	7	8	9	10
8·7	2·1633	1645	1656	1668	1679	1691	1702	1713	1725	1736	1	2	3	5	6	7	8	9	10
8·8	2·1748	1759	1770	1782	1793	1804	1815	1827	1838	1849	1	2	3	5	6	7	8	9	10
8·9	2·1861	1872	1883	1894	1905	1917	1928	1939	1950	1961	1	2	3	4	6	7	8	9	10
9·0	2·1972	1983	1994	2006	2017	2028	2039	2050	2061	2072	1	2	3	4	6	7	8	9	10
9·1	2·2083	2094	2105	2116	2127	2138	2148	2159	2170	2181	1	2	3	4	5	7	8	9	10
9·2	2·2192	2203	2214	2225	2235	2246	2257	2268	2279	2289	1	2	3	4	5	6	8	9	10
9·3	2·2300	2311	2322	2332	2343	2354	2364	2375	2386	2396	1	2	3	4	5	6	7	9	10
9·4	2·2407	2418	2428	2439	2450	2460	2471	2481	2492	2502	1	2	3	4	5	6	7	8	10
9·5	2·2513	2523	2534	2544	2555	2565	2576	2586	2597	2607	1	2	3	4	5	6	7	8	9
9·6	2·2618	2628	2638	2649	2659	2670	2680	2690	2701	2711	1	2	3	4	5	6	7	8	9
9·7	2·2721	2732	2742	2752	2762	2773	2783	2793	2803	2814	1	2	3	4	5	6	7	8	9
9·8	2·2824	2834	2844	2854	2865	2875	2885	2895	2905	2915	1	2	3	4	5	6	7	8	9
9·9	2·2925	2935	2946	2956	2966	2976	2986	2996	3006	3016	1	2	3	4	5	6	7	8	9

$\log_e 10 = 2{\cdot}3026.$ $\log_e 1000 = 6{\cdot}9078.$
$\log_e 100 = 4{\cdot}6052.$ $\log_e 10000 = 9{\cdot}2103.$

HYPERBOLIC FUNCTIONS

x	e^x	e^{-x}	sinh x	cosh x	x	e^x	e^{-x}	sinh x	cosh x
·00	1·0000	1·0000	0	1·0000	·50	1·6487	·6065	·5211	1·1276
·01	1·0101	·9900	·0100	1·0001	·51	1·6653	·6005	·5324	1·1329
·02	1·0202	·9802	·0200	1·0002	·52	1·6820	·5945	·5438	1·1383
·03	1·0305	·9704	·0300	1·0005	·53	1·6989	·5886	·5552	1·1438
·04	1·0408	·9608	·0400	1·0008	·54	1·7160	·5827	·5666	1·1494
·05	1·0513	·9512	·0500	1·0013	·55	1·7333	·5769	·5782	1·1551
·06	1·0618	·9418	·0600	1·0018	·56	1·7507	·5712	·5897	1·1609
·07	1·0725	·9324	·0701	1·0025	·57	1·7683	·5655	·6014	1·1669
·08	1·0833	·9231	·0801	1·0032	·58	1·7860	·5599	·6131	1·1730
·09	1·0942	·9139	·0901	1·0041	·59	1·8040	·5543	·6248	1·1792
·10	1·1052	·9048	·1002	1·0050	·60	1·8221	·5488	·6367	1·1855
·11	1·1163	·8958	·1102	1·0061	·61	1·8404	·5434	·6485	1·1919
·12	1·1275	·8869	·1203	1·0072	·62	1·8589	·5379	·6605	1·1984
·13	1·1388	·8781	·1304	1·0085	·63	1·8776	·5326	·6725	1·2051
·14	1·1503	·8694	·1405	1·0098	·64	1·8965	·5273	·6846	1·2119
·15	1·1618	·8607	·1506	1·0113	·65	1·9155	·5220	·6967	1·2188
·16	1·1735	·8521	·1607	1·0128	·66	1·9348	·5169	·7090	1·2258
·17	1·1853	·8437	·1708	1·0145	·67	1·9542	·5117	·7213	1·2330
·18	1·1972	·8353	·1810	1·0162	·68	1·9739	·5066	·7336	1·2402
·19	1·2092	·8270	·1911	1·0181	·69	1·9937	·5016	·7461	1·2476
·20	1·2214	·8187	·2013	1·0201	·70	2·0138	·4966	·7586	1·2552
·21	1·2337	·8106	·2115	1·0221	·71	2·0340	·4916	·7712	1·2628
·22	1·2461	·8025	·2218	1·0243	·72	2·0544	·4868	·7838	1·2706
·23	1·2586	·7945	·2320	1·0266	·73	2·0751	·4819	·7966	1·2785
·24	1·2712	·7866	·2423	1·0289	·74	2·0959	·4771	·8094	1·2865
·25	1·2840	·7788	·2526	1·0314	·75	2·1170	·4724	·8223	1·2947
·26	1·2969	·7711	·2629	1·0340	·76	2·1383	·4677	·8353	1·3030
·27	1·3100	·7634	·2733	1·0367	·77	2·1598	·4630	·8484	1·3114
·28	1·3231	·7558	·2837	1·0395	·78	2·1815	·4584	·8615	1·3199
·29	1·3364	·7483	·2941	1·0423	·79	2·2034	·4538	·8748	1·3286
·30	1·3499	·7408	·3045	1·0453	·80	2·2255	·4493	·8881	1·3374
·31	1·3634	·7334	·3150	1·0484	·81	2·2479	·4449	·9015	1·3464
·32	1·3771	·7261	·3255	1·0516	·82	2·2705	·4404	·9150	1·3555
·33	1·3910	·7189	·3360	1·0549	·83	2·2933	·4360	·9286	1·3647
·34	1·4049	·7118	·3466	1·0584	·84	2·3164	·4317	·9423	1·3740
·35	1·4191	·7047	·3572	1·0619	·85	2·3396	·4274	·9561	1·3835
·36	1·4333	·6977	·3678	1·0655	·86	2·3632	·4232	·9700	1·3932
·37	1·4477	·6907	·3785	1·0692	·87	2·3869	·4190	·9840	1·4029
·38	1·4623	·6839	·3892	1·0731	·88	2·4109	·4148	·9981	1·4128
·39	1·4770	·6771	·4000	1·0770	·89	2·4351	·4107	1·0122	1·4229
·40	1·4918	·6703	·4108	1·0811	·90	2·4596	·4066	1·0265	1·4331
·41	1·5068	·6637	·4216	1·0852	·91	2·4843	·4025	1·0409	1·4434
·42	1·5220	·6570	·4325	1·0895	·92	2·5093	·3985	1·0554	1·4539
·43	1·5373	·6505	·4434	1·0939	·93	2·5345	·3946	1·0700	1·4645
·44	1·5527	·6440	·4543	1·0984	·94	2·5600	·3906	1·0847	1·4753
·45	1·5683	·6376	·4653	1·1030	·95	2·5857	·3867	1·0995	1·4862
·46	1·5841	·6313	·4764	1·1077	·96	2·6117	·3829	1·1144	1·4973
·47	1·6000	·6250	·4875	1·1125	·97	2·6379	·3791	1·1294	1·5085
·48	1·6161	·6188	·4986	1·1174	·98	2·6645	·3753	1·1446	1·5199
·49	1·6323	·6126	·5098	1·1225	·99	2·6912	·3716	1·1598	1·5314

HYPERBOLIC FUNCTIONS

x	e^x	e^{-x}	sinh x	cosh x	x	e^x	e^{-x}	sinh x	cosh x
1·00	2·7183	·3679	1·1752	1·5431	3·50	33·115	·0302	16·543	16·573
1·05	2·8577	·3499	1·2539	1·6038	3·55	34·813	·0287	17·392	17·421
1·10	3·0042	·3329	1·3356	1·6685	3·60	36·598	·0273	18·285	18·313
1·15	3·1582	·3166	1·4208	1·7374	3·65	38·475	·0260	19·224	19·250
1·20	3·3201	·3012	1·5095	1·8107	3·70	40·447	·0247	20·211	20·236
1·25	3·4903	·2865	1·6019	1·8884	3·75	42·521	·0235	21·249	21·272
1·30	3·6693	·2725	1·6984	1·9709	3·80	44·701	·0224	22·339	22·362
1·35	3·8574	·2592	1·7991	2·0583	3·85	46·993	·0213	23·486	23·507
1·40	4·0552	·2466	1·9043	2·1509	3·90	49·402	·0202	24·691	24·711
1·45	4·2631	·2346	2·0143	2·2488	3·95	51·935	·0193	25·958	25·977
1·50	4·4817	·2231	2·1293	2·3524	4·00	54·598	·0183	27·290	27·308
1·55	4·7115	·2122	2·2496	2·4619	4·05	57·397	·0174	28·690	28·707
1·60	4·9530	·2019	2·3756	2·5775	4·10	60·340	·0166	30·162	30·178
1·65	5·2070	·1920	2·5075	2·6995	4·15	63·434	·0158	31·709	31·725
1·70	5·4739	·1827	2·6456	2·8283	4·20	66·686	·0150	33·336	33·351
1·75	5·7546	·1738	2·7904	2·9642	4·25	70·105	·0143	35·046	35·060
1·80	6·0496	·1653	2·9422	3·1075	4·30	73·700	·0136	36·843	36·857
1·85	6·3598	·1572	3·1013	3·2585	4·35	77·478	·0129	38·733	38·746
1·90	6·6859	·1496	3·2682	3·4177	4·40	81·451	·0123	40·719	40·732
1·95	7·0287	·1423	3·4432	3·5855	4·45	85·627	·0117	42·808	42·819
2·00	7·3891	·1353	3·6269	3·7622	4·50	90·017	·0111	45·003	45·014
2·05	7·7679	·1287	3·8196	3·9483	4·55	94·632	·0106	47·311	47·322
2·10	8·1662	·1225	4·0219	4·1443	4·60	99·484	·0101	49·737	49·747
2·15	8·5849	·1165	4·2342	4·3507	4·65	104·59	·00956	52·288	52·297
2·20	9·0250	·1108	4·4571	4·5679	4·70	109·95	·00910	54·969	54·978
2·25	9·4877	·1054	4·6913	4·7966	4·75	115·58	·00865	57·788	57·796
2·30	9·9742	·1003	4·9370	5·0372	4·80	121·51	·00823	60·751	60·759
2·35	10·486	·0954	5·1951	5·2905	4·85	127·74	·00783	63·866	63·874
2·40	11·023	·0907	5·4662	5·5570	4·90	134·29	·00745	67·141	67·149
2·45	11·588	·0863	5·7510	5·8373	4·95	141·17	·00708	70·584	70·591
2·50	12·182	·0821	6·0502	6·1323	5·00	148·41	·00674	74·203	74·210
2·55	12·807	·0781	6·3645	6·4426	5·05	156·02	·00641	78·008	78·014
2·60	13·464	·0743	6·6947	6·7690	5·10	164·02	·00610	82·008	82·014
2·65	14·154	·0707	7·0417	7·1123	5·15	172·43	·00580	86·213	86·219
2·70	14·880	·0672	7·4063	7·4735	5·20	181·27	·00552	90·633	90·639
2·75	15·643	·0639	7·7894	7·8533	5·25	190·57	·00525	95·280	95·286
2·80	16·445	·0608	8·1919	8·2527	5·30	200·34	·00499	100·17	100·17
2·85	17·288	·0578	8·6150	8·6728	5·35	210·61	·00475	105·30	105·31
2·90	18·174	·0550	9·0596	9·1146	5·40	221·41	·00452	110·70	110·71
2·95	19·106	·0523	9·5268	9·5791	5·45	232·76	·00430	116·38	116·38
3·00	20·086	·0498	10·018	10·068	5·50	244·69	·00409	122·34	122·35
3·05	21·115	·0474	10·534	10·581	5·55	257·24	·00389	128·62	128·62
3·10	22·198	·0450	11·076	11·122	5·60	270·43	·00370	135·21	135·22
3·15	23·336	·0429	11·647	11·689	5·65	284·29	·00352	142·14	142·15
3·20	24·533	·0408	12·246	12·287	5·70	298·87	·00335	149·43	149·44
3·25	25·790	·0388	12·876	12·915	5·75	314·19	·00318	157·09	157·10
3·30	27·113	·0369	13·538	13·575	5·80	330·30	·00303	165·15	165·15
3·35	28·503	·0351	14·234	14·269	5·85	347·23	·00288	173·62	173·62
3·40	29·964	·0334	14·965	14·999	5·90	365·04	·00274	182·52	182·52
3·45	31·500	·0317	15·734	15·766	5·95	383·75	·00261	191·88	191·90
					6·00	403·43	·00248	201·71	201·72